Marx Joyce
Abbott Hardy Machiavelli Montaigne Chesterton Cooper Emerson Austen
Defoe Melville Haggard Hugo
Christie Carroll Eliot Grimm
Stoker Molière
Wilde Maupassant Byron
Engels Schiller
Garnett Einstein Fitzgerald Smith
Hawthorne Kafka
Goethe Dostoyevsky
Cotton Hall
Baum Kipling Doyle Willis
Leslie Dumas Henry Nietzsche
Flaubert Turgenev Balzac
Stockton Vatsyayana Crane
Burroughs Verne
Whitman Gogol
Curtis Tocqueville Vinci
Homer Widger Tolstoy Busch
Darwin Thoreau
Potter Freud Twain Plato Scott
Zola Lawrence
Kant Jowett Stevenson Dickens
Andersen Burton Hesse Harte
London Descartes Cervantes
Poe Aristotle Wells Voltaire
Hale James Hastings Cooke
Bunner Shakespeare
Richter Chambers Irving
da Benedict
Doré Swift Dante Chekhov Shaw Wodehouse Alcott
Pushkin
Newton

tredition®

tredition was established in 2006 by Sandra Latusseck and Soenke Schulz. Based in Hamburg, Germany, tredition offers publishing solutions to authors and publishing houses, combined with worldwide distribution of printed and digital book content. tredition is uniquely positioned to enable authors and publishing houses to create books on their own terms and without conventional manufacturing risks.

For more information please visit: www.tredition.com

TREDITION CLASSICS

This book is part of the TREDITION CLASSICS series. The creators of this series are united by passion for literature and driven by the intention of making all public domain books available in printed format again - worldwide. Most TREDITION CLASSICS titles have been out of print and off the bookstore shelves for decades. At tredition we believe that a great book never goes out of style and that its value is eternal. Several mostly non-profit literature projects provide content to tredition. To support their good work, tredition donates a portion of the proceeds from each sold copy. As a reader of a TREDITION CLASSICS book, you support our mission to save many of the amazing works of world literature from oblivion. See all available books at www.tredition.com.

Project Gutenberg

The content for this book has been graciously provided by Project Gutenberg. Project Gutenberg is a non-profit organization founded by Michael Hart in 1971 at the University of Illinois. The mission of Project Gutenberg is simple: To encourage the creation and distribution of eBooks. Project Gutenberg is the first and largest collection of public domain eBooks.

The Cruise of the Betsey or, A Summer Ramble Among the Fossiliferous Deposits of the Hebrides. With Rambles of a Geologist or, Ten Thousand Miles Over the Fossiliferous Deposits of Scotland

Hugh Miller

Imprint

This book is part of TREDITION CLASSICS

Author: Hugh Miller
Cover design: Buchgut, Berlin – Germany

Publisher: tredition GmbH, Hamburg - Germany
ISBN: 978-3-8472-2565-2

www.tredition.com
www.tredition.de

THE

CRUISE OF THE BETSEY;

OR,

A SUMMER RAMBLE AMONG THE FOSSIL-IFEROUS DEPOSITS OF THE HEBRIDES.

WITH

RAMBLES OF A GEOLOGIST;

OR,

TEN THOUSAND MILES OVER THE FOSSIL-IFEROUS DEPOSITS OF SCOTLAND.

BY

HUGH MILLER, LL. D.,

AUTHOR OF "THE OLD RED SANDSTONE," "FOOTPRINTS OF THE CREATOR," "MY SCHOOLS AND SCHOOLMASTERS," "THE TESTIMONY OF THE ROCKS," ETC.

BOSTON:
GOULD AND LINCOLN,
59 WASHINGTON STREET.

NEW YORK: SHELDON AND COMPANY.
CINCINNATI: GEO. S. BLANCHARD.

1862.

AUTHORIZED EDITION.

By a special arrangement with the late Hugh Miller, Gould and Lincoln became the authorized American publishers of his works. By a similar arrangement made with the family since his decease, they will also publish his Posthumous Works, of which the present volume is the first.

ELECTROTYPED BY W. F. DRAPER, ANDOVER, MASS.

PRINTED BY GEO. C. RAND & AVERY, BOSTON.

PREFACE.

Naturalists of every class know too well how Hugh Miller died—the victim of an overworked brain; and how that bright and vigorous spirit was abruptly quenched forever.

During the month of May (1857) Mrs. Miller came to Malvern, after recovering from the first shock of bereavement, in search of health and repose, and evidently hoping to do justice, on her recovery, to the literary remains of her husband. Unhappily the excitement and anxiety naturally attaching to a revision of her husband's works proved over much for one suffering under such recent trial, and from an affection of the brain and spine which ensued; and, in consequence, Mrs. Miller has been forbidden, for the present, to engage in any work of mental labor.

Under these circumstances, and at Mrs. Miller's request, I have undertaken the editing of "The Cruise of the Betsey, or a Summer Ramble among the Fossiliferous Deposits of the Hebrides," as well as "The Rambles of a Geologist," hitherto unpublished, save as a series of articles in the "Witness" newspaper. The style and arguments of Hugh Miller are so peculiarly his own, that I have not presumed to alter the text, and have merely corrected some statements incidental to the condition of geological knowledge at the time this work was penned. "The Cruise of the Betsey" was written for that well-known paper the "Witness" during the period when a disputation productive of much bitter feeling waged between the Free and Established Churches of Scotland; but as the Disruption and its history possesses little interest to a large class of the readers of this work, who will rejoice to follow their favorite author among the isles and rocks of the "bonnie land," I have expunged *some* passages, which I am assured the author would have omitted had he lived to reprint this interesting narrative of his geological rambles. Hugh Miller battled nobly for his faith while living. The sword is in the scabbard: let it rest!

W.S. SYMONDS.

Pendock Rectory, April 1, 1858.

CONTENTS.

PART I.

THE CRUISE OF THE BETSEY.

CHAPTER IV.

CHAPTER V.

CHAPTER VI.

CHAPTER VII.

CHAPTER VIII.

CHAPTER IX.

ford — Rain — Island of Pabba — Description of the Island — Its Geological Structure — Astrea — Polypifers — *Gryphœa incurva* — Three Groups of Fossils in the Lias of Skye — Abundance of the Petrifactions of Pabba — Scenery — Pabba a "piece of smooth, level England" — Fossil Shells of Pabba — - Voyage resumed — Kyle Akin — Ruins of Castle Maoil — A "Thornback" Dinner — The Bunch of Deep Sea Tangle — The Caileach Stone — Kelp Furnaces — Escape of the Betsey from sinking.

CHAPTER X.

Isle Orusay — The Sabbath — A Sailor-minister's Sermon for Sailors — The Scuir Sermon — Loch Carron — Groups of Moraines — A sheep District — The Editor of the *Witness* and the Establishment Clergyman — Dingwall — Conon-side revisited — The Pond and its Changes — New Faces. — The Stonemason's Mark — The Burying-ground of Urquhart — An old Acquaintance — Property Qualification for Voting in Scotland — Montgerald Sandstone Quarries — Geological Science in Cromarty — The Danes at Cromarty — The Danish Professor and the "Old Red Sandstone" — Harmonizing Tendencies of Science.

CHAPTER XI.

Ichthyolite Beds — An interesting Discovery — Two Storeys of Organic Remains in the Old Red Sandstone — Ancient Ocean of Lower Old Red — Two great Catastrophes — Ancient Fish Scales — Their skilful Mechanism displayed by examples — Bone Lips — Arts of the Slater and Tiler as old as Old Red Sandstone — Jet Trinkets — Flint Arrow-heads — Vitrified Forts of Scotland — Style of grouping Lower Old Red Fossils — Illustration from Cromarty Fishing Phenomena — Singular Remains of Holoptychius — Ramble with Mr. Robert Dick — Color of the Planet Mars — Tombs never dreamed of by Hervey — Skeleton of the Bruce — Gigantic Holoptychius — "Coal money Currency" — Upper Boundary of Lower Old Red — Every one may add to the Store of Geological Facts — Discoveries of Messrs. Dick and Peach.

CHAPTER XII.

Ichthyolite Beds of Clune and Lethenbarn — Limestone Quarry — Destruction of Urns and Sarcophagi in the Lime-kiln — Nodules opened — Beautiful coloring of the Remains — Patrick Duff's Description — New Genus of Morayshire Ichthyolite described — Form and size of the Nodules or Stone Coffins — Illustration from Mrs. Marshall's Cements — Forest of Darnaway — The Hill of Berries — Sluie — Elgin — Outliers of the Weald and the Oölite — Description of the Weald at Linksfield — Mr. Duff's *Lepidotus minor* — Eccentric Types of Fish Scales — Visit to the Sandstones of Scat-Craig — Fine suit of Fossils at Scat-Craig — True graveyard Bones, not mere Impressions — Varieties of pattern — The Diker's "Carved Flowers" — *Stagonolepis*, a new Genus — Termination of the Ramble.

CHAPTER XIII.

SUPPLEMENTARY.

Supplementary — Isolated Reptile Remains in Eigg — Small Isles revisited — The Betsey again — Storm bound — Tacking — Becalmed — Medusæ caught and described — Rain — A Shoal of Porpoises — Change of Weather — The bed-ridden Woman — The Poor Law Act for Scotland — Geological Excursion — Basaltic Columns — Oölitic Beds — Abundance of Organic Remains — Hybodus Teeth — Discovery of reptile Remains *in situ* — Musical Sand of Laig re-examined — Explanation suggested — Sail for Isle Ornsay — Anchored Clouds — A Leak sprung — Peril of the Betsey — At work with Pump and Pails — Safe in Harbor — Return to Edinburgh.

PART II.

RAMBLES OF A GEOLOGIST.

CHAPTER I.

Embarkation — A foundered Vessel — Lateness of the Harvest dependent on the Geological character of the Soil—A Granite Harvest and an Old Red Harvest—Cottages of Redstone and of Granite—Arable Soil of Scotland the result of a Geological Grinding Agency—Locality of the Famine of 1846—Mr. Longmuir's Fossils—Geology necessary to a Theologian—Popularizers of Science when dangerous—"Constitution of Man," and "Vestiges of Creation"—Atop of the Banff Coach—A Geologist's Field Equipment—The trespassing "Stirk"—Silurian Schists inlaid with Old Red—Bay of Gamrie, how formed—Gardenstone—Geological Free-masonry illustrated—How to break an Ichthyolite Nodule—An old Rhyme mended—A raised Beach—Fossil Shells—Scotland under Water at the time of the Boulder-clays.

CHAPTER II.

Character of the Rocks near Gardenstone—A Defunct Father-lasher—A Geological Inference—Village of Gardenstone—The drunken Scot—Gardenstone Inn—Lord Gardenstone—A Tempest threatened—The Author's Ghost Story—The Lady in Green—Her Appearance and Tricks—The Rescued Children—The murdered Peddler and his Pack—Where the Green Dress came from—Village of Macduff—Peculiar Appearance of the Beach at the Mouth of the Deveron—Dr. Emslie's Fossils—*Pterichthys quadratus*—Argillaceous Deposits of Blackpots—Pipe-laying in Scotland—Fossils of Blackpots Clay—Mr. Longmuir's Description of them—Blackpots Deposit a Re-formation of a Liasic Patch—Period of its Formation.

CHAPTER III.

From Blackpots to Portsoy—Character of the Coast—Burn of Boyne—Fever Phantoms—Graphic Granite—Maupertuis and the Runic Inscription—Explanation of the *quo modo* of Graphic Granite—Portsoy Inn—Serpentine Beds—Portsoy Serpentine unrivalled for small ornaments—Description of it—Significance of the term *serpentine*—Elizabeth Bond and her "Letters"—From Portsoy to Cullen—Attritive Power of the Ocean illustrated—The Equinoctial—From Cullen to Fochabers—The Old Red again—The old

Pensioner—Fochabers—Mr. Joss, the learned Mail-guard—The Editor a sort of Coach-guard—On the Coach to Elgin—Geology of Banffshire—Irregular paging of the Geologic Leaves—Geologic Map of the County like Joseph's Coat—Striking Illustration.

CHAPTER IV.

Yellow-hued Houses of Elgin—Geology of the Country indicated by the coloring of the Stone Houses—Fossils of Old Red north of the Grampians different from those of Old Red south—Geologic Formations at Linksfield difficult to be understood—Ganoid Scales of the Wealden—Sudden Reaction, from complex to simple, in the Scales of Fishes—Pore-covered Scales—Extraordinary amount of Design exhibited in Ancient Ganoid Scales—Holoptychius Scale illustrated by Cromwell's "fluted pot"—Patrick Duff's Geological Collection—Elgin Museum—Fishes of the Ganges—Armature of Ancient Fishes—Compensatory Defences—-The Hermit-crab—Spines of the Pimelodi—Ride to Campbelton—Theories of the formation of Ardersier and Fortrose Promontories—Tradition of their construction by the Wizard, Michael Scott—A Region of Legendary Lore.

CHAPTER V.

Rosemarkie and its Scaurs—Kaes' Craig—A Jackdaw Settlement—"Rosemarkie Kaes" and "Cromarty Cooties"—"The Danes," a Group of Excavations—At Home in Cromarty—The Boulder-clay of Cromarty "begins to tell its story"—One of its marked Scenic Peculiarities—Hints to Landscape Painters—"Samuel's Well"—A Chain of Bogs geologically accounted for—Another Scenic Peculiarity—"*Ha-has* of Nature's digging"—The Author's earliest Field of Hard Labor—Picturesque Cliff of Boulder-clay—Scratchings on the Sandstone—Invariable Characteristic of true Boulder-clay—Scratchings on Pebbles in the line of the longer axis—Illustration from the Boulder-clay of Banff.

CHAPTER VI.

Organisms of the Boulder-clay not unequivocal—First Impressions of the Boulder-clay—Difficulty of accounting for its barrenness of Remains—Sir Charles Lyell's reasoning—A Fact to the contrary—Human Skull dug from a Clay-bank—The Author's Change of Belief respecting Organic Remains of the Boulder-clay—Shells from the Clay at Wick—Questions respecting them settled—Conclusions confirmed by Mr. Dick's Discoveries at Thurso—Sir John Sinclair's Discovery of Boulder-clay Shells in 1802—Comminution of the Shells illustrated—*Cyprina islandica*—Its Preservation in larger Proportions than those of other Shells accounted for—Boulder-clays of Scotland reformed during the existing Geological Epoch—Scotland in the Period of the Boulder-clay "merely three detached groups of Islands"—Evidence of the Subsidence of the Land in Scotland—Confirmed by Rev. Mr. Cumming's conclusion—High-lying Granite Boulders—Marks of a succeeding elevatory Period—Scandinavia now rising—Autobiography of a Boulder desirable—A Story of the Supernatural.

CHAPTER VII.

Relation of the deep red stone of Cromarty to the Ichthyolite Beds of the System—Ruins of a Fossil-charged Bed—Journey to Avoch—Red Dye of the Boulder-clay distinct from the substance itself—Variation of Coloring in the Boulder-clay Red Sandstone accounted for—Hard-pan how formed—A reformed Garden—An ancient Battle-field—Antiquity of Geologic and Human History compared—Burn of Killein—Observation made in boyhood confirmed—Fossil-nodules—Fine Specimen of *Coccosteus decipiens*—Blank strata of Old Red—New View respecting the Rocks of Black Isle—A Trip up Moray and Dingwall Friths—Altered color of the Boulder-clay—Up the Auldgrande River—Scenery of the great Conglomerate—Graphic Description—Laidlaw's Boulder—*Vaccinium myrtillus*—Profusion of Travelled Boulders—The Boulder *Clach Malloch*—Its zones of Animal and Vegetable life.

CHAPTER VIII.

Imaginary Autobiography of the *Clach Malloch* Boulder — Its Creation — Its Long Night of unsummed Centuries — Laid open to light on a desert Island — Surrounded by an Arctic Vegetation — Undermined by the rising Sea — Locked up and floated off on an Ice-field — At rest on the Sea-bottom — Another Night of unsummed Years — The Boulder raised again above the waves by the rising of the Land — Beholds an Altered Country — Pine Forests and Mammals — Another Period of Ages passes — The Boulder again floated off by an Iceberg — Finally at rest on the Shore of Cromarty Bay — Time and Occasion of naming it — Strange Phenomena accounted for by Earthquakes — How the Boulder of Petty Bay was moved — The Boulder of Auldgrande — The old Highland Paupers — The little Parsi Girl — Her Letter to her Papa — But one Human Nature on Earth — Journey resumed — Conon Burying Ground — An aged Couple — Gossip.

CHAPTER IX.

The Great Conglomerate — Its Undulatory and Rectilinear Members — Knock Farril and its Vitrified Fort — The old Highlanders an observant race — The Vein of Silver — Summit of Knock Farril — Mode of accounting for the Luxuriance of Herbage in the ancient Scottish Fortalices — The green Graves of Culloden — Theories respecting the Vitrification of the Hill-forts — Combined Theories of Williams and Mackenzie probably give the correct account — The Author's Explanation — Transformations of Fused Rocks — Strathpetlier — The Spa — Permanent Odoriferous Qualities of an ancient Sea-bottom converted into Rock — Mineral Springs of the Spa — Infusion of the powdered rock a substitute — Belemnite Water — The lively young Lady's Comments — A befogged Country seen from a hill-top — Ben-Wevis — Journey to Evanton — A Geologist's Night-mare — The Route Home — Ruins of Craig house — Incompatibility of Tea and Ghosts — End of the Tour.

CHAPTER X.

Recovered Health — Journey to the Orkneys — Aboard the Steamer at Wick — Mr. Bremner — Masonry of the Harbor of Wick — The greatest Blunders result from good Rules misapplied—Mr. Bremner's Theory about sea-washed Masonry—Singular Fracture of the Rock near Wick—The Author's mode of accounting for it—"Simple but not obvious" Thinking—Mr. Bremner's mode of making stone Erections under Water—His exploits in raising foundered Vessels—Aspect of the Orkneys—The ungracious Schoolmaster—In the Frith of Kirkwall—Cathedral of St Magnus—Appearance of Kirkwall—Its "perished suppers"—Its ancient Palaces—Blunder of the Scotch Aristocracy—The patronate Wedge—Breaking Ground in Orkney—Minute Gregarious Coccosteus—True Position of the Coccosteus' Eyes—Ruins of one of Cromwell's Forts—Antiquities of Orkney—The Cathedral—Its Sculptures—The Mysterious Cell—Prospect from the Tower—Its Chimes—Ruins of Castle Patrick.

CHAPTER XI.

The Bishop's Palace at Orkney—Haco the Norwegian—Icelandic Chronicle respecting his Expedition to Scotland—His Death—Removal of his Remain to Norway—Why Norwegian Invasion ceased—Straw-plaiting—The Lassies of Orkney—Orkney Type of Countenance—Celtic and Scandinavian—An accomplished Antiquary—Old Manuscripts—An old Tune book—Manuscript Letter of Mary Queen of Scots—Letters of General Monck—The fearless Covenanter—Cave of the Rebels—Why the tragedy of "Gustavus Vasa" was prohibited—Quarry of Pickoquoy—Its Fossil Shells—Journey to Stromness—Scenery—Birth-place of Malcolm, the Poet—His History—One of his Poems—His Brother a Free Church Minister—New Scenery.

CHAPTER XII.

Hills of Orkney—Their Geologic Composition—Scene of Scott's "Pirate"—Stromness—Geology of the District—"Seeking beasts"—Conglomerate in contact with Granite—A palæozoic Hudson's Bay—Thickness of Conglomerate of Orkney—Oldest Vertebrate yet discovered in Orkney—Its Size—Figure of a characteristic plate of the Asterolepis—Peculiarity of Old Red Fishes—Length of the Asterolepis—A rich Ichthyolite Bed—Arrangement of the Layers—Queries as to the Cause of it—Minerals—An abandoned Mine—A lost Vessel—Kelp for Iodine—A dangerous Coast—Incidents of Shipwreck—Hospitality—Stromness Museum—Diplopterus mistaken for Dipterus—Their Resemblances and Differences—Visit to a remarkable Stack—Paring the Soil for Fuel, and consequent Barrenness—Description of the Stack—Wave-formed Caves—Height to which the Surf rises.

CHAPTER XIII.

Detached Fossils—Remains of the Pterichthys—Terminal Bones of the Coccosteus, etc., preserved—Internal Skeleton of Coccosteus—The shipwrecked Sailor in the Cave—Bishop Grahame—His Character, as drawn by Baillie—His Successor—Ruins of the Bishop's Country-house—Sub-aërial Formation of Sandstone—Formation near New Kaye—Inference from such Formation—Tour resumed—Loch of Stennis—Waters of the Loch fresh, brackish, and salt—Vegetation varied accordingly—Change produced in the Flounder by fresh water—The Standing Stones, second only to Stonehenge—Their Purpose—Their Appearance and Situation—Diameter of the Circle—What the Antiquaries say of it—Reference to it in the "Pirate"—Dr. Hibbert's Account.

CHAPTER XIV.

On Horseback—A pared Moor—Small Landholders—Absorption of small holdings in England and Scotland—Division of Land favorable to Civil and Religious Rights—Favorable to social Elevation—An inland Parish—The Landsman and Lobster—Wild Flowers of Orkney—Law of Compensation illustrated by the Tobacco Plant—Poverty tends to Productiveness—Illustrated in Ireland—Profusion of Ichthyolites—Orkney a land of Defunct Fishes—Sandwick—A Collection of Coccostean Flags—A Quarry full of Heads of Dipteri—The Bergil, or Striped Wrasse—Its Resemblance to the Dipterus—Poverty of the Flora of the Lower Old Red—No true Coniferous Wood in the Orkney Flagstones—Departure for Hoy—The intelligent Boatman—Story of the Orkney Fisherman.

CHAPTER XV.

Hoy—Unique Scenery—The Dwarfie Stone of Hoy—Sir Walter Scott's Account of it—Its Associations—Inscription of Names—George Buchanan's Consolation—The mythic Carbuncle of the Hill of Hoy—No Fossils at Hoy—Striking Profile of Sir Walter Scott on the Hill of Hoy—Sir Walter, and Shetland and Orkney—Originals of two Characters in "The Pirate"—Bessie Millie—Garden of Gow, the "Pirate"—Childhood's Scene of Byron's "Torquil"—The Author's Introduction to his Sister—A German Visitor—German and Scotch Sabbath-keeping habits contrasted—Mr. Watt's Specimens of Fossil Remains—The only new Organism found in Orkney—Back to Kirkwall—to Wick—Vedder's Ode to Orkney.

[Pg 15]

THE CRUISE OF THE BETSEY.

CHAPTER I.

Preparation—Departure—Recent and Ancient Monstrosities—A Free Church Yacht—Down the Clyde—Jura—Prof. Walker's Experiment—Whirlpool near Scarba—Geological Character of the Western Highlands—An Illustration—Different Ages of Outer and Inner Hebrides—Mt. Blanc and the Himalayas "mere upstarts"—Esdaile Quarries—Oban—A Section through Conglomerate and Slate examined—M'Dougal's Dog-stone—Power of the Ocean to move Rocks—Sound of Mull—The Betsey—The Minister's Cabin—Village of Tobermory—The "Florida," a Wreck of the Invincible Armada—Geologic Exploration and Discovery—At Anchor.

The pleasant month of July had again come round, and for full five weeks I was free. Chisels and hammers, and the bag for specimens, were taken from their corner in the dark closet, and packed up with half a stone weight of a fine *soft* Conservative Edinburgh newspaper, valuable for a quality of preserving old things entire. At noon on St. Swithin's day (Monday the 15th), I was speeding down the Clyde in the Toward Castle steamer, for Tobermory in Mull. In the previous season I had intended passing direct from the Oölitic deposits of the eastern coast of Scotland, to the Oölitic deposits of the Hebrides. But the weeks glided all too quickly away among the ichthyolites of Caithness and Cromarty, and the shells and lignites of Sutherland and Ross. My friend, too, the Rev. Mr. Swanson, of Small Isles, on whose assistance I had reckoned, was [Pg 16] in the middle of his troubles at the time, with no longer a home in his parish, and not yet provided with one elsewhere; and I concluded he would have but little heart, at such a season, for breaking into rocks, or for passing from the too pressing monstrosities of an existing state of things, to the old lapidified monstrosities of the past. And so my design on the Hebrides had to be postponed for a twelvemonth. But my friend, now afloat in his Free Church yacht, had got a home on the sea beside his island charge, which, if not very secure when nights were dark and winds loud, and the little vessel tilted high to the long roll of the Atlantic, lay at least beyond the reach of man's intolerance, and not beyond the protecting care of the Almighty. He had written me that he would run down his vessel from Small Isles to meet me at Tobermory, and in consequence of the arrangement I was now on my way to Mull.

St. Swithin's day, so important in the calendar of our humbler meteorologists, had in this part of the country its alternate fits of sunshine and shower. We passed gaily along the green banks of the Clyde, with their rich flat fields glittering in moisture, and their lines of stately trees, that, as the light flashed out, threw their shadows over the grass. The river expanded into the estuary, the estuary into the open sea; we left behind us beacon, and obelisk, and rock-perched castle;—

"Merrily down we drop
Below the church, below the tower,
Below the light house top,"

and, as the evening fell, we were ploughing the outer reaches of the Frith, with the ridgy table-land of Ayrshire stretching away, green, on the one side, and the serrated peaks of Arran rising dark and high on the other. At sunrise next morning our boat lay, unloading a portion of her cargo, in one of the ports of Islay, and we could see the [Pg 17] Irish coast resting on the horizon to the south and west, like a long undulating bank of thin blue cloud; with the island of Rachrin—famous for the asylum it had afforded the Bruce when there was no home for him in Scotland,—presenting in front its mass of darker azure. On and away! We swept past Islay, with its low fertile hills of mica-schist and slate; and Jura, with its flat dreary moors, and its far-seen gigantic paps, on one of which, in the last age, Professor Walker, of Edinburgh, set water a-boil with six degrees of heat less than he found necessary for the purpose on the plain below. The Professor describes the view from the summit, which includes in its wide circle at once the Isle of Skye and the Isle of Man, as singularly noble and imposing; two such prospects more, he says, would bring under the eye the whole island of Great Britain, from the Pentland Frith to the English Channel. We sped past Jura. Then came the Gulf of Coryvrekin, with the bare mountain island of Scarba overlooking the fierce, far-famed whirlpool, that we could see from the deck, breaking in long lines of foam, and sending out its waves in wide rings on every side, when not a speck of white was visible elsewhere in the expanse of sea around us. And then came an opener space, studded with smaller islands,—mere hill-tops rising out of the sea, with here and there insulated groups of pointed rocks, the skeletons of perished hills, amid which the tide chafed and fretted, as if laboring to complete on the broken remains their work of denudation and ruin.

The disposition of land and water on this coast suggests the idea that the Western Highlands, from the line in the interior, whence the rivers descend to the Atlantic, with the islands beyond to the outer Hebrides, are all parts of one great mountainous plane, inclined slantways into the sea. First, the long withdrawing valleys of the main land, with their brown mossy streams, change their character as [Pg 18] they clip beneath the sea-level, and become salt-water lochs. The lines of hills that rise over them jut out as promontories, till cut off by some transverse valley, lowered still more deeply into the brine, and that exists as a kyle, minch, or sound, swept twice every tide by powerful currents. The sea deepens as the plain slopes downward; mountain-chains stand up out of the water as larger islands, single mountains as smaller ones, lower eminences as mere groups of pointed rocks; till at length, as we pass outwards, all trace of the submerged land disappears, and the wide ocean stretches out and away its unfathomable depths. The model of some Alpine country raised in plaster on a flat board, and tilted slantways, at a low angle, into a basin of water, would exhibit, on a minute scale, an appearance exactly similar to that presented by the western coast of Scotland and the Hebrides. The water would rise along the hollows, longitudinal and transverse, forming sounds and lochs, and surround, island-like, the more deeply submerged eminences. But an examination of the geology of the coast, with its promontories and islands, communicates a different idea. These islands and promontories prove to be of very various ages and origin. The *outer* Hebrides may have existed as the inner skeleton of some ancient country, contemporary with the main land, and that bore on its upper soils the productions of perished creations, at a time when by much the larger portion of the *inner* Hebrides,—Skye, and Mull, and the Small Isles,—existed as part of the bottom of a wide sound, inhabited by the Cephalopoda and Enaliosaurians of the Lias and the Oölite. Judging from its components, the Long Island, like the Lammermoors and the Grampians, may have been smiling to the sun when the Alps and the Himalaya Mountains lay buried in the abyss; whereas the greater part of Skye and Mull must have been, like these vast mountain-chains of [Pg 19] the Continent, an oozy sea-floor, over which the ligneous productions of the neighboring lands, washed down by the streams, grew heavy and sank, and on which the belemnite dropped its spindle and the ammonite its shell. The idea imparted of *old* Scotland to the geologist here,—of Scotland, proudly, aristocratically, supereminently old,—for it can call Mont Blanc a mere upstart, and Dhawalageri, with its twenty-eight thousand feet of elevation, a heady fellow of yesterday,—is not that of a land settling down by the head, like a foundering vessel, but of a land whose hills and islands, like its great aristocratic families, have arisen from the level in very various ages, and under the operation of circumstances essentially diverse.

We left behind us the islands of Lunga, Luing, and Seil, and entered the narrow Sound of Kerrera, with its border of Old Red conglomerate resting on the clay-slate of the district. We had passed Esdaile near enough to see the workmen employed in the quarries of the island, so extensively known in commerce for their roofing slate, and several small vessels beside them, engaged in loading; and now we had got a step higher in the geological scale, and could mark from the deck the peculiar character of the conglomerate, which, in cliffs washed by the sea, when the binding matrix is softer than the pebbles which it encloses, roughens, instead of being polished, by the action of the waves, and which, along the eastern side of the Sound here, seems as if formed of cannon-shot, of all sizes, embedded in cement. The Sound terminates in the beautiful bay of Oban, so quiet and sheltered, with its two island breakwaters in front,—its semi-circular sweep of hill behind,—its long white-walled village, bent like a bow, to conform to the inflection of the shore,—its mural precipices behind, tapestried with ivy,—its rich patches of green pasture,—its bosky dingles of shrub and tree,—and, perched on the [Pg 20] seaward promontory, its old, time-eaten keep. "In one part of the harbor of Oban," says Dr. James Anderson, in his "Practical Treatise on Peat Moss," (1794), "where the depth of the sea is about twenty fathoms, the bottom is found to consist of quick peat, which affords no safe anchorage." I made inquiry at the captain of the steamer, regarding this submerged deposit, but he had never heard of it. There are, however, many such on the coasts of both Britain and Ireland. We staid at Oban for several hours, waiting the arrival of the Fort William steamer; and, taking out hammer and chisel from my bag, I stepped ashore to question my ancient acquaintance, the Old Red conglomerate, and was fortunate enough to meet on the pier-head, as I landed, one of the best of companions for assisting in such work, Mr. Colin Elder, of Isle Ornsay,—the gentleman who had so kindly furnished my friend Mr. Swanson with an asylum for his family, when there was no longer a home for them in Small Isles. "You are much in luck," he said, after our first greeting: "one of the villagers, in improving his garden, has just made a cut for some fifteen or twenty yards along the face of the precipice behind the village, and laid open the line of junction between the conglomerate and the clay-slate. Let us go and see it."

I found several things worthy of notice in the chance section to which I was thus introduced. The conglomerate lies uncomfortably along the edges of the slate strata, which present under it an appearance exactly similar to that which they exhibit under the rolled stones and shingle of the neighboring shore, where we find them laid bare beside the harbor, for several hundred yards. And, mixed with the pebbles of various character and origin of which the conglomerate is mainly composed, we see detached masses of the slate, that still exhibit on their edges the [Pg 21] identical lines of fracture characteristic of the rock, which they received, when torn from the mass below, myriads of ages before. In the incalculably remote period in which the conglomerate base of the Old Red Sandstone was formed, the clay-slate of this district had been exactly the same sort of rock that it is now. Some long anterior convulsion had upturned its strata, and the sweep of water, mingled with broken fragments of stone, had worn smooth the exposed edges, just as a similar agency wears the edges exposed at the present time. Quarries might have been opened in this rock, as now, for a roofing-slate, had there been quarriers to open them, or houses to roof over; it was in every respect as ancient a looking stone then as in the present late age of the world. There are no sermons that seem stranger or more impressive to one who has acquired just a little of the language in which they are preached, than those which, according to the poet, are to be found in stones; a bit of fractured slate, embedded among a mass of rounded pebbles, proves voluble with ideas of a kind almost too large for the mind of man to grasp. The eternity that hath passed is an ocean without a further shore, and a finite conception may in vain attempt to span it over. But from the beach, strewed with wrecks, on which we stand to contemplate it, we see far out towards the cloudy horizon, many a dim islet and many a pinnacled rock, the sepulchres of successive eras,—the monuments of consecutive creations: the entire prospect is studded over with these landmarks of a hoar antiquity, which, measuring out space from space, constitute the vast whole a province of time; nor can the eye reach to the open, shoreless infinitude beyond, in which only God existed; and, as in a sea-scene in nature, in which headland stretches dim and blue beyond headland, and islet beyond islet, the distance seems not lessened, but increased, by the crowded [Pg 22] objects—we borrow a larger, not a smaller idea of the distant eternity, from the vastness of the measured periods that occur between.

Over the lower bed of conglomerate, which here, as on the east coast, is of great thickness, we find a bed of gray stratified clay, containing a few calcareo-argillaceous nodules. The conglomerate cliffs to the north of the village present appearances highly interesting to the geologist. Rising in a long wall within the pleasure-grounds of Dunolly castle, we find them wooded atop and at the base; while immediately at their feet there stretches out a grassy lawn, traversed by the road from the village to the castle, which sinks with a gradual slope into the existing sea-beach, but which ages ago must have been a sea-beach itself. We see the bases of the precipices hollowed and worn, with all their rents and crevices widened into caves; and mark, at a picturesque angle of the rock, what must have been once an insulated sea-stack, some thirty or forty feet in height, standing up from amid the rank grass, as at one time it stood up from amid the waves. Tufts of fern and sprays of ivy bristle from its sides, once roughened by the serrated kelp-weed and the tangle. The Highlanders call it M'Dougal's Dog-stone, and say that the old chieftains of Lorne made use of it as a post to which to fasten their dogs,—animals wild and gigantic as themselves,—when the hunters were gathering to rendezvous, and the impatient beagles struggled to break away and begin the chase on their own behalf. It owes its existence as a stack—for the precipice in which it was once included has receded from around it for yards—to an immense boulder in its base—by far the largest stone I ever saw in an Old Red conglomerate. The mass is of a rudely rhomboidal form, and measures nearly twelve feet in the line of its largest diagonal. A second huge pebble in the same [Pg 23] detached spire measures four feet by about three. Both have their edges much rounded, as if, ere their deposition in the conglomerate, they had been long exposed to the wear of the sea; and both are composed of an earthy amygdaloidal trap. I have stated elsewhere ["Old Red Sandstone," Chapter XII.], that I had scarce ever seen a stone in the Old Red conglomerate which I could not raise from the ground; and ere I said so I had examined no inconsiderable extent of this deposit, chiefly, however, along the eastern coast of Scotland, where its larger pebbles rarely exceed two hundred weight. How account for the occurrence of pebbles of so gigantic a size here? We can but guess at a solution, and that very vaguely. The islands of Mull and Kerrera form, in the present state of things, inner and outer breakwaters between what is now the coast of Oban and the waves of the Atlantic; but Mull, in the times of even the Oölite, must have existed as a mere sea-bottom; and Kerrera, composed mainly of trap, which has brought with it to the surface patches of the conglomerate, must, when the conglomerate was in forming, have been a mere sea-bottom also. Is it not possible, that when the breakwaters *were not*, the Atlantic *was*, and that its tempests, which in the present time can transport vast rocks for hundreds of yards along the exposed coasts of Shetland and Orkney, may have been the agent here in the transport of these huge pebbles of the Old Red conglomerate? "Rocks that two or three men could not lift," say the Messrs. Anderson of Inverness, in describing the storms of Orkney, "are washed about even on the tops of cliffs, which are between sixty and a hundred feet above the surface of the sea, when smooth; and detached masses of rock, of an enormous size, are well known to have been carried a considerable distance between low and high-water mark.' "A little way from the Brough," says Dr. Patrick [Pg 24] Neill, in his 'Tour through Orkney and Shetland,' "we saw the prodigious effects of a late winter storm: many great stones, one of them of several tons weight, had been tossed up a precipice twenty or thirty feet high, and laid fairly on the green sward." There is something farther worthy of notice in the stone of which the two boulders of the Dog-stack are composed. No species of rock occurs more abundantly in the embedded pebbles of this ancient conglomerate than rocks of the trap family. We find in it trap-porphyries, greenstones, clink-stones, basalts, and amygdalolds, largely mingled with fragments of the granitic, clay-slate, and quartz rocks. The Plutonic agencies must have been active in the locality for periods amazingly protracted; and many of the masses protruded at a very early time seem identical in their composition with rocks of the trap family, which in other parts of the country we find referred to much later eras. There occur in this deposit rolled pebbles of a basalt, which in the neighborhood of Edinburgh would be deemed considerably more modern than the times of the Mountain Limestone, and in the Isle of Skye, considerably more modern than the times of the Oölite.

The sunlight was showering its last slant rays on island and loch, and then retreating upwards along the higher hills, chased by the shadows, as our boat quitted the bay of Oban, and stretched northwards, along the end of green Lismore, for the Sound of Mull. We had just enough of day left, as we reached mid sea, to show us the gray fronts of the three ancient castles,-- which at this point may be at once seen from the deck,—Dunolly, Duart, and Dunstaffnage; and enough left us as we entered the Sound, to show, and barely show, the Lady Rock, famous in tradition, and made classic by the pen of Campbell, raising its black back amid the tides, like a [Pg 25] belated porpoise. And then twilight deepened into night, and we went snorting through the Strait with a stream of green light curling off from either bow in the calm, towards the high dim land, that seemed standing up on both sides like tall hedges over a green lane. We entered the Bay of Tobermory about midnight, and cast anchor amid a group of little vessels. An exceedingly small boat shot out from the side of a yacht of rather diminutive proportions, but tautly rigged for her size, and bearing an outrigger astern. The water this evening was full of phosphoric matter, and it gleamed and sparkled around the little boat like a northern aurora around a dark cloudlet. There was just light enough to show that the oars were plied by a sailor-like man in a Guernsey frock, and that another sailor-like man,—the skipper, mayhap,—attired in a cap and pea-jacket, stood in the stern. The man in the Guernsey frock was John Stewart, sole mate and half the crew of the Free Church yacht Betsey; and the skipper-like man in the pea-jacket was my friend the minister of the Protestants of Small Isles. In five minutes more I was sitting with Mr. Elder beside the little iron stove in the cabin of the Betsey; and the minister, divested of his cap and jacket, but still looking the veritable skipper to admiration, was busied in making us a rather late tea.

The cabin,—my home for the greater part of the three following weeks, and that of my friend for the greater part of the previous twelvemonth,—I found to be an apartment about twice the size of a common bed, and just lofty enough under the beams to permit a man of five feet eleven to stand erect in his night-cap. A large table, lashed to the floor, furnished with tiers of drawers of all sorts and sizes, and bearing a writing desk bound to it a-top, occupied the middle space, leaving just room enough [Pg 26] for a person to pass between its edges and the narrow coffin-like beds in the sides, and space enough at its fore-end for two seats in front of the stove. A jealously barred skylight opened above; and there depended from it this evening a close lantern-looking lamp, sufficiently valuable, no doubt, in foul weather, but dreary and dim on the occasions when all one really wished from it was light. The peculiar furniture of the place gave evidence to the mixed nature of my friend's employment. A well-thumbed chart of the Western Islands lay across an equally well-thumbed volume of Henry's "Commentary." There was a Polyglot and a spy-glass in one corner, and a copy of Calvin's "Institutes," with the latest edition of "The Coaster's Sailing Directions," in another; while in an adjoining state-room, nearly large enough to accommodate an arm-chair, if the chair could have but contrived to get into it, I caught a glimpse of my friend's printing press and his case of types, canopied overhead by the blue ancient of the vessel, bearing, in stately six-inch letters of white bunting, the legend, "Free Church Yacht." A door opened, which communicated with the forecastle, and John Stewart, stooping very much, to accommodate himself to the low-roofed passage, thrust in a plate of fresh herrings, splendidly toasted, to give substantiality and relish to our tea. The little rude forecastle, a considerably smaller apartment than the cabin, was all a-glow with the bright fire in the coppers, itself invisible; we could see the chain-cable dangling from the hatchway to the floor, and John Stewart's companion, a powerful-looking, handsome young man, with broad bare breast, and in his shirt-sleeves, squatted full in front of the blaze, like the household goblin described by Milton, or the "Christmas Present" of Dickens. Mr. Elder left us for the steamer, in which he prosecuted his voyage next morning to Skye; and we [Pg 27] tumbled in, each to his narrow bed,—comfortable enough sort of resting places, though not over soft; and slept so soundly, that we failed to mark Mr. Elder's return for a few seconds, a little after daybreak. I found at my bedside, when I awoke, a fragment of rock which he had brought from the shore, charged with Liasic fossils; and a note he had written, to say that the deposit to which it belonged occurred in the trap immediately above the village-mill; and further, to call my attention to a house near the middle of the village, built of a mouldering red sandstone, which had been found *in situ* in digging the foundations. I had but little time for the work of exploration in Mull, and the information thus kindly rendered enabled me to economize it.

The village of Tobermory resembles that of Oban. A quiet bay has its secure island-breakwater in front; a line of tall, well-built houses, not in the least rural in their aspect, but that seem rather as if they had been transported from the centre of some stately city entire and at once, sweeps round its inner inflection, like a bent bow; and an amphitheatre of mingled rock and wood rises behind. With all its beauty, however, there hangs about the village an air of melancholy. Like some of the other western coast villages, it seems not to have grown, piece-meal, as a village ought, but to have been made wholesale, as Frankenstein made his man; and to be ever asking, and never more incessantly than when it is at its quietest, why it should have been made at all? The remains of the Florida, a gallant Spanish ship, lie off its shores, a wreck of the Invincible Armada, "deep whelmed," according to Thompson,

> "What time,
> Snatched sudden by the vengeful blast,
> [Pg 28] The scattered vessels drove, and on blind shelve,
> And pointed rock that marks th' indented shore,

> Relentless dashed, where loud the northern main
> Howls through the fractured Caledonian isles."

Macculloch relates, that there was an attempt made, rather more than a century ago, to weigh up the Florida, which ended in the weighing up of merely a few of her guns, some of them of iron greatly corroded; and that, on scraping them, they became so hot under the hand that they could not be touched, but that they lost this curious property after a few hours' exposure to the air. There have since been repeated instances elsewhere, he adds, of the same phenomenon, and chemistry has lent its solution of the principles on which it occurs; but, in the year 1740, ere the riddle was read, it must have been deemed a thoroughly magical one by the simple islanders of Mull. It would seem as if the guns, heated in the contest with Drake, Hawkins, and Frobisher, had again kindled, under some supernatural influence, with the intense glow of the lost battle.

The morning was showery; but it cleared up a little after ten, and we landed to explore. We found the mill a little to the south of the village, where a small stream descends, all foam and uproar, from the higher grounds along a rocky channel half-hidden by brushwood; and the Liasic bed occurs in an exposed front directly over it, coped by a thick bed of amygdaloidal trap. The organisms are numerous; and, when we dig into the bank beyond the reach of the weathering influences, we find them delicately preserved, though after a fashion that renders difficult their safe removal. Originally the bed must have existed as a brown argillaceous mud, somewhat resembling that which forms in the course of years, under a scalp of muscles; and it has hardened into a more silt-like clay, in which the fossils occur, not as petrifactions, [Pg 29] but as shells in a state of decay, except in some rare cases, in which a calcareous nodule has formed within or around them. Viewed in the group, they seem of an intermediate character, between the shells of the Lias and the Oölite. One of the first fossils I disinterred was the Gryphæa obliquata,—a shell characteristic of the Liasic formation; and the fossil immediately after, the Pholadomy æqualis, a shell of the

Oölitic one. There occurs in great numbers a species of small Pecten,—some of the specimens scarce larger than a herring scale; a minute Ostrea, a sulcated Terebratula, an Isocardia, a Pullastra, and groups of broken serpulæ in vast abundance. The deposit has also its three species of Ammonite, existing as mere impressions in the clay; and at least two species of Belemnite,—one of the two somewhat resembling the Belemnites abbreviatus, but smaller and rather more elongated: while the other, of a spindle form, diminishing at both ends, reminds one of the Belemnites minimus of the Gault. The Red Sandstone in the centre of the village occurs detached, like this Liasic bed, amid the prevailing trap, and may be seen *in situ* beside the southern gable of the tall, deserted looking house at the hill-foot, that has been built of it. It is a soft, coarse-grained, mouldering stone, ill fitted for the purposes of the architect; and more nearly resembles the New Red Sandstone of England and Dumfriesshire, than any other rock I have yet seen in the north of Scotland. I failed to detect in it aught organic.

We weighed anchor about two o'clock, and beat gallantly out the Sound, in the face of an intermittent baffling wind and a heavy swell from the sea. I would fain have approached nearer the precipices of Ardnamurchan, to trace along their inaccessible fronts the strange reticulations of trap figured by Macculloch; but prudence and the skipper forbade our trusting even the docile little [Pg 30] Betsey, on one of the most formidable lee shores in Scotland, in winds so light and variable, and with the swell so high. We could hear the deep roar of the surf for miles, and see its undulating strip of white flickering under stack and cliff. The scenery here seems rich in legendary association. At one tack we bore into Bloody Bay, on the Mull coast,—the scene of a naval battle between two island chiefs; at another, we approached, on the mainland, a cave inaccessible save from the sea, long the haunt of a ruthless Highland pirate. Ere we rounded the headland of Ardnamurchan, the slant light of evening was gleaming athwart the green acclivities of Mull, barring them with long horizontal lines of shadow, where the trap terraces rise step beyond step, in the characteristic stair-like arrangement to which the rock owes its name; and the sun set as we were bearing down in one long tack on the Small Isles. We passed the Isle of Muck, with its one low hill; saw the pyramidal mountains of Rum looming tall in the offing; and then, running along the Isle of Eigg, with its colossal Scuir rising between us and the sky, as if it were a piece of Babylonian wall, or of the great wall of China, only vastly larger, set down on the ridge of a mountain, we entered the channel which separates the island from one of its dependencies, Eilean Chaisteil, and cast anchor in the tideway, about fifty yards from the rocks. We were now at home,—the only home which the proprietor of the island permits to the islanders' minister; and, after getting warm and comfortable over the stove and a cup of tea, we did what all sensible men do in their own homes when the night wears late,—got into bed.

[Pg 31]

CHAPTER II.

The Minister's Larder—No Harbor—Eigg Shoes—*Tormentilla erecta*—For the *Witness*' Sake—Eilean Chaisteil—Appearance of Eigg—Chapel of St. Donan—Shell-sand—Origin of Secondary Calcareous Rock suggested—Exploration of Eigg—Pitchstone Veins—A Bone Cave—Massacre at Eigg—Grouping of Human Bones in the Cave—Relics—The Horse's Tooth—A Copper Sewing Needle—Teeth found—Man a worse Animal than his Teeth show him to have been designed for—Story of the Massacre—Another Version—Scuir of Eigg—The Scuir a Giant's Causeway—Character of the Columns—Remains of a Prostrate Forest.

We had rich tea this morning. The minister was among his people; and our first evidence of the fact came in the agreeable form of three bottles of fine fresh cream from the shore. Then followed an ample baking of nice oaten cakes. The material out of which the cakes were manufactured had been sent from the minister's store aboard,—for oatmeal in Eigg is rather a scarce commodity in the middle of July; but they had borrowed a crispness and flavor from the island, that the meal, left to its own resources, could scarcely have communicated; and the golden-colored cylinder of fresh butter which accompanied them was all the island's own. There was an ample supply of eggs too, as one not quite a conjuror might have expected from a country bearing such a name,—eggs with the milk in them; and, with cream, butter, oaten cakes, eggs, and tea, all of the best, and with sharp-set sea-air appetites to boot, we fared sumptuously. There is properly no harbor in the island. We lay in a narrow channel, through which, twice every twenty-four hours, the tides sweep powerfully in one direction, and then as powerfully [Pg 32] in the direction opposite; and our anchors had a trick of getting foul, and canting stock downwards in the loose sand, which, with pointed rocks all around us, over which the current ran races, seemed a very shrewd sort of trick indeed. But a kedge and halser, stretched thwartwise to a neighboring crag, and jammed fast in a crevice, served in moderate weather to keep us tolerably right. In the severer seasons, however, the kedge is found inadequate, and the minister has to hoist sail and make out for the open sea, as if served with a sudden summons of ejectment.

Among the various things brought aboard this morning, there was a pair of island shoes for the minister's cabin use, that struck my fancy not a little. They were all around of a deep madder red color, soles, welts and uppers; and, though somewhat resembling in form the little yawl of the Betsey, were sewed not unskilfully with thongs; and their peculiar style of tie seemed of a kind suited to furnish with new idea a fashionable shoemaker of the metropolis. They were altogether the production of Eigg, from the skin out of which they had been cut, with the lime that had prepared it for the tan, and the root by which the tan had been furnished, down to the last on which they had been moulded, and the artisan that had cast them off, a pair of finished shoes. There are few trees, and, of course, no bark to spare, in the island; but the islanders find a substitute in the astringent lobiferous root of the *Tormentilla erecta,* which they dig out for the purpose among the heath, at no inconsiderable expense of time and trouble. I was informed by John Stewart, an adept in all the multifarious arts of the island, from the tanning of leather and the tilling of land, to the building of a house or the working of a ship, that the infusion of root had to be thrice changed for every skin, and that it [Pg 33] took a man nearly a day to gather roots enough for a single infusion. I was further informed that it was not unusual for the owner of a skin to give it to some neighbor to tan, and that, the process finished, it was divided equally between them, the time and trouble bestowed on it by the one being deemed equivalent to the property held in it by the other. I wished to call a pair of these primitive-looking shoes my own, and no sooner was the wish expressed, than straightway one islander furnished me with leather, and another set to work upon the shoes. When I came to speak of remuneration, however, the islanders shook their heads. "No, no, not from the *Witness*: there are not many that take our part, and the *Witness* does." I hold the shoes, therefore, as my first retainer, determined, on all occasions of just quarrel, to make common cause with the poor islanders.

The view from the anchoring ground presents some very striking features. Between us and the sea lies Eilean Chaisteil, a rocky trap islet, about half a mile in length by a few hundred yards in breadth; poor in pastures, but peculiarly rich in sea-weed, of which John Stewart used, he informed me, to make finer kelp, ere the trade was put down by act of Parliament, than could be made elsewhere in Eigg. This islet bore, in the remote past, its rude fort or dun, long since sunk into a few grassy mounds; and hence its name. On the landward side rises the island of Eigg proper, resembling in outline two wedges, placed point to point on a board. The centre is occupied by a deep angular gap, from which the ground slopes upward on both sides, till, attaining its extreme height at the opposite ends of the island, it drops suddenly on the sea. In the northern rising ground the wedge-like outline is complete; in the southern one it is somewhat modified by the gigantic Scuir, which rises direct on the apex of the height, *i.e.*, the [Pg 34] thick part of the wedge; and which, seen bows-on from this point of view, resembles some vast donjon keep, taller, from base to summit, by about a hundred feet, than the dome of St. Paul's. The upper slopes of the island are brown and moory, and present little on which the eye may rest, save a few trap terraces, with rudely columnar fronts; its middle space is mottled with patches of green, and studded with dingy cottages, each of which this morning, just a little before the breakfast hour, had its own blue cloudlet of smoke diffused around it; while along the beach, patches of level sand, alternated with tracts of green bank, or both, give place to stately ranges of basaltic columns, or dingy groups of detached rocks. Immediately in front of the central hollow, as if skilfully introduced, to relieve the tamest part of the prospect, a noble wall of semi-circular columns rises some eighty or a hundred feet over the shore; and on a green slope, directly above, we see the picturesque ruins of the Chapel of St. Donan, one of the disciples of Columba, and the Culdee saint and apostle of the island.

One of the things that first struck me, as I got on deck this morning, was the extreme whiteness of the sand. I could see it gleaming bright through the transparent green of the sea, three fathoms below our keel, and, in a little flat bay directly opposite, it presented almost the appearance of pulverized chalk. A stronger contrast to the dingy trap-rocks around which it lies could scarce be produced, had contrast for effect's sake been the object. On landing on the exposed shelf to which we had fastened our halser, I found the origin of the sand interestingly exhibited. The hollows of the rock, a rough trachyte, with a surface like that of a steel rasp, were filled with handfuls of broken shells thrown up by the surf from the sea-banks beyond: fragments of echini, bits of the valves of razor-fish, the [Pg 35] island cyprina, mactridæ, buccinidæ, and fractured periwinkles, lay heaped together in vast abundance. In hollow after hollow, as I passed shorewards, I found the fragments more and more comminuted, just as, in passing along the successive vats of a paper-mill, one finds the linen rags more and more disintegrated by the cylinders; and immediately beyond the inner edge of the shelf, which is of considerable extent, lies the flat bay, the ultimate recipient of the whole, filled to the depth of several feet, and to the extent of several hundred yards, with a pure shell-sand, the greater part of which had been thus washed ashore in handfuls, and ground down by the blended agency of the trachyte and the surf. Once formed, however, in this way it began to receive accessions from the exuviæ of animals that love such localities,—the deep arenaceous bed and soft sand-beach; and these now form no inconsiderable proportion of the entire mass. I found the deposit thickly inhabited by spatangi, razor-fish, gapers, and large, well-conditioned cockles, which seemed to have no idea whatever that they were living amid the debris of a charnel house. Such has been the origin here of a bed of shell-sand, consisting of many thousand tons, and of which at least eighty per cent. was once associated with animal life. And such, I doubt not, is the history of many a calcareous rock in the later secondary formations. There are strata, not a few, of the Cretaceous and Oölitic groups, that would be found—could we but trace their beginnings with a certainty and clearness equal to that with which we can unravel the story of this deposit—to be, like it, elaborations from dead matter, made through the agency of animal secretion.

We set out on our first exploratory ramble in Eigg an hour before noon. The day was bracing and breezy, and a clear sun looked cheerily down on island, and strait, and [Pg 36] blue open sea. We rowed southwards in our little boat, through the channel of Eilean Chaisteil, along the trap-rocks of the island, and landed under the two pitchstone veins of Eigg, so generally known among mineralogists, and of which specimens may be found in so many cabinets. They occur in an earthy, greenish-black amygdaloid, which forms a range of sea-cliffs varying in height from thirty to fifty feet, and that, from their sad hue and dull fracture, seem to absorb the light; while the veins themselves, bright and glistening, glitter in the sun, as if they were streams of water traversing the face of the rock. The first impression they imparted, in viewing them from the boat, was, that the inclosing mass was a pitch caldron, rather of the roughest and largest, and much begrimmed by soot, that had cracked to the heat, and that the fluid pitch was forcing its way outwards through the rents. The veins expand and contract, here diminishing to a strip a few inches across, there widening into a comparatively broad belt, some two or three feet over; and, as well described by M'Culloch, we find the inclosed pitchstone changing in color, and assuming a lighter or darker hue, as it nears the edge or recedes from it. In the centre it is of a dull olive green, passing gradually into blue, which in turn deepens into black; and it is exactly at the point of contact with the earthy amygdaloid that the black is most intense, and the fracture of the stone glassiest and brightest. I was lucky enough to detach a specimen, which, though scarce four inches across, exhibits the three colors characteristic of the vein,—its bar of olive green on the one side, of intense black on the other, and of blue, like that of imperfectly fused bottle-glass, in the centre. This curious rock,—so nearly akin in composition and appearance to obsidian,—a mineral which, in its dense form, closely resembles the coarse dark-colored glass of which common bottles are [Pg 37] made, and which, in its lighter form, exists as pumice,—constitutes one of the links that connect the trap with the unequivocally volcanic rocks. The one mineral may be seen beside smoking crater, as in the Lipari Isles, passing into pumice; while the other may be converted into a substance almost identical with pumice, by the chemist. "It is stated by the Honorable George Knox, of Dublin," says Mr. Robert Allan, in his valuable mineralogical work, "that the pitchstone of Newry, on being exposed to a high temperature, loses its bitumen and water, and is converted into a light substance in every respect resembling pumice." But of pumice in connection with the pitchstones of Eigg, more anon.

Leaving our boat to return to the Betsey at John Stewart's leisure, and taking with us his companion, to assist us in carrying such specimens as we might procure, we passed westwards for a few hundred yards under the crags, and came abreast of a dark angular opening at the base of the precipice, scarce two feet in height, and in front of which there lies a little sluggish, ankle-deep pool, half mud, half water, and matted over with grass and rushes. Along the mural face of the rock of earthy amygdaloid there runs a nearly vertical line, which in one of the stratified rocks one might perhaps term the line of a fault, but which in a trap rock may merely indicate where two semi-molten masses had pressed against each other without uniting—just as currents of cooling lead, poured by the plumber from the opposite end of a groove, sometimes meet and press together, so as to make a close, polished joint, without running into one piece. The little angular opening forms the lower termination of the line, which, hollowing inwards, recedes near the bottom into a shallow cave, roughened with tufts of fern and bunches of long silky grass, here and there enlivened by the delicate flowers of [Pg 38] the lesser rock-geranium. A shower of drops patters from above among the weeds and rushes of the little pool. My friend the minister stopped short. "There," he said, pointing to the hollow, "you will find such a bone cave as you never saw before. Within that opening there lie the remains of an entire race, palpably destroyed, as geologists in so many other cases are content merely to imagine, by one great catastrophe. That is the famous cave of Frances (*Uamh Fraingh*), in which the whole people of Eigg were smoked to death by the M'Leods."

We struck a light, and, worming ourselves through the narrow entrance, gained the interior,—a true rock gallery, vastly more roomy and lofty than one could have anticipated from the mean vestibule placed in front of it. Its extreme length we found to be two hundred and sixty feet; its extreme breadth twenty-seven feet; its height, where the roof rises highest, from eighteen to twenty feet. The cave seems to have owed its origin to two distinct causes. The trap-rocks on each side of the vertical fault-like crevice which separates them are greatly decomposed, as if by the moisture percolating from above; and directly in the line of the crevice must the surf have charged, wave after wave, for ages ere the last upheaval of the land. When the Dog-stone at Dunolly existed as a sea-stack, skirted with algæ, the breakers on this shore must have dashed every tide through the narrow opening of the cavern, and scooped out by handfuls the decomposing trap within. The process of decomposition, and consequent enlargement, is still going on inside, but there is no longer an agent to sweep away the disintegrated fragments. Where the roof rises highest, the floor is blocked up with accumulations of bulky decaying masses, that have dropped from above; and it is covered over its entire area by a stratum of earthy rubbish, which has fallen from the sides and ceiling in such [Pg 39] abundance, that it covers up the straw beds of the perished islanders, which still exist beneath as a brown mouldering felt, to the depth of from five to eight inches. Never yet was tragedy enacted on a gloomier theatre. An uncertain twilight glimmers gray at the entrance, from the narrow vestibule; but all within, for full two hundred feet, is black as with Egyptian darkness. As we passed onward with our one feeble light, along the dark mouldering walls and roof, which absorbed every straggling ray that reached them, and over the dingy floor, ropy and damp, the place called to recollection that hall in Roman story, hung and carpeted with black, into which Domitian once thrust his senate, in a frolic, to read their own names on the coffin-lids placed against the wall. The darkness seemed to press upon us from every side, as if it were a dense jetty fluid, out of which our light had scooped a pailful or two, and that was rushing in to supply the vacuum; and the only objects we saw distinctly visible were each other's heads and faces, and the lighter parts of our dress.

The floor, for about a hundred feet inwards from the narrow vestibule, resembles that of a charnel-house. At almost every step we came upon heaps of human bones grouped together, as the Psalmist so graphically describes, "as when one cutteth and cleaveth wood upon the earth." They are of a brownish, earthy hue, here and there tinged with green; the skulls, with the exception of a few broken fragments, have disappeared; for travellers in the Hebrides have of late years been numerous and curious; and many a museum,—that at Abbotsford among the rest,—exhibits, in a grinning skull, its memorial of the Massacre at Eigg. We find, too, further marks of visitors in the single bones separated from the heaps and scattered over the area; but enough still remains to show, in the general disposition of the remains, that the hapless islanders died under the walls [Pg 40] in families, each little group separated by a few feet from the others. Here and there the remains of a detached skeleton may be seen, as if some robust islander, restless in his agony, had stalked out into the middle space ere he fell; but the social arrangement is the general one. And beneath every heap we find, at the depth, as has been said, of a few inches, the remains of the straw-bed upon which the family had lain, largely mixed with the smaller bones of the human frame, ribs and vertebræ, and hand and feet bones; occasionally, too, with fragments of unglazed pottery, and various other implements of a rude housewifery. The minister found for me, under one family heap, the pieces of a half-burned, unglazed earthen jar, with a narrow mouth, that, like the sepulchral urns of our ancient tumuli, had been moulded by the hand, without the assistance of the potter's wheel; and to one of the fragments there stuck a minute pellet of gray hair. From under another heap he disinterred the handle-stave of a child's wooden porringer (*bicker*), perforated by a hole still bearing the mark of the cord that had hung it to the wall; and beside the stave lay a few of the larger, less destructible bones of the child, with what for a time puzzled us both not a little,—one of the grinders of a horse. Certain it was, no horse could have got there to have dropped a tooth,—a foal of a week old could not have pressed itself through the opening; and how the single grinder, evidently no recent introduction into the cave, could have got mixed up in the straw with the human bones, seemed an enigma somewhat of the class to which the reel in the bottle belongs. I found in Edinburgh an unexpected commentator on the mystery, in the person of my little boy,—an experimental philosopher in his second year. I had spread out on the floor the curiosities of Eigg,—among the rest, the relics of the cave, including the pieces of earthern jar, and the fragment of [Pg 41] the porringer; but the horse's tooth seemed to be the only real curiosity among them in the eyes of little Bill. He laid instant hold of it; and, appropriating it as a toy, continued playing with it till he fell asleep. I have now little doubt that it was first brought into the cave by the poor child amid whose mouldering remains Mr. Swanson found it. The little pellet of gray hair spoke of feeble old age involved in this wholesale massacre with the vigorous manhood of the island; and here was a story of unsuspecting infancy amusing itself on the eve of destruction with its toys. Alas, for man! "Should not I spare Nineveh, that great city," said God to the angry prophet, "wherein are more than six score thousand persons that cannot discern between their right hand and their left?" God's image must have been sadly defaced in the murderers of the poor inoffensive children of Eigg, ere they could have heard their feeble wailings, raised, no doubt, when the stifling atmosphere within began first to thicken, and yet ruthlessly persist in their work of indiscriminate destruction.

Various curious things have from time to time been picked up from under the bones. An islander found among them, shortly before our visit, a sewing needle of copper, little more than an inch in length; fragments of Eigg shoes, of the kind still made in the island, are of comparatively common occurrence; and Mr. James Wilson relates, in the singularly graphic and powerful description of *Uamh Fraingh*, which occurs in his "Voyage round the Coasts of Scotland" (1841), that a sailor, when he was there, disinterred, by turning up a flat stone, a "buck-tooth" and a piece of money,—the latter a rusty copper coin, apparently of the times of Mary of Scotland. I also found a few teeth; they were sticking fast in a fragment of jaw; and, taking it for granted, as I suppose I may, that the dentology of the murderous M'Leods outside the cave must have very [Pg 42] much resembled that of the murdered M'Donalds within, very harmless looking teeth they were for being those of an animal so maliciously mischievous as man. I have found in the Old Red Sandstone the strong-based tusks of the semi-reptile Holoptychius; I have chiselled out of the limestone of the Coal Measures the sharp, dagger-like incisors of the Megalichthys; I have picked up in the Lias and Oölite the cruel spikes of the Crocodile and the Ichthyosaurus; I have seen the trenchant, saw-edged teeth of gigantic Cestracions and Squalidæ that had been disinterred from the Chalk and the London Clay; and I have felt, as I examined them, that there could be no possibility of mistake regarding the nature of the creatures to which they had belonged;—they were teeth made for hacking, tearing, mangling,—for amputating limbs at a bite, and laying open bulky bodies with a crunch; but I could find no such evidence in the human jaw, with its three inoffensive looking grinders, that the animal it had belonged to,—far more ruthless and cruel than reptile-fish, crocodiles, or sharks,—was of such a nature that it could destroy creatures of even its own kind by hundreds at a time, when not in the least incited by hunger, and with no ultimate intention of eating them. Man must surely have become an immensely worse animal than his teeth show him to have been designed for; his teeth give no real evidence regarding his real character. Who, for instance, could gather from the dentology of the M'Leods the passage in their history to which the cave of Frances bears evidence?

We quitted the cave, with its stagnant damp atmosphere and its mouldy unwholesome smells, to breathe the fresh sea-air on the beach without. Its story, as recorded by Sir Walter in his "Tales of a Grandfather," and by Mr. Wilson, in his "Voyage," must be familiar to the reader; and I learned from my friend, versant in all the various [Pg 43] island traditions regarding it, that the less I inquired into its history on the spot, the more was I likely to feel satisfied that I knew something about it. There seem to have been no chroniclers, in this part of the Hebrides, in the rude age of the unglazed pipkin and the copper needle; and many years seem to have elapsed ere the story of their hapless possessors was committed to writing; and so we find it existing in various and somewhat conflicting editions. "Some hundred years ago," says Mr. Wilson, "a few of the M'Leods landed in Eigg from Skye, where, having greatly misconducted themselves, the Eiggites strapped them to their own boats, which they sent adrift into the ocean. They were, however, rescued by some clansmen; and, soon after, a strong body of the M'Leods set sail from Skye, to revenge themselves on Eigg. The natives of the latter island feeling they were not of sufficient force to offer resistance, went and hid themselves (men, women, and children) in this secret cave, which is narrow, but of great subterranean length, with an exceedingly small entrance. It opens from the broken face of a steep bank along the shore; and, as the whole coast is cavernous, their particular retreat would have been sought for in vain by strangers. So the Skye-men, finding the island uninhabited, presumed the natives had fled, and satisfied their revengeful feelings by ransacking and pillaging the empty houses. Probably the *movables* were of no great value. They then took their departure and left the island, when the sight of a solitary human being among the cliffs awakened their suspicion, and induced them to return. Unfortunately a slight sprinkling of snow had fallen, and the footsteps of an individual were traced to the mouth of the cave. Not having been there ourselves at the period alluded to, we cannot speak with certainty as to the nature of the parley which ensued, or the terms [Pg 44] offered by either party; but we know that those were not the days of protocols. The ultimatum was unsatisfactory to the Skye-men, who immediately proceeded to 'adjust the preliminaries' in their own way, which adjustment consisted in carrying a vast collection of heather, ferns, and other combustibles, and making a huge fire just in the very entrance of the *Uamh Fraingh*. which they kept up for a length of time; and thus, by 'one fell smoke,' they smothered the entire population of the island."

Such is Mr. Wilson's version of the story, which, in all its leading circumstances, agrees with that of Sir Walter. According, however, to at least one of the Eigg versions, it was the M'Leod himself who had landed on the island, driven there by a storm. The islanders, at feud with the M'Leod's at the time, inhospitably rose upon him, as he bivouacked on the shores of the Bay of Laig; and in a fray, in which his party had the worse, his back was broken, and he was forced off half dead to sea. Several months after, on his partial recovery, he returned, crook-backed and infirm, to wreak his vengeance on the inhabitants, all of whom, warned of his coming by the array of his galleys in the offing, hid themselves in the cave, in which, however, they were ultimately betrayed—as narrated by Sir Walter and Mr. Wilson—by the track of some footpaths in a sprinkling of snow; and the implacable chieftain, giving orders on the discovery, to unroof the houses in the neighborhood, raised high a pile of rafters against the opening, and set it on fire. And there he stood in front of the blaze, hump-backed and grim, till the wild, hollow cry from the rock within had sunk into silence, and there lived not a single islander of Eigg, man, woman, or child. The fact that their remains should have been left to moulder in the cave is proof enough, of itself, that none survived to bury the [Pg 45] dead. I am inclined to believe, from the appearance of the place, that smoke could scarcely have been the real agent of destruction; then, as now, it would have taken a great deal of pure smoke to smother a Highlander. It may be perhaps deemed more probable, that the huge fire of rafter and roof-tree piled close against the opening, and rising high over it, would draw out the oxygen within as its proper food, till at length all would be exhausted; and life would go out for want of it, like the flame of a candle under an upturned jar. Sir Walter refers the date of the event to some time "about the close of the sixteenth century;" and the coin of Queen Mary, mentioned by Mr. Wilson, points at a period at least not much earlier; but the exact time of its occurrence is so uncertain, that a Roman Catholic priest of the Hebrides, in lately showing his people what a very bad thing Protestantism is, instanced, as a specimen of its average morality, the affair of the cave. The *Protestant* M'Leods of Skye, he said, full of hatred in their hearts, had murdered, wholesale, their wretched brethren, the *Protestant* M'Donalds of Eigg, and sent them off to perdition before their time.

Quitting the beach, we ascended the breezy hill-side on our way to the Scuir,—an object so often and so well described, that it might be perhaps prudent, instead of attempting one description more, to present the reader with some of the already existing ones. "The Scuir of Eigg," says Professor Jamieson, in his 'Mineralogy of the Western Islands,' "is perfectly mural, and extends for upwards of a mile and a half, and rises to a height of several hundred feet. It is entirely columnar, and the columns rise in successive ranges, until they reach the summit, where, from their great height, they appear, when viewed from below, diminutive. Staffa is an object of the [Pg 46] greatest beauty and regularity; the pillars are as distinct as if they had been reared by the hand of art; but it has not the extent or sublimity of the Scuir of Eigg. The one may be compared with the greatest exertions of human power; the other is characteristic of the wildest and most inimitable works of nature." "The height of this extraordinary object is considerable," says M'Culloch, dashing off his sketch with a still bolder hand; "yet its powerful effect arises rather from its peculiar form, and the commanding elevation which it occupies, than from its positive altitude. Viewed in one direction, it presents a long irregular wall, crowning the summit of the highest hill, while in the other it resembles a huge tower. Thus it forms no natural combination of outline with the surrounding land, and hence acquires that independence in the general landscape which increases its apparent magnitude, and produces that imposing effect which it displays. From the peculiar position of the Scuir, it must also inevitably be viewed from a low station. Hence it everywhere towers high above the spectator; while, like other objects on the mountain outline, its apparent dimensions are magnified, and its dark mass defined on the sky, so as to produce all the additional effects arising from strong oppositions of light and shadow. The height of this rock is sufficient in this stormy country frequently to arrest the passage of the clouds, so as to be further productive of the most brilliant effects in landscape. Often they may be seen hovering on its summit, and adding ideal dimensions to the lofty face, or, when it is viewed on the extremity, conveying the impression of a tower, the height of which is such as to lie in the regions of the clouds. Occasionally they sweep along the base, leaving its huge and black mass involved in additional gloom, and resembling the castle of some Arabian enchanter, built on the clouds, and suspended [Pg 47] in air." It might be perhaps deemed somewhat invidious to deal with pictures such as these in the style the connoisseur in the "Vicar of Wakefield" dealt with the old painting, when, seizing a brush, he daubed it over with brown varnish, and then asked the spectators whether he had not greatly improved the tone of the coloring. And yet it is just possible, that in the case of at least M'Culloch's picture, the brown varnish might do no manner of harm. But a homelier sketch, traced out on almost the same leading lines, with just a little less of the aërial in it, may have nearly the same subduing effect; I have, besides, a few curious touches to lay in, which seem hitherto to have escaped observation and the pencil; and in these several circumstances must lie my apology for adding one sketch more to the sketches existing already.

The Scuir of Eigg, then, is a veritable Giant's Causeway, like that on the coast of Antrim, taken and magnified rather more than twenty times in height, and some five or six times in breadth, and then placed on the ridge of a hill nearly nine hundred feet high. Viewed sideways, it assumes, as described by M'Culloch, the form of a perpendicular but ruinous rampart, much gapped above, that runs for about a mile and a quarter along the top of a lofty sloping talus. Viewed endways, it resembles a tall massy tower,—such a tower as my friend, Mr. D.O. Hill, would delight to draw, and give delight by drawing,—a tower three hundred feet in breadth by four hundred and seventy feet in height, perched on the apex of a pyramid, like a statue on a pedestal. This strange causeway is columnar from end to end; but the columns, from their great altitude and deficient breadth, seem mere rodded shafts in the Gothic style; they rather resemble bundles of rods than well-proportioned pillars. Few of them exceed eighteen inches in diameter, and many of them fall short of half a [Pg 48] foot; but, though lost in the general mass of the Scuir as independent columns, when we view it at an angle sufficiently large to take in its entire bulk, they yet impart to it that graceful linear effect which we see brought out in tasteful pencil sketches and good line engravings. We approached it this day from the shore in the direction in which the eminence it stands upon assumes the pyramidal form, and itself the tower-like outline. The acclivity is barren and stony,—a true desert foreground, like those of Thebes and Palmyra; and the huge square shadow of the tower stretched dark and cold athwart it. The sun shone out clearly. One half the immense bulk before us, with its delicate vertical lining, lay from top to bottom in deep shade, massive and gray; one half presented its many-sided columns to the light, here and there gleaming with tints of extreme brightness, where the pitchstones presented their glassy planes to the sun; its general outline, whether pencilled by the lighter or darker tints, stood out sharp and clear; and a stratum of white fleecy clouds floated slowly amid the delicious blue behind it. But the minuter details I must reserve for my next chapter. One fact, however, anticipated just a little out of its order, may heighten the interest of the reader. There are massive buildings,—bridges of noble span, and harbors that abut far into the waves,—founded on wooden piles; and this hugest of hill-forts we find founded on wooden piles also. It is built on what a Scotch architect would perhaps term a pile-*brander* of the *Pinites Eiggensis*, an ancient tree of the Oölite. The gigantic Scuir of Eigg rests on the remains of a prostrate forest.

[Pg 49]

CHAPTER III.

Structure of the Scuir—A stray Column—The Piazza—A buried Pine Forest the Foundation of the Scuir—Geological Poachers in a Fossil Preserve—*Pinites Eiggensis*—Its Description—Witham's Experiments on Fossil Pine of Eigg—Rings of the Pine—Ascent of the Scuir—Appearance of the Top—White Pitchstone—Mr. Greig's Discovery of Pumice—A Sunset Scene—The Manse and the Yacht—The Minister's Story—A Cottage Repast—American Timber drifted to the Hebrides—Agency of the Gulf Stream—The Minister's Sheep.

As we climbed the hill-side, and the Shinar-like tower before us rose higher over the horizon at each step we took, till it seemed pointing at the middle sky, we could mark peculiarities in its structure which escape notice in the distance. We found it composed of various beds, each of which would make a Giant's Causeway entire, piled over each other like stories in a building, and divided into columns, vertical, or nearly so, in every instance except in one bed near the base, in which the pillars incline to a side, as if losing footing under the superincumbent weight. Innumerable polygonal fragments,—single stones of the building,—lie scattered over the slope, composed, like almost all the rest of the Scuir, of a peculiar and very beautiful stone, unlike any other in Scotland—a dark pitchstone-porphyry, which, inclosing crystals of glassy feldspar, resembles in the hand-specimen, a mass of black sealing-wax, with numerous pieces of white bugle stuck into it. Some of the detached polygons are of considerable size; few of them larger and bulkier, however, than a piece of column of this characteristic porphyry, about ten feet in length by two feet in diameter, which lies a full mile away from any of the others, in the [Pg 50] line of the old burying-ground, and distant from it only a few hundred yards. It seems to have been carried there by man: we find its bearing from the Scuir lying nearly at right angles with the direction of the drift-boulders of the western coast, which are, besides, of rare occurrence in the Hebrides; nor has it a single neighbor; and it seems not improbable, as a tradition of the island testifies, that it was removed thus far for the purpose of marking some place of sepulture, and that the catastrophe of the cave arrested its progress after by far the longer and rougher portion of the way had been passed. The dry-arm bones of the charnel-house in the rock may have been tugging around it when the galleys of the M'Leod hove in sight. The traditional history of Eigg, said my friend the minister, compared with that of some of the neighboring islands, presents a decapitated aspect: the M'Leods cut it off by the neck. Most of the present inhabitants can tell which of their ancestors, grandfather, or great-grandfather, or great-great-grandfather, first settled in the place, and where they came from; and, with the exception of a few vague legends about St. Donan and his grave, which were preserved apparently among the people of the other Small Isles, the island has no early traditional history.

We had now reached the Scuir. There occur, intercalated with the columnar beds, a few bands of a buff-colored non-columnar trap, described by M'Culloch as of a texture intermediate between a greenstone and a basalt, and which, while the pitchstone around it seems nearly indestructible, has weathered so freely as to form horizontal grooves along the face of the rock, from two to five yards in depth. One of these runs for several hundred feet along the base of the Scuir, just at the top of the talus, and greatly resembles a piazza, lacking the outer pillars. It is from ten to twelve feet in height, by from fifteen to twenty in depth; the [Pg 51] columns of the pitch stone-bed immediately above it seem perilously hanging in mid air; and along their sides there trickles, in even the driest summer weather,—for the Scuir is a condenser on an immense scale—minute runnels of water, that patter ceaselessly in front of the long deep hollow, like rain from the eaves of a cottage during a thunder shower. Inside, however, all is dry, and the floor is covered to the depth of several inches with the dung of sheep and cattle, that find, in this singular mountain piazza, a place of shelter. We had brought a pickaxe with us; and the dry and dusty floor, composed mainly of a gritty conglomerate, formed the scene of our labors. It is richly fossiliferous, though the organisms have no specific variety; and never, certainly, have I found the remains of former creations in a scene in which they more powerfully addressed themselves to the imagination. A stratum of peat-moss, mixed with fresh-water shells, and resting on a layer of vegetable mould, from which the stumps and roots of trees still protruded, was once found in Italy, buried beneath an ancient tesselated pavement; and the whole gave curious evidence of a kind fitted to picture to the imagination a background vista of antiquity, all the more remotely ancient in aspect from the venerable age of the object in front. Dry ground covered by wood, a lake, a morass, and then dry ground again, had all taken precedence, on the site of the tesselated pavement, in this instance, of an old Roman villa. But what was antiquity in connection with a Roman villa, to antiquity in connection with the Scuir of Eigg? Under the old foundations of this huge wall we find the remains of a pine forest, that, long ere a single bed of the porphyry had burst from beneath, had sprung up and decayed on hill and beside stream in some nameless land,—had then been swept to the sea,—had been entombed deep at the bottom in a grit of Oölite,—had been heaved up to the surface, and high [Pg 52] over it, by volcanic agencies working from beneath,—and had finally been built upon, as moles are built upon piles, by the architect that had laid down the masonry of the gigantic Scuir, in one fiery layer after another. The mountain wall of Eigg, with its dizzy elevation of four hundred and seventy feet, is a wall founded on piles of pine laid crossways; and, strange as the fact may seem, one has but to dig into the floor of this deep-hewn piazza, to be convinced that at least it *is* a fact.

Just at this interesting stage, however, our explorations bade fair to be interrupted. Our man who carried the pickaxe had lingered behind us for a few hundred yards, in earnest conversation with an islander; and he now came up, breathless and in hot haste, to say that the islander, a Roman Catholic tacksman in the neighborhood, had peremptorily warned him that the Scuir of Eigg was the property of Dr. M'Pherson of Aberdeen, not ours, and that the Doctor would be very angry at any man who meddled with it. "That message," said my friend, laughing, but looking just a little sad through the laugh, "would scarce have been sent us when I was minister of the Establishment here; but it seems allowable in the case of a poor Dissenter, and is no bad specimen of the thousand little ways in which the Roman Catholic population of the island try to annoy me, now that they see my back to the wall." I was tickled with the idea of a fossil preserve, which coupled itself in my mind, through a trick of the associative faculty, with the idea of a great fossil act for the British empire, framed on the principles of the game-laws; and, just wondering what sort of disreputable vagabonds geological poachers would become under its deteriorating influence, I laid hold of the pickaxe and broke into the stonefast floor; and thence I succeeded in abstracting,—feloniously, I dare say, though the crime has not yet got [Pg 53] into the statute-book—some six or eight pieces of the *Pinites Eiggensis*, amounting in all to about half a cubic foot of that very ancient wood—value unknown. I trust, should the case come to a serious bearing, the members of the London Geological Society will generously subscribe half-a-crown a-piece to assist me in feeing counsel. There are more interests than mine at stake in the affair. If I be cast and committed,—I, who have poached over only a few miserable districts in Scotland,—pray, what will become of some of them,—the Lyells, Bucklands, Murchisons and Sedgwicks,—who have poached over whole continents?

We were successful in procuring several good specimens of the Eigg pine, at a depth, in the conglomerate, of from eight to eighteen inches. Some of the upper pieces we found in contact with the decomposing trap out of which the hollow piazza above had been scooped; but the greater number, as my set of specimens abundantly testify, lay embedded in the original Oölitic grit in which they had been locked up, in, I doubt not, their present fossil state, ere their upheaval, through Plutonic agency, from their deep-sea bottom. The annual rings of the wood, which are quite as small as in a slow-growing Baltic pine, are distinctly visible in all the better pieces I this day transferred to my bag. In one fragment I reckon sixteen rings in half an inch, and fifteen in the same space in another. The trees to which they belonged seem to have grown on some exposed hill-side, where, in the course of half a century, little more than from two or three inches were added to their diameter. The *Pinites Eiggensis*, or Eigg pine, was first introduced to the notice of the scientific world by the late Mr. Witham, in whose interesting work on "The Internal Structure of Fossil Vegetables" the reader may find it figured and described. The specimen in which he studied its peculiarities "was found," he says, "at the base of the magnificent [Pg 54] mural escarpment named the Scuir of Eigg,—not, however, *in situ*, but among fragments of rocks of the Oölitic series." The authors of the "Fossil Flora," where it is also figured, describe it as differing very considerably in structure from any of the coniferæ of the Coal

Measures. "Its medullary rays," says Messrs. Lindley and Hutton, "appear to be more numerous, and frequently are not continued through one zone of wood to another, but more generally terminate at the concentric circles. It abounds also in turpentine vessels, or lacunæ, of various sizes, the sides of which are distinctly defined." Viewed through the microscope, in transparent slips, longitudinal and transverse, it presents, within the space of a few lines, objects fitted to fill the mind with wonder. We find the minutest cells, glands, fibres, of the original wood preserved uninjured. *There* still are those medullary rays entire that communicated between the pith and the outside,—*there* still the ring of thickened cells that indicated the yearly check which the growth received when winter came on,—*there* the polygonal reticulations of the cross section, without a single broken mesh,—*there*, too, the elongated cells in the longitudinal one, each filled with minute glands that take the form of double circles,—*there* also, of larger size and less regular form, the lacunæ in which the turpentine lay: every nicely organized speck, invisible to the naked eye, we find in as perfect a state of keeping in the incalculably ancient pile-work on which the gigantic Scuir is founded, as in the living pines that flourish green on our hill-sides. A net-work, compared with which that of the finest lace ever worn by the fair reader would seem a net-work of cable, has preserved entire, for untold ages, the most delicate peculiarities of its pattern. There is not a mesh broken, nor a circular dot away!

The experiments of Mr. Witham on the Eigg fossil, furnish [Pg 55] an interesting example of the light which a single, apparently simple, discovery may throw on whole departments of fact. He sliced his specimen longitudinally and across, fastened the slices on glass, ground them down till they became semi-transparent, and then, examining them under reflected light by the microscope, marked and recorded the specific peculiarities of their structure. And we now know, in consequence, that the ancient Eigg pine, to which the detached fragment picked up at the base of the Scuir belonged,—a pine alike different from those of the earlier carboniferous period and those which exist contemporary with ourselves,—was, some *three creations* ago, an exceedingly common tree in the country now called Scotland,—as much so, perhaps, as the Scotch fir is at the present day. The fossil trees found in such abundance in the neighborhood of Helmsdale that they are burnt for lime,—the fossil wood of Eathie, in Cromartyshire, and that of Shandwick, in Ross,—all belong to the *Pinites Eiggensis*. It seems to have been a straight and stately tree, in most instances, as in the Eigg specimens, of slow growth. One of the trunks I saw near Navidale measured two feet in diameter, but a full century had passed ere it attained to a bulk so considerable; and a splendid specimen in my collection, from the same locality, which measures twenty-one inches, exhibits even *more* than a hundred annual rings. In one of my specimens, and one only, the rings are of great breadth. They differ from those of all the others in the proportion in which I have seen the annual rings of a young, vigorous fir that had sprung up in some rich, moist hollow, differ from the annual rings of trees of the same species that had grown in the shallow, hard soil of exposed hill-sides. And this one specimen furnishes curious evidence that the often-marked but little understood law, which gives us our better and worse seasons in alternate groups, various in number and [Pg 56] uncertain in their time of recurrence, obtained as early as the age of the Oölite. The rings follow each other in groups of lesser and larger breadth. One group of four rings measures an inch and a quarter across, while an adjoining group of five rings measures only five-eighth parts; and in a breadth of six inches there occur five of these alternate groups. For some four or five years together, when this pine was a living tree, the springs were late and cold, and the summers cloudy and chill, as in that group of seasons which intervened between 1835 and 1841; and then, for four or five years, more springs were early and summers genial, as in the after group of 1842, 1843 and 1844. An arrangement in nature,—first observed, as we learn from Bacon, by the people of the Low Countries, and which has since formed the basis of meteoric tables, and of predictions and elaborate cycles of the weather,—bound together the twelvemonths of the Oölitic period in alternate bundles of better and worse: vegetation throve vigorously during the summers of one group, and languished, in those of another, in a state of partial development.

Sending away our man shipwards, laden with a bag of fossil wood, we ascended by a steep broken ravine to the top of the Scuir. The columns, as we pass on towards the west, diminish in size, and assume in many of the beds considerable variety of direction and form. In one bed they belly over with a curve, like the ribs of some wrecked vessel from which the planking has been torn away; in another they project in a straight line, like muskets planted slantways on the ground to receive a charge of cavalry; in others the inclination is inwards, like that of ranges of stakes placed in front of a sea-dyke, to break the violence of the waves; while yet in others they present, as in the eastern portion of the Scuir, the common vertical direction. The ribbed appearance of every crag and cliff, imparts to the [Pg 57] scene a peculiar character; every larger mass of light and shadow is corded with minute stripes; and the feeling experienced among the more shattered peaks, and in the more broken recesses, seems near akin to that which it is the tendency of some magnificent ruin to excite, than that which awakens amid the sublime of nature. We feel as if the pillared rocks around us were like the Cyclopean walls of Southern Italy,—the erections of some old gigantic race passed from the earth forever. The feeling must have been experienced on former occasions, amid the innumerable pillars of the Scuir; for we find M'Culloch, in his description, ingeniously analyzing it. "The resemblance to architecture here is much increased," he says, "by the columnar structure, which is sufficiently distinguishable, even from a distance, and produces a strong effect of artificial regularity when seen near at hand. To this vague association in the mind of the efforts of art with the magnitude of nature, is owing much of that sublimity of character which the Scuir presents. The sense of power is a fertile source of the sublime; and as the appearance of power exerted, no less than that of simplicity, is necessary to confer this character on architecture, so the mind, insensibly transferring the operations of nature to the efforts of art where they approximate in character, becomes impressed with a feeling rarely excited by her more ordinary forms, where these are even more stupendous."

The top of the Scuir, more especially towards its eastern termination, resembles that of some vast mole not yet levelled over by the workmen; the pavement has not yet been laid down, and there are deep gaps in the masonry, that run transversely, from side to side, still to fill up. Along one of these ditch-like gaps, which serves to insulate the eastern and highest portion of the Scuir from all its other portions, we find fragments of a rude wall of uncemented [Pg

58] stones, the remains of an ancient hill-fort; which, with its natural rampart of rock on three of its four sides, more than a hundred yards in sheer descent, and with its deep ditch and rude wall on the fourth, must have formed one of the most inaccessible in the kingdom. The masses of pitchstone a-top, though so intensely black within, are weathered on the surface into almost a pure white; and we found lying detached among them, fragments of common amygdaloid and basalt, and minute slaty pieces of chalcedony that had formed apparently in fissures of the trap. We would have scrutinized more narrowly at the time had we expected to find anything more rare; but I did not know until full four months after, that aught more rare was to be found. Had we examined somewhat more carefully, we might possibly have done what Mr. Woronzow Greig did on the Scuir about eighteen years previous,—picked up on it a piece of *bona fide* Scotch pumice. This gentleman, well known through his exertions in statistical science, and for his love of science in general, and whose tastes and acquirements are not unworthy the son of Mrs. Somerville, has kindly informed me by letter regarding his curious discovery. "I visited the island of Eigg," he says, "in 1825 or 1826, for the purpose of shooting, and remained in it several days; and as there was a great scarcity of game, I amused myself in my wanderings by looking about for natural curiosities. I knew little about Geology at the time, but, collecting whatever struck my eye as uncommon, I picked up from the sides of the Scuir, among various other things, a bit of fossil wood, and, nearly at the summit of the eminence, a piece of pumice of a deep brownish-black color, and very porous, the pores being large and round, and the substance which divided them of a uniform thickness. This last specimen I gave to Mr. Lyell, who said that it could not originally have belonged to Eigg, though it [Pg 59] might possibly have been washed there by the sea,—a suggestion, however, with which its place on the top of the Scuir seems ill to accord. I may add, that I have since procured a larger specimen from the same place." This seems a curious fact, when we take into account the identity, in their mineral components, of the pumice and obsidian of the recent volcanoes; and that pitchstone, the obsidian of the trap-rocks, is resolvable into a pumice by the art of the chemist. If pumice was to be found anywhere in Scotland, we might *a priori* expect to find it in connection with by far the largest mass of pitchstone in the kingdom. It is just possible, however, that Mr. Greig's two specimens may not date farther back, in at least their existing state, than the days of the hill-fort. Powerful fires would have been required to render the exposed summit of the Scuir at all comfortable; there is a deep peat-moss in its immediate neighborhood, that would have furnished the necessary fuel; the wind must have been sufficiently high on the summit to fan the embers into an intense white heat; and if it was heat but half as intense as that which was employed in fusing into one mass the thick vitrified ramparts of Craig Phadrig and Knock Farril, on the east coast, it could scarce have failed to anticipate the experiment of the Hon. Mr. Knox, of Dublin, by converting some of the numerous pitchstone fragments that lie scattered about, "into a light substance in every respect resembling pumice."

It was now evening, and rarely have I witnessed a finer. The sun had declined half-way adown the western sky, and for many yards the shadow of the gigantic Scuir lay dark beneath us along the descending slope. All the rest of the island, spread out at our feet as in a map, was basking in yellow sunshine; and with its one dark shadow thrown from its one mountain-elevated wall of rock, it seemed some [Pg 60] immense fantastical dial, with its gnomon rising tall in the midst. Far below, perched on the apex of the shadow, and half lost in the line of the penumbra, we could see two indistinct specks of black, with a dim halo around each,—specks that elongated as we arose, and contracted as we sat, and went gliding along the line as we walked. The shadows of two gnats disporting on the edge of an ordinary gnomon would have seemed vastly more important, in proportion, on the figured plane of the dial, than these, our ghostly representatives, did here. The sea, spangled in the wake of the sun with quick glancing light, stretched out its blue plain around us; and we could see included in the wide prospect, on the one hand, at once the hill-chains of Morven and Kintail, with the many intervening lochs and bold jutting headlands that give variety to the mainland; and, on the other, the variously complexioned Hebrides, from the Isle of Skye to Uist and Barra, and from Uist and Barra to Tiree and Mull. The contiguous Small Isles, Muck and Rum, lay moored immediately beside us, like vessels of the same convoy that in some secure roadstead drop anchor within hail of each other. I could willingly have lingered on the top of the Scuir until after sunset; but the minister, who, ever and anon, during the day, had been conning over some notes jotted on a paper of wonderfully scant dimensions, reminded me that this was the evening of his week-day discourse, and that we were more than a particularly rough mile from the place of meeting, and within, half an hour of the time. I took one last look of the scene ere we commenced our descent. There, in the middle of the ample parish glebe, that looked richer and greener in the light of the declining sun than at any former period during the day,—rose the snug parish manse; and yonder,—in an open island channel, with a strip of dark rocks fringing the land within, and another dark strip [Pg 61] fringing the barren Eilean Chaisteil outside,—lay the Betsey, looking wonderfully diminutive, but evidently a little thing of high spirit, taut-masted, with a smart rake aft, and a spruce outrigger astern, and flaunting her triangular flag of blue in the sun. I pointed first to the manse, and then to the yacht. The minister shook his head.

"'Tis a time of strange changes," he said; "I thought to have lived and died in that house, and found a quiet grave in the burying-ground yonder beside the ruin; but my path was a clear though a rugged one; and from almost the moment that it opened up to me, I saw what I had to expect. It has been said that I might have lain by here in this out-of-the-way corner, and suffered the Church question to run its course, without quitting my hold of the Establishment. And so I perhaps might. It is easy securing one's own safety, in even the worst of times, if one look no higher; and I, as I had no opportunity of mixing in the contest, or of declaring my views respecting it, might be regarded as an unpledged man. But the principles of the Evangelical party were my principles; and it would have been consistent with neither honor nor religion to have hung back in the day of battle, and suffered the men with whom in heart I was at one to pay the whole forfeit of our common quarrel. So I attended the Convocation, and pledged myself to stand or fall with my brethren. On my return I called my people together, and told them how the case stood, and that in May next I bade fair to be a dependent for a home on the proprietor of Eigg. And so they petitioned the proprietor that he might give me leave to build a house among them,—exactly the same sort of favor granted to the Roman Catholics of the island. But month after month passed, and they got no reply to their petition; and I was left in suspense, not knowing whether I was to have a home among them or no. I did feel the case [Pg 62] a somewhat hard one. The father of Dr. M'Pherson of Eigg had been, like myself, a humble Scotch minister; and the Doctor, however indifferent to his people's wishes in such a matter, might have just thought that a man in his father's station in life, with a wife and family dependent on him, was placed by his silence in cruel circumstances of uncertainty. Ere the Disruption took place, however, I came to know pretty conclusively what I had to expect. The Doctor's factor came to Eigg, and, as I was informed, told the Islanders that it was not likely the Doctor would permit a *third* place of worship on the Island: the Roman Catholics had one, and the Establishment had a kind of one, and there was to be no more. The factor, an active messenger-at-arms, useful in raising rents in these parts, has always been understood to speak the mind of his master; but the congregation took heart in the emergency, and sent off a second petition to Dr. M'Pherson, a week or so previous to the Disruption. Ere *it* received an answer, the Disruption took place; and, laying the whole circumstances before my brethren in Edinburgh, who, like myself, interpreted the silence of the Doctor into a refusal, I suggested to them the scheme of the Betsey, as the only scheme through which I could keep up unbroken my connection with my people. So the trial is now over, and here we are, and yonder is the Betsey."

We descended the Scuir together for the place of meeting, and entered, by the way, the cottage of a worthy islander, much attached to his minister. "We are both very hungry," said my friend: "we have been out among the rocks since breakfast-time, and are wonderfully disposed to eat. Do not put yourself about, but give us anything you have at hand." There was a bowl of rich milk brought us, and a splendid platter of mashed potatoes, and we dined like princes. I observed, for the first time, in the interior [Pg 63] of this cottage, what I had frequent occasion to remark afterwards, that much of the wood used in building in the smaller and outer islands of the Hebrides must have drifted across the Atlantic, borne eastwards and northwards by the great Gulf-stream. Many of the beams and boards, sorely drilled by the *Teredo navalis*, are of American timber, that, from time to time, has been cast upon the shore,—a portion of it, apparently, from timber-laden vessels unfortunate in their voyage, but a portion of it, also, with root and branch still attached, bearing mark of having been swept to the sea by transatlantic rivers. Nuts and seeds of tropical plants are occasionally picked up on the beach. My friend gave me a bean or nut of the *Dolichos urens*, or cow-itch shrub, of the West Indies, which an islander had found on the shore sometime in the previous year, and given to one of the manse children as a toy; and I attach some little interest to it, as a curiosity of the same class with the large canes and the fragment of carved wood found floating near the shores of Madeira by the brother-in-law of Columbus, and which, among other pieces of circumstantial evidence, led the great navigator to infer the existence of a western continent. Curiosities of this kind seem still more common in the northern than in the western islands of Scotland. "Large exotic nuts or seeds," says Dr. Patrick Neill, in his interesting "Tour," quoted in a former chapter, "which in Orkney are known by the name of Molucca beans, are occasionally found among the *rejectamenta* of the sea, especially after westerly winds. There are two kinds commonly found: the larger (of which the fishermen very generally make snuff-boxes) seem to be seeds from the great pod of the *Mimosa scandens* of the West Indies; the smaller seeds, from the pod of the *Dolichos urens*, also a native of the same region. It is probable that the currents of the ocean, and particularly that great current which issues from the [Pg 64] Gulf of Florida, and is hence denominated the Gulf Stream, aid very much in transporting across the mighty Atlantic these American products. They are generally quite fresh and entire, and afford an additional proof how impervious to moisture, and how imperishable, nuts and seeds generally are."

The evening was fast falling ere the minister closed his discourse; and we had but just light enough left, on reaching the Betsey, to show us that there lay a dead sheep on the deck. It had been sent aboard to be killed by the minister's factotum, John Stewart; but John was at the evening preaching at the time, and the poor sheep, in its attempts to set itself free, had got itself entangled among the cords, and strangled itself. "Alas, alas!" exclaimed the minister, "thus ends our hope of fresh mutton for the present, and my hapless speculation as a sheep farmer for evermore." I learned from him, afterwards, over our tea, that shortly previous to the Convocation he had got his glebe,—one of the largest in Scotland,—well stocked with sheep and cattle, which he had to sell, immediately on the Disruption, in miserably bad condition, at a loss of nearly fifty per cent. He had a few sheep, however, that would not sell at all, and that remained on the glebe, in consequence, until his successor entered into possession. And he, honest man, straightway impounded them, and got them incarcerated in a dark, dirty hole, somewhat in the way Giant Despair incarcerated the pilgrims,—a thing he had quite a legal right to do, seeing that the mile-long glebe, with its many acres of luxuriant pasture, was now as much his property as it had been Mr. Swanson's a few months before, and seeing Mr. Swanson's few sheep had no right to crop his grass. But a worthy neighbor interfered,—Mr. M'Donald, of Keil, the principal tenant in the island. Mr. M'Donald,—a practical commentator on the law of kindness,—was [Pg 65] sorely scandalized at what he deemed the new minister's gratuitous unkindness to a brother in calamity; and, relieving the sheep, he brought them to his own farm, where he found them board and lodging on my friend's behalf, till they could be used up at leisure. And it was one of the last of this unfortunate lot that now contrived to escape from us by anticipating John Stewart. "A black beginning makes a black ending," said Gouffing Jock, an ancient border shepherd, when his only sheep, a black ewe, the sole survivor of a flock smothered in a snow-storm, was worried to death by his dogs. Then, taking down his broadsword, he added, "Come awa, my auld friend; thou and I maun e'en stock Bowerhope-Law ance mair!" Less warlike than Gouffing Jock, we were content to repeat over the dead, on this occasion, simply the first portion of his speech; and then, betaking ourselves to our cabin, we forgot all our sorrows over our tea.

[Pg 66]

CHAPTER IV.

An Excursion—The Chain of Crosses—Bay of Laig—Island of Rum—Description of the Island—Superstitions banished by pure Religion—Fossil Shells—Remarkable Oyster Bed—New species of Belemnite—Oölitic Shells—White Sandstone Precipices—Gigantic

Petrified Mushrooms—"Christabel" in Stone—Musical Sand—*Jabel Nakous*, or Mountain of the Bell—Experiments of Travellers at *Jabel Nakous*—Welsted's Account—*Reg-Rawan*, or the Moving Sand—The Musical Sounds inexplicable—Article on the subject in the North British Review.

There had been rain during the night; and when I first got on deck, a little after seven, a low stratum of mist, that completely enveloped the Scuir, and truncated both the eminence on which it stands and the opposite height, stretched like a ruler across the flat valley which indents so deeply the middle of the island. But the fogs melted away as the morning rose, and ere our breakfast was satisfactorily discussed, the last thin wreath had disappeared from around the columned front of the rock-tower of Eigg, and a powerful sun looked down on moist slopes and dank hollows, from which there arose in the calm a hazy vapor, that, while it softened the lower features of the landscape, left the bold outline relieved against a clear sky. Accompanied by our attendant of the previous day, bearing bag and hammer, we set out a little before eleven for the north-western side of the island, by a road which winds along the central hollow. My friend showed me as we went, that on the edge of an eminence, on which the traveller journeying westwards catches the last glimpse of the chapel of St. Donan, there had once been a rude cross erected, and another rude cross on an eminence on which he catches the last glimpse of the [Pg 67] first; and that there had thus been a chain of stations formed from sea to sea, like the sights of a land-surveyor, from one of which a second could be seen, and a third from the second, till, last of all, the emphatically holy point of the island,—the burial-place of the old Culdee,—came full in view. The unsteady devotion, that journeyed, fancy-bound, along the heights, to gloat over a dead man's bones, had its clue to carry it on in a straight line. Its trail was on the ground; it glided snake-like from cross to cross, in quest of dust; and, without its finger-posts to guide it, would have wandered devious. It is surely a better devotion that, instead of thus creeping over the earth to a mouldy sepulchre, can at once launch into the sky, secure of finding Him who once arose from one. In less than an hour we were descending on the Bay of Laig, a semi-circular indentation of the coast, about a mile in length, and, where it opens to the main sea, nearly two miles in breadth; with the noble island of Rum rising high in front, like some vast breakwater; and a meniscus of comparatively level land, walled in behind by a semi-circular rampart of continuous precipice, sweeping round its shores. There are few finer scenes in the Hebrides than that furnished by this island bay and its picturesque accompaniments,—none that break more unexpectedly on the traveller who descends upon it from the east; and rarely has it been seen, to greater advantage than on the delicate day, so soft, and yet so sunshiny and clear, on which I paid it my first visit.

The island of Rum, with its abrupt sea-wall of rock, and its steep-pointed hills, that attain, immediately over the sea, an elevation of more than two thousand feet, loomed bold and high in the offing, some five miles away, but apparently much nearer. The four tall summits of the island rose clear against the sky like a group of pyramids; its lower slopes and precipices, variegated and relieved by graceful [Pg 68] alternations of light and shadow, and resting on their blue basement of sea, stood out with equal distinctness; but the entire middle space from end to end was hidden in a long horizontal stratum of gray cloud, edged atop with a lacing of silver. Such was the aspect of the noble breakwater in front. Fully two-thirds of the semi-circular rampart of rock which shuts in the crescent-shaped plain directly opposite lay in deep shadow; but the sun shone softly on the plain itself, brightening up many a dingy cottage, and many a green patch of corn; and the bay below stretched out, sparkling in the light. There is no part of the island so thickly inhabited as this flat meniscus. It is composed almost entirely of Oölitic rocks, and bears atop, especially where an ancient oyster-bed of great depth forms the subsoil, a kindly and fertile mould. The cottages lie in groups; and, save where a few bogs, which it would be no very difficult matter to drain, interpose their rough shag of dark green, and break the continuity, the plain around them waves with corn. Lying fair, green and populous within the sweep of its inaccessible rampart of rock, at least twice as lofty as the ramparts of Babylon of old, it reminds one of the suburbs of some ancient city lying embosomed, with all its dwellings and fields, within some roomy crescent of the city wall. We passed, ere we entered on the level, a steep-sided narrow dell, through which a small stream finds its way from the higher grounds, and which terminates at the upper end in an abrupt precipice, and a lofty but very slim cascade. "One of the few superstitions that still linger on the island," said my friend the minister, "is associated with that wild hollow. It is believed that shortly before a death takes place among the inhabitants, a tall withered female may be seen in the twilight, just yonder where the rocks open, washing a shroud in the stream. John, there, will perhaps tell you how she was spoken to on one occasion, by [Pg 69] an over-bold, over-inquisitive islander, curious to know whose shroud she was preparing; and how she more than satisfied his curiosity, by telling him it was his own. It is a not uninteresting fact," added the minister, "that my poor people, since they have become more earnest about their religion, think very little about ghosts and spectres: their faith in the realities of the unseen world seems to have banished from their minds much of their old belief in its phantoms."

In the rude fences that separate from each other the little farms in this plain, we find frequent fragments of the oyster bed, hardened into a tolerably compact limestone. It is seen to most advantage, however, in some of the deeper cuttings in the fields, where the surrounding matrix exists merely as an incoherent shale; and the shells may be picked out as entire as when they lay, ages before, in the mud, which we still see retaining around them its original color. They are small, thin, triangular, much resembling in form some specimens of the *Ostrea deltoidea,* but greatly less in size. The nearest resembling shell in Sowerby is the *Ostrea acuminata,*—an oyster of the clay that underlies the great Oölite of Bath. Few of the shells exceed an inch and a half in length, and the majority fall short of an inch. What they lack in bulk, however, they make up in number. They are massed as thickly together, to the depth of several feet, as shells on the heap at the door of a Newhaven fisherman, and extend over many acres. Where they lie open we can still detect the triangular disc of the hinge, with the single impression of the abductor muscle; and the foliaceous character of the shell remains in most instances as distinct as if it had undergone no mineral change. I have seen nowhere in Scotland, among the secondary formations, so unequivocal an oyster-bed; nor do such beds seem to be at all common in formations older than the Tertiary in England, though the oyster itself is sufficiently so. We find [Pg 70] Mantell stating, in his recent work ("Medals of Creation"), after first describing an immense oyster bed of the London Basin, that underlies the city (for what is now London was once an oyster-bed), that in the chalk below, though it contains several species of Ostrea, the shells are diffused promiscuously throughout the general mass. Leaving, however, these oysters of the Oölite, which never net inclosed nor drag disturbed, though they must have formed the food of many an extinct order of fish,—mayhap reptile,—we pass on in a south-western direction, descending in the geological scale as we go, until we reach the southern side of the Bay of Laig. And there, far below tide-mark, we find a dark-colored argillaceous shale of the Lias, greatly obscured by boulders of trap,—the only deposit of the Liasic formation in the island.

A line of trap-hills that rises along the shore seems as if it had strewed half its materials over the beach. The rugged blocks lie thick as stones in a causeway, down to the line of low ebb,—memorials of a time when the surf dashed against the shattered bases of the trap-hills, now elevated considerably beyond its reach; and we can catch but partial glimpses of the shale below. Wherever access to it can be had, we find it richly fossiliferous; but its organisms, with the exception of its Belemnites, are very imperfectly preserved. I dug up from under the trap-blocks some of the common Liasic Ammonites of the north-eastern coast of Scotland, a few of the septa of a large Nautilus, broken pieces of wood, and half-effaced casts of what seems a branched coral; but only minute portions of the shells have been converted into stone; here and there a few chambers in the whorls of an Ammonite or Nautilus, though the outline of the entire organism lies impressed in the shale; and the ligneous and polyparious fossils we find in a still greater state of decay. The Belemnite alone, as [Pg 71] is common with this robust fossil,—so often the sole survivor of its many contemporaries,—has preserved its structure entire. I disinterred from the shale good specimens of the Belemnite *sulcatus* and Belemnite *elongatus*, and found, detached on the surface of the bed, a fragment of a singularly large Belemnite, a full inch and a quarter in diameter, the species of which I could not determine.

Returning by the track we came, we reach the bottom of the bay, which we find much obscured with sand and shingle; and pass northwards along its side, under a range of low sandstone precipices, with interposing grassy slopes, in which the fertile Oölitic meniscus descends to the beach. The sandstone, white and soft, and occurring in thick beds, much resembles that of the Oölite of Sutherland. We detect in it few traces of fossils; now and then a carbonaceous marking, and now and then what seems a thin vein of coal, but which proves to be merely the bark of some woody stem, converted into a glossy bituminous lignite, like that of Brora. But in beds of a blue clay, intercalated with the sandstone, we find fossils in abundance, of a character less obscure. We spent a full half-hour in picking out shells from the bottom of a long dock-like hollow among the rocks, in which a bed of clay has yielded to the waves, while the strata on either side stand up over it like low wharfs on the opposite side of a river. The shells, though exceedingly fragile,—for they partake of the nature of the clayey matrix in which they are imbedded,—rise as entire as when they had died among the mud, years, mayhap ages, ere the sandstone had been deposited over them; and we were enabled at once to detect their extreme dissimilarity, as a group, to the shells of the Liasic deposit we had so lately quitted. We did not find in this bed a single Ammonite, Belemnite, or Nautilus; but chalky Bivalves, resembling our existing Tellina, in vast abundance, [Pg 72] mixed with what seem to be a small Buccinum and a minute Trochus, with numerous rather equivocal fragments of a shell resembling an Oiliva. So thickly do they lie clustered together in this deposit, that in some patches where the sad-colored argillaceous ground is washed bare by the sea, it seems marbled with them into a light gray tint. The group more nearly resembles in type a recent one than any I have yet seen in a secondary deposit, except perhaps in the Weald of Moray, where we find in one of the layers a Planorbis scarce distinguishable from those of our ponds and ditches, mingled with a Paludina that seems as nearly modelled after the existing form. From the absence of the more characteristic shells of the Oölite, I am inclined to deem the deposit one of estuary origin. Its clays were probably thrown down, like the silts of so many of our rivers, in some shallow bay, where the waters of a descending stream mingled with those of the sea, and where, though shells nearly akin to our existing periwinkles and whelks congregate thickly, the Belemnite, seared by the brackish water, never plied its semi-cartilaginous fins, or the Nautilus or Ammonite hoisted its membranaceous sail.

We pass on towards the north. A thick bed of an extremely soft white sandstone presents here, for nearly half a mile together, its front to the waves, and exhibits, under the incessant wear of the surf, many singularly grotesque combinations of form. The low precipices, undermined at the base, beetle over like the sides of stranded vessels. One of the projecting promontories we find hollowed through and through by a tall rugged archway; while the outer pier of the arch,—if pier we may term it,—worn to a skeleton, and jutting outwards with a knee-like angle, presents the appearance of a thin ungainly leg and splay foot, advanced, as if in awkward courtesy, to the breakers. [Pg 73] But in a winter or two, judging from its present degree of attenuation, and the yielding nature of its material, which resembles a damaged mass of arrow-root, consolidated by lying in the leaky hold of a vessel, its persevering courtesies will be over, and pier and archway must lie in shapeless fragments on the beach. Wherever the surf has broken into the upper surface of this sandstone bed, and worn it down to nearly the level of the shore, what seem a number of double ramparts, fronting each other, and separated by deep square ditches exactly parallel in the sides, traverse the irregular level in every direction. The ditches vary in width from one to twelve feet; and the ramparts, rising from three to six feet over them, are perpendicular as the walls of houses, where they front each other, and descend on the opposite sides in irregular slopes. The iron block, with square groove and projecting ears, that receives the bar of a railway, and connects it with the stone below, represents not inadequately a section of one of these ditches, with its ramparts. They form here the sole remains of dykes of an earthy trap, which, though at one time in a state of such high fusion that they converted the portions of soft sandstone in immediate contact with them into the consistence of quartz rock, have long since mouldered away, leaving but the hollow rectilinear rents which they had occupied, surmounted by the indurated walls which they had baked. Some of the most curious appearances, however, connected with the sandstone, though they occur chiefly in an upper bed, are exhibited by what seem fields of petrified mushrooms, of a gigantic size, that spread out in some places for hundreds of yards under the high-water level. These apparent mushrooms stand on thick squat stems, from a foot to eighteen inches in height; the heads are round like those of toad-stools, and vary from one foot to nearly two yards in diameter. In some specimens we [Pg 74] find two heads joined together in a form resembling a squat figure of *eight*, of what printers term the Egyptian type, or, to borrow the illustration of M'Culloch, "like the ancient military projectile known by the name of double-headed shot;" in other specimens three heads have coalesced in a trefoil shape, or rather in a shape like that of an ace of clubs divested of the stem. By much the greater number, however, are spherical. They are composed of concretionary masses, consolidated, like the walls of the dykes, though under some different process, into a hard siliceous stone, that has resisted those disintegrating influences of the weather and the surf, under which the yielding matrix in which they were embedded has worn from around them. Here and there we find them lying detached on the beach, like huge shot, compared with which the greenstone balls of Mons Meg are but marbles for children to play with; in other cases they project from the mural front of rampart-like precipices, as if they had been showered into them by the ordnance of some besieging battery, and had stuck fast in the mason-work. Abbotsford has been described as a romance in stone and lime; we have here, on the shores of Laig, what seems a wild but agreeable tale, of the extravagant cast of "Christabel," or the "Rhyme of the Ancient Mariner," fretted into sandstone. But by far the most curious part of the story remains to be told.

The hollows and fissures of the lower sandstone bed we find filled with a fine quartzose sand, which, from its pure white color, and the clearness with which the minute particles reflect the light, reminds one of accumulations of potato-flour drying in the sun. It is formed almost entirely of disintegrated particles of the soft sandstone; and as we at first find it occurring in mere handfuls, that seem as if they had been detached from the mass during the last few [Pg 75] tides, we begin to marvel to what quarter the missing materials of the many hundred cubic yards of rock, ground down along the shore in this bed during the last century or two, have been conveyed away. As we pass on northwards, however, we see the white sand occurring in much larger quantities,—here heaped up in little bent-covered hillocks above the reach of the tide,—there stretching out in level, ripple-marked wastes into the waves,—yonder rising in flat narrow spits among the shallows. At length we reach a small, irregularly-formed bay, a few hundred feet across, floored with it from side to side; and see it, on the one hand, descending deep into the sea, that exhibits over its whiteness a lighter tint of green, and, on the other, encroaching on the land, in the form of drifted banks, covered with the plants common to our tracts of sandy downs. The sandstone bed that has been worn down to form it contains no fossils, save here and there a carbonaceous stem; but in an underlying harder stratum we occasionally find a few shells; and, with a specimen in my hand charged with a group of bivalves resembling the existing conchifera of our sandy beaches, I was turning aside this sand of the Oölite, so curiously reduced to its original state, and marking how nearly the recent shells that lay embedded in it resembled the extinct ones that had lain in it so long before, when I became aware of a peculiar sound that it yielded to the tread, as my companions paced over it. I struck it obliquely with my foot, where the surface lay dry and incoherent in the sun, and the sound elicited was a shrill, sonorous note, somewhat resembling that produced by a waxed thread, when tightened between the teeth and the hand, and tipped by the nail of the forefinger. I walked over it, striking it obliquely at each step, and with every blow the shrill note was repeated. My companions joined me; and we performed a concert, in which, if we could [Pg 76] boast of but little variety in the tones produced, we might at least challenge all Europe for an instrument of the kind which produced them. It seemed less wonderful that there should be music in the granite of Memnon, than in the loose Oölitic sand of the Bay of Laig. As we marched over the drier tracts, an incessant *woo, woo, woo,* rose from the surface, that might be heard in the calm some twenty or thirty yards away; and we found that where a damp semi-coherent stratum lay at the depth of three or four inches beneath, and all was dry and incoherent above, the tones were loudest and sharpest, and most easily evoked by the foot. Our discovery,—for I trust I may regard it as such,—adds a third locality to two previously known ones, in which what may be termed the musical sand,—no unmeet counterpart to the "singing water" of the tale,—has now been found. And as the island of Eigg is considerably more accessible than *Jabel Nakous,* in Arabia Petræa, or *Reg-Rawan*, in the neighborhood of Cabul, there must be facilities presented through the discovery which did not exist hitherto, for examining the phenomenon in acoustics which it exhibits,—a phenomenon, it may be added, which some of our greatest masters of the science have confessed their inability to explain.

Jabel Nakous, or the "Mountain of the Bell," is situated about three miles from the shores of the Gulf of Suez, in that land of wonders which witnessed for forty years the journeyings of the Israelites, and in which the granite peaks of Sinai and Horeb overlook an arid wilderness of rock and sand. It had been known for many ages by the wild Arab of the desert, that there rose at times from this hill a strange, inexplicable music. As he leads his camel past in the heat of the day, a sound like the first low tones of an Æolian harp stirs the hot breezeless air. It swells louder and louder in progressive undulations, till at length [Pg 77] the dry baked earth seems to vibrate under foot, and the startled animal snorts and rears, and struggles to break away. According to the Arabian account of the phenomenon, says Sir David Brewster, in his "Letters on Natural Magic," there is a convent miraculously preserved in the bowels of the hill; and the sounds are said to be those of the "*Nakous*, a long metallic ruler, suspended horizontally, which the priest strikes with a hammer, for the purpose of assembling the monks to prayer." There exists a tradition that on one occasion a wandering Greek saw the mountain open, and that, entering by the gap, he descended into the subterranean convent, where he found beautiful gardens and fountains of delicious water, and brought with him to the upper world, on his return, fragments of consecrated bread. The first European traveller who visited *Jabel Nakous*, says Sir David, was M. Seetzen, a German. He journeyed for several hours over arid sands, and under ranges of precipices inscribed by mysterious characters, that tell, haply, of the wanderings of Israel under Moses. And reaching, about noon, the base of the musical fountain, he found it composed of a white friable sandstone, and presenting on two of its sides sandy declivities. He watched beside it for an hour and a quarter, and then heard, for the first time, a low undulating sound, somewhat resembling that of a humming top, which rose and fell, and ceased and began, and then ceased again; and in an hour and three quarters after, when in the act of climbing along the declivity, he heard the sound yet louder and more prolonged. It seemed as if issuing from under his knees, beneath which the sand, disturbed by his efforts, was sliding downwards along the surface of the rock. Concluding that the sliding sand was the cause of the sounds, not an effect of the vibrations which they occasioned, he climbed to the top of one of the declivities, and, sliding downwards, exerted himself [Pg 78] with hands and feet to set the sand in motion. The effect produced far exceeded his expectations; the incoherent sand rolled under and around in a vast sheet; and so loud was the noise produced, that "the earth seemed to tremble beneath him to such a degree, that he states he should certainly have been afraid if he had been ignorant of the cause." At the time Sir David Brewster wrote (1832), the only other European who had visited *Jabel Nakous* was Mr. Gray, of University College, Oxford. This gentleman describes the noises he heard, but which he was unable to trace to their producing cause, as "beginning with a low continuous murmuring sound, which seemed to rise beneath his feet," but "which gradually changed into pulsations as it became louder, so as to resemble the striking of a clock, and became so strong at the end of five minutes *as to detach the sand*." The Mountain of the Bell has been since carefully explored by Lieutenant J. Welsted, of the Indian navy; and the reader may see it exhibited in a fine lithograph, in his travels, as a vast irregularly conical mass of broken stone, somewhat resembling one of our Highland cairns, though, of course, on a scale immensely more huge, with a steep, angular slope of sand resting in a hollow in one of its sides, and rising to nearly its apex. "It forms," says Lieutenant Welsted, "one of a ridge of low, calcareous hills, at a distance of three and a half miles from the beach, to which a sandy plain, extending with a gentle rise to their base, connects them. Its height, about four hundred feet, as well as the material of which it is composed,—a light-colored friable sandstone,—is about the same as the rest of the chain; but an inclined plane of almost impalpable sand rises at an angle of forty degrees with the horizon, and is bounded by a semi-circle of rocks, presenting broken, abrupt, and pinnacled forms, and extending to the base of this remarkable hill. Although their shape and arrangement [Pg 79] in some respects may be said to resemble a whispering gallery, yet I determined by experiment that their irregular surface renders them but ill adapted for the production of an echo. Seated at a rock at the base of the sloping eminence, I directed one of the Bedouins to ascend; and it was not until he had reached some distance that I perceived the sand in motion, rolling down the hill to the depth of a foot. It did not, however, descend in one continued stream; but, as the Arab scrambled up, it spread out laterally and upwards, until a considerable portion of the surface was in motion. At their commencement the sounds might be compared to the faint strains of an Æolian harp when its strings first catch the breeze: as the sand became more violently agitated by the increased velocity of the descent, the noise more nearly resembled that produced by drawing the moistened fingers over glass. As it reached the base, the reverberations attained the loudness of distant thunder, causing the rock on which we were seated to vibrate; and our camels,—animals not easily frightened,—became so alarmed that it was with difficulty their drivers could restrain them."

"The hill of *Reg-Rawan* or the 'Moving Sand,'" says the late Sir Alexander Burnes, by whom the place was visited in the autumn of 1837, and who has recorded his visit in a brief paper, illustrated by a rude lithographic view, in the "Journal of the Asiatic Society" for 1838, "is about forty miles north of Cabul, towards Hindu-kush, and near the base of the mountains." It rises to the height of about four hundred feet, in an angle formed by the junction of two ridges of hills; and a sheet of sand, "pure as that of the sea-shore," and which slopes in an angle of forty degrees, reclines against it from base to summit. As represented in the lithograph, there projects over the steep sandy slope on each side, as in the "Mountain of the Bell," still steeper [Pg 80] barriers of rock; and we are told by Sir Alexander, that though "the mountains here are generally composed of granite or mica, at *Reg-Rawan* there is sandstone and lime." The situation of the sand is curious, he adds: it is seen from a great distance; and as there is none other in the neighborhood, "it might almost be imagined, from its appearance, that the hill had been cut in two, and that the sand had gushed forth as from a sand-bag." "When set in motion by a body of people who slide down it, a sound is emitted. On the first trial we distinctly heard two loud hollow sounds, such as would be given by a large drum;"—"there is an echo in the place; and the inhabitants have a belief that the sounds are only heard on Friday, when the saint of *Reg-Rawan*, who is interred hard by, permits." The phenomenon, like the resembling one in Arabia, seems to have attracted attention among the inhabitants of the country at an early period; and the notice of an eastern annalist, the Emperor Baber, who flourished late in the fifteenth century, and, like Cæsar, conquered and recorded his conquests, still survives. He describes it as the *Khwaja Reg-Rawan*, "a small hill, in which there is a line of sandy ground reaching from the top to the bottom," from which there "issues in the summer season the sound of drums and nagarets." In connection with the fact that the musical sand of Eigg is composed of a disintegrated sandstone of the Oölite, it is not quite unworthy of notice that sandstone and lime enter into the composition of the hill of *Reg-Rawan*,—that the district in which the hill is situated is not a sandy one,—and that its slope of sonorous sand seems as if it had issued from its side. These various circumstances, taken together, lead to the inference that the sand may have originated in the decomposition of the rock beneath. It is further noticeable, that the *Jabel Nakous* is composed of a white friable sandstone, resembling that of the white friable [Pg 81] bed of the Bay of Laig, and that it belongs to nearly the same geological era. I owe to the kindness of Dr. Wilson of Bombay, two specimens which he picked up in Arabia Petræa, of spines of Cidarites of the mace-formed type so common in the Chalk and Oölite, but so rare in the older formations. Dr. Wilson informs me that they are of frequent occurrence in the desert of Arabia Petræa, where they are termed by the Arabs petrified olives; that nummulites are also abundant in the district; and that the various secondary rocks he examined in his route through it seem to belong to the Cretaceous group. It appears not improbable, therefore, that all the sonorous sand in the world yet discovered is formed, like that of Eigg, of disintegrated sandstone; and at least two-thirds of it of the disintegrated sandstone of secondary formations, newer than the Lias. But how it should be at all sonorous, whatever its age or origin, seems yet to be discovered. There are few substances that appear worse suited than sand to communicate to the atmosphere those vibratory undulations that are the producing causes of sound: the grains, even when sonorous individually, seem, from their inevitable contact with each other, to exist under the influence of that simple law in acoustics which arrests the tones of the ringing glass or struck bell, immediately as they are but touched by some foreign body, such as the hand or finger. The one grain, ever in contact with several other grains, is a glass or bell on which the hand always rests. And the difficulty has been felt and acknowledged. Sir John Herschel, in referring to the phenomenon of the *Jabel Nakous*, in his "Treatise on Sound," in the "Encyclopædia Metropolitana," describes it as to him "utterly inexplicable;" and Sir David Brewster, whom I had the pleasure of meeting in December last, assured me it was not less a puzzle to him than to Sir John. An eastern traveller, who attributes its [Pg 82] production to "a reduplication of impulse setting air in vibration in a focus of echo," means, I suppose, saying nearly the same thing as the two philosophers, and merely conveys his meaning in a less simple style.

I have not yet procured what I expect to procure soon,—sand enough from the musical bay at Laig to enable me to make its sonorous qualities the subject of experiment at home. It seems doubtful whether a small quantity set in motion on an artificial slope will serve to evolve the phenomena which have rendered the Mountain of the Bell so famous. Lieutenant Welsted informs us, that when his Bedouin first set the sand in motion, there was scarce any perceptible sound heard;—it was rolling downwards for many yards around him to the depth of a foot, ere the music arose; and it is questionable whether the effect could be elicited with some fifty or sixty pounds weight of the sand of Eigg, on a slope of but at most a few feet, which it took many hundred weight of sand of *Jabel Nakous*, and a slope of many yards, to produce. But in the stillness of a close room, it is just possible that it may. I have, however, little doubt, that from small quantities the sound evoked by the foot on the shore may be reproduced: enough will lie within the reach of experiment to demonstrate the strange difference which exists between this sonorous sand of the Oölite, and the common unsonorous sand of our sea-beaches; and it is certainly worth while examining into the nature and producing causes of a phenomenon so curious in itself, and which has been characterized by one of the most distinguished of living philosophers as "the most celebrated of all the acoustic wonders which the natural world presents to us." In the forthcoming number of the "North British Review,"—which appears on Monday first, [1]—the reader will find the sonorous sand of Eigg referred to, in an [Pg 83] article the authorship of which will scarcely be mistaken. "We have here," says the writer, after first describing the sounds of *Jabel Nakous*, and then referring to those of Eigg, "the phenomenon in its simple state, disembarrassed from reflecting rocks, from a hard bed beneath, and from cracks and cavities that might be supposed to admit the sand; and indicating as its cause, either the accumulated vibration of the air when struck by the driven sand, or the accumulated sounds occasioned by the mutual impact of the particles of sand against each other. If a musket-ball passing through the air emits a whistling note, each individual particle of sand must do the same, however faint be the note which it yields; and the accumulation of these infinitesimal vibrations must constitute an audible sound, varying with the number and velocity of moving particles. In like manner, if two plates of silex or quartz, which are but large crystals of sand, give out a musical sound when mutually struck, the impact or collision of two minute crystals or particles of sand must do the same, in however inferior a degree; and the union of all these sounds, though singly imperceptible, may constitute the musical notes of the Bell Mountain, or the lesser sounds of the trodden sea-beach at Eigg."

Here is a vigorous effort made to unlock the difficulty. I should, however, have mentioned to the philosophic writer,—what I inadvertently failed to do,—that the sounds elicited from the sand of Eigg seem as directly evoked by the slant blow dealt it by the foot, as the sounds similarly evoked from a highly waxed floor, or a board strewed over with ground rosin. The sharp shrill note follows the stroke, altogether independently of the grains driven into the air. My omission may serve to show how much safer it is for those minds of the observant order, that serve as hands and eyes to the reflective ones, to prefer incurring the risk of [Pg 84] being even tediously minute in their descriptions, to the danger of being inadequately brief in them. But, alas! for purposes of exact science, rarely are verbal descriptions otherwise than inadequate. Let us look, for example, at the various accounts given us of *Jabel Nakous*. There are strange sounds heard proceeding from a hill in Arabia, and various travellers set themselves to describe them. The tones are those of the convent *Nakous*, says the wild Arab;—there must be a convent buried under the hill. More like the sounds of a humming-top, remarks a phlegmatic German traveller. Not quite like them, says an English one in an Oxford gown;—they resemble rather the striking of a clock. Nay, listen just a little longer and more carefully, says a second Englishman, with epaulettes on his shoulder: "the sounds at their commencement may be compared to the faint strains of an Æolian harp when its strings first catch the breeze," but anon, as the agitation of the sand increases, they "more nearly resemble those produced by drawing the moistened fingers over glass." Not at all, exclaims the warlike Zahor Ed-din Muhammed Baber, twirling his whiskers: "I know a similar hill in the country towards Hindu-kush: it is the sound of drums and nagarets that issues from the sand." All we really know of this often-described music of the desert, after reading all the descriptions, is, that its tones bear certain analogies to certain other tones,—analogies that seem stronger in one direction to one ear, and stronger in another direction to an ear differently constituted, but which do not exactly resemble any other sounds in nature. The strange music of *Jabel Nakous*, as a combination of tones, is essentially unique.

[Pg 85]

CHAPTER V.

Trap-Dykes—"Cotton Apples"—Alternation of Lacustrine with Marine Remains—Analogy from the Beds of Esk—Aspect of the Island on its narrow Front—The Puffin—Ru-Stoir—Development of Old Red Sandstone—Striking Columnar character of Ru-Stoir—Discovery of Reptilian Remains—John Stewart's wonder at the Bones in the Stones—Description of the Bones—"Dragons, Gorgons, and Chimeras"—Exploration and Discovery pursued—The Midway Shieling—A Celtic Welcome—Return of the Yacht—"Array of Fossils new to Scotch Geology"—A Geologist's Toast—Hoffman and his Fossil.

We leave behind us the musical sand, and reach the point of the promontory which forms the northern extremity of the Bay of Laig. Wherever the beach has been swept bare, we see it floored with trap-dykes worn down to the level, but in most places accumulations of huge blocks of various composition cover it up, concealing the nature of the rock beneath. The long semi-circular wall of precipice which, sweeping inwards at the bottom of the bay, leaves to the inhabitants between its base and the beach their fertile meniscus of land, here abuts upon the coast. We see its dark forehead many hundred feet overhead, and the grassy platform beneath, now narrowed to a mere talus, sweeping upwards to its base from the shore,—steep, broken, lined thick with horizontal pathways, mottled over with ponderous masses of rock.

Among the blocks that load the beach, and render our onward progress difficult and laborious, we detect occasional fragments of an amygdaloidal basalt, charged with a white zeolite, consisting of crystals so extremely slender that the balls, with their light fibrous contents, remind us of cotton [Pg 86] apples divested of the seeds. There occur, though more rarely, masses of a hard white sandstone, abounding in vegetable impressions, which, from their sculptured markings, recalled to memory the Sigillaria of the Coal Measures. Here and there, too, we find fragments of a calcareous stone, so largely charged with compressed shells, chiefly bivalves, that it may be regarded as a shell breccia. There occur, besides, slabs of fibrous limestone, exactly resembling the limestone of the ichthyolite beds of the Lower Old Red; and blocks of a hard gray stone, of silky lustre in the fresh fracture, thickly speckled with carbonaceous markings. These fragmentary masses,—all of them, at least, except the fibrous limestone, which occurs in mere plank-like bands,—represent distinct beds, of which this part of the island is composed, and which present their edges, like courses of ashlar in a building, in the splendid section that stretches from the tall brow of the precipice to the beach; though in the slopes of the talus, where the lower beds appear in but occasional protrusions and land-slips, we find some difficulty in tracing their order of succession.

Near the base of the slope, where the soil has been undermined and the rock laid bare by the waves, there occur beds of a bituminous black shale,—resembling the dark shales so common in the Coal Measures,—that seem to be of fresh water or estuary origin. Their fossils, though numerous, are ill preserved; but we detect in them scales and plates of fishes, at least two species of minute bivalves, one of which very much resembles a Cyclas; and in some of the fragments, shells of Cypris lie embedded in considerable abundance. After all that has been said and written by way of accounting for those alternations of lacustrine with marine remains, which are of such frequent occurrence in the various formations, secondary and tertiary, from the Coal Measures downwards, it does seem strange enough [Pg 87] that the estuary, or fresh-water lake, should so often in the old geologic periods have changed places with the sea. It is comparatively easy to conceive that the inner Hebrides should have once existed as a broad ocean sound, bounded on one or either side by Oölitic islands, from which streams descended, sweeping with them, to the marine depths, productions, animal and vegetable, of the land. But it is less easy to conceive, that in that sound, the area covered by the ocean one year should have been covered by a fresh-water lake in perhaps the next, and then by the ocean again a few years after. And yet among the Oölitic deposits of the Hebrides evidence seems to exist that changes of this nature actually took place. I am not inclined to found much on the apparently fresh-water character of the bituminous shales of Eigg;—the embedded fossils are all too obscure to be admitted in evidence; but there can exist no doubt that fresh water, or at least estuary formations, do occur among the marine Oölites of the Hebrides. Sir R. Murchison, one of the most cautious, as he is certainly one of the most distinguished, of living geologists, found in a northern district of Skye, in 1826, a deposit containing Cyclas, Paludina, Neritina,—all shells of unequivocally fresh-water origin,—which must have been formed, he concludes, in either a lake or estuary. What had been sea at one period had been estuary or lake at another. In every case, however, in which these intercalated deposits are restricted to single strata of no great thickness, it is perhaps safer to refer their formation to the agency of temporary land-floods, than to that of violent changes of level, now elevating and now depressing the surface. There occur, for instance, among the marine Oölites of Brora,—the discovery of Mr. Robertson, of Inverugie,—two strata containing fresh-water fossils in abundance; but the one stratum is little more than an inch in thickness,—the other [Pg 88] little more than a foot; and it seems considerably more probable, that such deposits should have owed their existence to extraordinary land-floods, like those which in 1829 devastated the province of Moray, and covered over whole miles of marine beach with the spoils of land and river, than that a sea-bottom should have been elevated for their production, into a fresh-water lake, and then let down into a sea-bottom again. We find it recorded in the "Shepherd's Calendar," that after the thaw which followed the great snow-storm of 1794, there were found on a part of the sands of the Solway Frith known as the Beds of Esk, where the tide disgorges much of what is thrown into it by the rivers, "one thousand eight hundred and forty sheep, nine black cattle, three horses, two men, one woman, forty-five dogs, and one hundred and eighty hares, beside a number of meaner animals." A similar storm in an earlier time, with a soft sea-bottom prepared to receive and retain its spoils, would have formed a fresh-water stratum intercalated in a marine deposit.

Rounding the promontory, we lose sight of the Bay of Laig, and find the narrow front of the island that now presents itself exhibiting the appearance of a huge bastion. The green talus slopes upwards, as its basement, for full three hundred feet: and a noble wall of perpendicular rock, that towers over and beyond for at least four hundred feet more, forms the rampart. Save towards the sea, the view is of but limited extent; we see it restricted, on the landward side, to the bold face of the bastion; and in a narrow and broken dell that runs nearly parallel to the shore for a few hundred yards between the top of the talus and the base of the rampart,—a true covered way,—we see but the rampart alone. But the dizzy front of black basalt, dark as night, save where a broad belt of light-colored sandstone traverses it in an angular direction, like a white [Pg 89] sash thrown across a funeral robe,—the fantastic peaks and turrets in which the rock terminates atop,—the masses of broken ruins, roughened with moss and lichen, that have fallen from above, and lie scattered at its base,—the extreme loneliness of the place, for we have left behind us every trace of the human family,—and the expanse of solitary sea which it commands,—all conspire to render the scene a profoundly imposing one. It is one of those scenes in which a man feels that he is little, and that nature is great. There is no precipice in the island in which the puffin so delights to build as among the dark pinnacles overhead, or around which the silence is so frequently broken by the harsh scream of the eagle. The sun had got far adown the sky ere we had reached the covered way at the base of the rock. All lay dark below; and the red light atop, half absorbed by the dingy hues of the stone, shone with a gleam so faint and melancholy, that it served but to deepen the effect of the shadows.

The puffin, a comparatively rare bird in the inner Hebrides, builds, I was told, in great numbers in the continuous line of precipice which, after sweeping for a full mile round the Bay of Laig, forms the pinnacled rampart here, and then, turning another angle of the island, runs on parallel to the coast for about six miles more. In former times the puffin furnished the islanders, as in St. Kilda, with a staple article of food, in those hungry months of summer in which the stores of the old crop had begun to fail, and the new crop had not yet ripened; and the people of Eigg, taught by their necessities, were bold cragsmen. But men do not peril life and limb for the mere sake of a meal, save when they cannot help it; and the introduction of the potato has done much to put out the practice of climbing for the bird, except among a few young lads, who find excitement enough in the work to pursue it for its own sake, as an [Pg 90] amusement. I found among the islanders what was said to be a piece of the natural history of the puffin, sufficiently apocryphal to remind one of the famous passage in the history of the barnacle, which traced the lineage of the bird to one of the pedunculated cirripedes, and the lineage of the cirripede to a log of wood. The puffin feeds its young, say the islanders, on an oily scum of the sea, which renders it such an unwieldy mass of fat, that about the time when it should be beginning to fly, it becomes unable to get out of its hole. The parent bird, not in the least puzzled, however, treats the case medicinally, and,—like mothers of another two-legged genus, who, when their daughters get over stout, put them through a course of reducing acids to bring them down,—feeds it on sorrel leaves for several days together, till, like a boxer under training, it gets thinned to the proper weight, and becomes able, not only to get out of its cell, but also to employ its wings.

We pass through the hollow, and, reaching the farther edge of the bastion, towards the east, see a new range of prospect opening before us. There is first a long unbroken wall of precipice,—a continuation of the tall rampart overhead,—relieved along its irregular upper line by the blue sky. We mark the talus widening at its base, and expanding, as on the shores of the Bay of Laig, into an irregular grassy platform, that, sinking midway into a ditch-like hollow, rises again towards the sea, and presents to the waves a perpendicular precipice of redstone. The sinking sun shone brightly this evening;

and the warm hues of the precipice, which bears the name of *Ru-Stoir*,—the Red Head,—strikingly contrasted with the pale and dark tints of the alternating basalts and sandstones in the taller cliff behind. The ditch-like hollow, which seems to indicate the line of a fault, cuts off this red headland from all the other rocks of the island, from which it appears to differ as considerably [Pg 91] in texture as in hue. It consists mainly of thick beds of a pale red stone, which M'Culloch regarded as a trap, and which, intercalated with here and there a thin band of shale, and presenting not a few of the mineralogical appearances of what geologists of the school of the late Mr. Cunningham term Primary Old Red Sandstone, in some cases has been laid down as a deposit of Old Red proper, abutting in the line of a fault on the neighboring Oölites and basalts. In the geological map which I carried with me,—not one of high authority however,—I found it actually colored as a patch of this ancient system. The Old Red Sandstone is largely developed in the neighboring island of Rum, in the line of which the *Ru-Stoir* seems to have a more direct bearing than any of the other deposits of Eigg; and yet the conclusion regarding this red headland merely adds one proof more to the many furnished already, of the inadequacy of mineralogical testimony, when taken in evidence regarding the eras of the geologist. The hard red beds of *Ru-Stoir* belong, as I was fortunate enough this evening to ascertain, not to the ages of the Coccosteus and Pterichthys, but to the far later ages of the Plesiosaurus and the fossil crocodile. I found them associated with more reptilian remains, of a character more unequivocal than have been yet exhibited by any other deposit in Scotland.

What first strikes the eye, in approaching the *Ru-Stoir* from the west, is the columnar character of the stone. The precipices rise immediately over the sea, in rude colonnades of from thirty to fifty feet in height; single pillars, that have fallen from their places in the line, and exhibit a tenacity rare among the trap-rocks,—for they occur in unbroken lengths of from ten to twelve feet,—lie scattered below; and in several places where the waves have joined issue with the precipices in the line on which the [Pg 92] base of the columns rest, and swept away the supporting foundation, the colonnades open into roomy caverns, that resound to the dash of the sea. Wherever the spray lashes, the pale red hue of the stone prevails, and the angles of the polygonal shafts are rounded; while higher up all is sharp-edged, and the unweathered surface is covered by a gray coat of lichens. The tenacity of the prostrate columns first drew my attention. The builder scant of materials would have experienced no difficulty in finding among them sufficient lintels for apertures from eight to twelve feet in width. I was next struck with the peculiar composition of the stone; it much rather resembles an altered sandstone, in at least the weathered specimens, than a trap, and yet there seemed nothing to indicate that it was an *Old Red* Sandstone. Its columnar structure bore evidence to the action of great heat; and its pale red color was exactly that which the Oölitic sandstones of the island, with their slight ochreous tinge, would assume in a common fire. And so I set myself to look for fossils. In the columnar stone itself I expected none, as none occur in vast beds of the unaltered sandstones, out of some one of which I supposed it might possibly have been formed; and none I found: but in a rolled block of altered shale of a much deeper red than the general mass, and much more resembling Old Red Sandstone, I succeeded in detecting several shells, identical with those of the deposit of blue clay described in a former chapter. There occurred in it the small univalve resembling a Trochus, together with the oblong bivalve, somewhat like a Tellina; and, spread thickly throughout the block, lay fragments of coprolitic matter, and the scales and teeth of fishes. Night was coming on, and the tide had risen on the beach; but I hammered lustily, and laid open in the dark red shale a vertebral joint, a rib, and a parallelogramical fragment of solid bone, none [Pg 93] of which could have belonged to any fish. It was an interesting moment for the curtain to drop over the promontory of *Ru-Stoir*; I had thus already found in connection with it well nigh as many reptilian remains as had been found in all Scotland before,—for there could exist no doubt that the bones I laid open were such; and still more interesting discoveries promised to await the coming morning, and a less hasty survey. We found a hospitable meal awaiting us at a picturesque old two-story house, with, what is rare in the island, a clump of trees beside it, which rises on the northern angle of the Oölitic meniscus; and after our day's hard work in the fresh sea-air, we did ample justice to the viands. Dark night had long set in ere we reached our vessel.

Next day was Saturday; and it behooved my friend, the minister,—as scrupulously careful in his pulpit preparations for the islanders of Eigg as if his congregation were an Edinburgh one,—to remain on board, and study his discourse for the morrow. I found, however, no unmeet companion for my excursion in his trusty mate John Stewart. John had not very much English, and I had no Gaelic; but we contrived to understand one another wonderfully well; and ere evening I had taught him to be quite as expert in hunting dead crocodiles as myself. We reached the *Ru-Stoir*, and set hard to work with hammer and chisel. The fragments of red shale were strewed thickly along the shore for at least three quarters of a mile; wherever the red columnar rock appeared, there lay the shale, in water-worn blocks, more or less indurated; but the beach was covered over with shingle and detached masses of rock, and we could nowhere find it *in situ*. A winter storm powerful enough to wash the beach bare might do much to assist the explorer. There is a piece of shore on the eastern coast of Scotland, on which for years [Pg 94] together I used to pick up nodular masses of lime containing fish of the Old Red Sandstone; but nowhere in the neighborhood could I find the ichthyolite bed in which they had originally formed. The storm of a single night swept the beach; and in the morning the ichthyolites lay revealed *in situ* under a stratum of shingle which I had a hundred times examined, but which, though scarce a foot in thickness, had concealed from me the ichthyolite bed for five twelvemonths together!

Wherever the altered shale of *Ru-Stoir* has been thrown high on the beach, and exposed to the influences of the weather, we find it fretted over with minute organisms, mostly the scales, plates, bones, and teeth of fishes. The organisms, as is frequently the case, seem indestructible, while the hard matrix in which they are embedded has weathered from around them. Some of the scales present the rhomboidal outline, and closely resemble those of the *Lepidotus Minor* of the Weald; others approximate in shape to an isosceles triangle. The teeth are of various forms: some of them, evidently palatal, are mere blunted protuberances glittering with enamel,—some of them present the usual slim, thorn-like type common in the teeth of the existing fish of our coasts,—some again are squat and angular, and rest on rectilinear bases, prolonged considerably on each side of the body of the tooth, like the rim of a hat or the flat head of a scupper nail. Of the occipital plates, some present a smooth enamelled surface, while some are thickly tuberculated,—each tubercle bearing a minute depression in its apex, like a crater on the summit of a rounded hill. We find reptilian bones in abundance,—a thing new to Scotch geology,—and in a state of keeping peculiarly fine. They not a little puzzled John Stewart: he could not resist the evidence of his senses: they were bones, he said, real bones,—there could be no doubt of that: *there* were the joints of [Pg 95] a backbone, with the hole the brain-marrow had passed through; and *there* were shank-bones and ribs, and fishes' teeth; but how, he wondered, had they all got into the very heart of the hard red stones? He had seen what was called wood, he said, dug out of the side of the Scuir, without being quite certain whether it was wood or no; but there could be no uncertainty here. I laid open numerous vertebræ of various forms,—some with long spinous processes rising over the body or *centrum* of the bone,—which I found in every instance, unlike that of the Ichthyosaurus, only moderately concave on the articulating faces; in others the spinous process seemed altogether wanting. Only two of the number bore any mark of the suture which unites, in most reptiles, the annular process to the centrum; in the others both centrum and process seemed anchylosed, as in quadrupeds, into one bone; and there remained no scar to show that the suture had ever existed. In some specimens the ribs seem to have been articulated to the sides of the centrum; in others there is a transverse process, but no marks of articulation. Some of the vertebræ are evidently dorsal, some cervical, one apparently caudal; and almost all agree in showing in front two little eyelets, to which the great descending artery seems to have sent out blood-vessels in pairs. The more entire ribs I was lucky enough to disinter have, as in those of crocodileans, double heads; and a part of a fibula, about four inches in length, seems also to belong to this ancient family. A large proportion of the other bones are evidently Plesiosaurian. I found the head of the flat humerus so characteristic of the extinct order to which the Plesiosaurus has been assigned, and two digital bones of the paddle, that, from their comparatively slender and slightly curved form, so unlike the digitals of its cogener the Ichthyosaurus, could have belonged evidently to no other reptile. I observed, too, in the slightly curved articulations [Pg 96] of not a few of the vertebræ, the gentle convexity in the concave centre, which, if not peculiar to the Plesiosaurus, is at least held to distinguish it from most of its contemporaries. Among the various nondescript organisms of the shale, I laid open a smooth angular bone, hollowed something like a grocer's scoop; a three-pronged caltrop-looking bone, that seems to have formed part of a pelvic arch; another angular bone, much massier than the first, regarding the probable position of which I could not form a conjecture, but which some of my geological friends deem cerebral; an extremely dense bone, imperfect at each end, which presents the appearance of a cylinder slightly flattened; and various curious fragments, which, with what our Scotch museums have not yet acquired,—entire reptilian fossils for the purposes of comparison,—might, I doubt not, be easily assigned to their proper places. It was in vain that, leaving John to collect the scattered pieces of shale in which the bones occurred, I set myself again and again to discover the bed from which they had been detached. The tide had fallen, and a range of skerries lay temptingly off, scarce a hundred yards from the water's edge: the shale beds might be among them, with Plesiosauri and crocodiles stretching entire; and fain would I have swum off to them, as I had done oftener than once elsewhere, with my hammer in my teeth, and with shirt and drawers in my hat; but a tall brown forest of kelp and tangle in which even a seal might drown, rose thick and perilous round both shore and skerries; a slight swell was felting the long fronds together; and I deemed it better, on the whole, that the discoveries I had already made should be recorded, than that they should be lost to geology, mayhap for a whole age, in the attempt to extend them.

The water, beautifully transparent, permitted the eye to penetrate into its green depths for many fathoms around, [Pg 97] though every object presented, through the agitated surface, an uncertain and fluctuating outline. I could see, however, the pink-colored urchin warping himself up, by his many cables, along the steep rock-sides; the green crab stalking along the gravelly bottom; a scull of small rock-cod darting hither and thither among the tangle-roots; and a few large medusæ slowly flapping their continuous fins of gelatine in the opener spaces, a few inches under the surface. Many curious families had their representatives within the patch of sea which the eye commanded; but the strange creatures that had once inhabited it by thousands, and whose bones still lay sepulchred on its shores, had none. How strange, that the identical sea heaving around stack and skerry in this remote corner of the Hebrides should have once been thronged by reptile shapes more strange than poet ever imagined,—dragons, gorgons and chimeras! Perhaps of all the extinct reptiles, the Plesiosaurus was the most extraordinary. An English geologist has described it, grotesquely enough, and yet most happily, as a snake *threaded* through a tortoise. And here on this very spot, must these monstrous dragons have disported and fed; here must they have raised their little reptile heads and long swan-like necks over the surface, to watch an antagonist or select a victim; here must they have warred and wedded, and pursued all the various instincts of their unknown natures. A strange story, surely, considering it is a true one! I may mention in the passing, that some of the fragments of the shale in which the remains are embedded have been baked by the intense heat into an exceedingly hard, dark-colored stone, somewhat resembling basalt. I must add further, that I by no means determine the rock with which we find it associated to be in reality an altered sandstone. Such is the appearance which it presents where weathered; but its general aspect is that of a porphyritic [Pg 98] trap. Be it what it may, the fact is not at all affected, that the shores, wherever it occurs on this tract of insular coast, are strewed with reptilian remains of the Oölite.

The day passed pleasantly in the work of exploration and discovery; the sun had already declined far in the west; and, bearing with us our better fossils, we set out, on our return, by the opposite route to that along the Bay of Laig, which we had now thrice walked over. The grassy talus so often mentioned continues to run on the eastern side of the island for about six miles, between the sea and the inaccessible rampart of precipice behind. It varies in breadth from about two to four hundred yards; the rampart rises over it from three to five hundred feet; and a noble expanse of sea, closed in the distance by a still nobler curtain of blue hills, spreads away from its base: and it was along this grassy talus that our homeward road lay. Let the Edinburgh reader imagine the fine walk under Salisbury Crags lengthened some twenty times,—the line of precipices above heightened some five or six times,—the gravelly slope at the base not much increased in altitude, but developed transversely into a green undulating belt of hilly pasture, with here and there a sunny slope level enough for the plough, and here and there a rough wilderness of detached crags and broken banks; let him further imagine the sea sweeping around the base of this talus, with the nearest opposite land—bold, bare and undulating atop—some six or eight miles distant; and he will have no very inadequate idea of the peculiar and striking scenery through which, this evening, our homeward route lay. I have scarce ever walked over a more solitary tract. The sea shuts it in on the one hand, and the rampart of rocks on the other; there occurs along its entire length no other human dwelling than a lonely summer shieling; for full one-half the way we saw no trace of man; and the wildness of the few cattle which [Pg 99] we occasionally startled in the hollows showed us that man was no very frequent visitor among them. About half an hour before sunset we reached the midway shieling.

Rarely have I seen a more interesting spot, or one that, from its utter loneliness, so impressed the imagination. The shieling, a rude low-roofed erection of turf and stone, with a door in the centre some five feet in height or so, but with no window, rose on the grassy slope immediately in front of the vast continuous rampart. A slim pillar of smoke ascends from the roof, in the calm, faint and blue within the shadow of the precipice, but it caught the sunlight in its ascent, and blushed, ere it melted into the ether, a ruddy brown. A streamlet came pouring from above in a long white thread, that maintained its continuity unbroken for at least two-thirds of the way; and then, untwisting into a shower of detached drops, that pattered loud and vehemently in a rocky recess, it again gathered itself up into a lively little stream, and, sweeping past the shieling, expanded in front into a circular pond, at which a few milch cows were leisurely slaking their thirst. The whole grassy talus, with a strip mayhap a hundred yards wide, of deep green sea, lay within the shadow of the tall rampart; but the red light fell, for many a mile beyond, on the glassy surface; and the distant Cuchullin Hills, so dark at other times, had all their prominent slopes and jutting precipices tipped with bronze; while here and there a mist streak, converted into bright flame, stretched along their peaks or rested on their sides. Save the lonely shieling, not a human dwelling was in sight. An island girl of eighteen, more than merely good-looking, though much embrowned by the sun, had come to the door to see who the unwonted visitors might be, and recognized in John Stewart an old acquaintance. John informed her in her own language that I was Mr. Swanson's sworn friend, and not a *Moderate*, but one of their own people, and that I [Pg 100] had fasted all day, and had come for a drink of milk. The name of her minister proved a strongly recommendatory one: I have not yet seen the true Celtic interjection of welcome,—the kindly "O o o,"—attempted on paper; but I had a very agreeable specimen of it on this occasion, *viva voce*. And as she set herself to prepare for us a rich bowl of mingled milk and cream, John and I entered the shieling. There was a turf fire at the one end, at which there sat two little girls, engaged in keeping up the blaze under a large pot, but sadly diverted from their work by our entrance; while the other end was occupied by a bed of dry straw, spread on the floor from wall to wall, and fenced off at the foot by a line of stones. The middle space was occupied by the utensils and produce of the dairy,—flat wooden vessels of milk, a butter-churn, and a tub half-filled with curd; while a few cheeses, soft from the press, lay on a shelf above. The little girls were but occasional visitors, who had come, out of a juvenile frolic, to pass the night in the place; but I was informed by John that the shieling had two other inmates, young women, like the one so hospitably engaged in our behalf, who were out at the milking, and that they lived here all alone for several months every year, when the pasturage was at its best, employed in making butter and cheese for their master, worthy Mr. M'Donald of Keill. They must often feel lonely when night has closed darkly over mountain and sea, or in those dreary days of mist and rain so common in the Hebrides, when nought may be seen save the few shapeless crags that stud the nearer hillocks around them, and nought heard save the moaning of the wind in the precipices above, or the measured dash of the wave on the wild beach below. And yet they would do ill to exchange their solitary life and rude shieling for the village dwellings and gregarious habits of the females who ply their rural labors in bands among the [Pg 101] rich fields of the Lowlands, or for the unwholesome backroom and weary task-work of the city seamstress. The sunlight was fading from the higher hill-tops of Skye and Glenelg as we bade farewell to the lonely shieling and the hospitable island girl.

The evening deepened as we hurried southwards along the scarce visible pathway, or paused for a few seconds to examine some shattered block, bulky as a Highland cottage, that had fallen from the precipice above. Now that the whole landscape lay equally in shadow, one of the more picturesque peculiarities of the continuous rampart came out more strongly as a feature of the scene than when a strip of shade rested along the face of the rock, imparting to it a retiring character, and all was sunshine beyond. A thick bed of white sandstone, as continuous as the rampart itself, runs nearly horizontally about midway in the precipice for mile after mile, and, standing out in strong contrast with the dark-colored trap above and below, reminds one of a belt of white hewn work in a basalt house front, or rather,—for there occurs above a second continuous strip, of an olive hue, the color assumed, on weathering by a bed of amygdaloid,—of a piece of dingy old-fashioned furniture, inlaid with one stringed belt of bleached holly, and another of faded green-wood. At some of the more accessible points I climbed to the line of white belting, and found it to consist of the same soft quartzy sandstone that in the Bay of Laig furnishes the musical sand. Lower down there occur, alternating with the trap, beds of shale and of blue clay, but they are lost mostly in the talus. Ill adapted to resist the frosts and rains of winter, their exposed edges have mouldered into a loose soil, now thickly covered over with herbage; and, but for the circumstance that we occasionally find them laid bare by a water-course, we would scarce be aware of their existence at all. The shale exhibits everywhere, [Pg 102] as on the opposite side of the *Ru-Stoir*, faint impressions of a minute shell resembling a Cyclas, and ill-preserved fragments of fish-scales. The blue clay I found at one spot where the pathway had cut deep into the hill-side, richly charged with bivalves of the species I had seen so abundant in the resembling clay of the Bay of Laig; but the closing twilight prevented me from ascertaining whether it also contained the characteristic univalves of the deposit, and whether its shells,—for they seem identical with those of the altered shales of the *Ru-Stoir*,—might not be associated, like these, with reptilian remains. Night fell fast, and the streaks of mist that had mottled the hills at sunset began to spread gray over the heavens in a continuous curtain; but there was light enough left to show me that the trap became more columnar as we neared our journey's end. One especial jutting in the rock presented in the gloom the appearance of an ancient portico, with pediment and cornice, such as the traveller sees on the hill-sides of Petræa in front of some old tomb; but it may possibly appear less architectural by day. At length, passing from under the long line of rampart, just as the stars that had begun to twinkle over it were disappearing, one after one, in the thickening vapor, we reached the little bay of Kildonan, and found the boat waiting us on the beach. My friend the minister, as I entered the cabin, gathered up his notes from the table, and gave orders for the tea-kettle; and I spread out before him—a happy man—an array of fossils new to Scotch Geology. No one not an enthusiastic geologist or a zealous Roman Catholic can really know how vast an amount of interest may attach to a few old bones. Has the reader ever heard how fossil relics once saved the dwelling of a monk, in a time of great general calamity, when all his other relics proved of no avail whatever?

[Pg 103]

Thomas Campbell, when asked for a toast in a society of authors, gave the memory of Napoleon Bonaparte; significantly adding, "he once hung a bookseller." On a nearly similar principle I would be disposed to propose among geologists a grateful bumper in honor of the revolutionary army that besieged Maestricht. That city, some seventy-five or eighty years ago, had its zealous naturalist in the person of M. Hoffmann, a diligent excavator in the quarries of St. Peter's mountain, long celebrated for its extraordinary fossils. Geology, as a science, had no existence at the time; but Hoffmann was doing, in a quiet way, all he could to give it a beginning;—he was transferring from the rock to his cabinet, shells, and corals, and crustacea, and the teeth and scales of fishes, with now and then the vertebræ, and now and then the limb-bone, of a reptile. And as he honestly remunerated all the workmen he employed, and did no manner of harm to any one, no one heeded him. On one eventful morning, however, his friends the quarriers laid bare a most extraordinary fossil,—the occipital plates of an enormous saurian, with jaws four and a half feet long, bristling over with teeth, like *chevaux de frise*; and after Hoffmann, who got the block in which it lay embedded, cut out entire, and transferred to his house, had spent week after week in painfully relieving it from the mass, all Maestricht began to speak of it as something really wonderful.

There is a cathedral on St. Peter's mountain,—the mountain itself is church-land; and the lazy canon, awakened by the general talk, laid claim to poor Hoffmann's wonderful fossil as *his* property. He was lord of the manor, he said, and the mountain and all that it contained belonged to him. Hoffmann defended his fossil as he best could in an expensive lawsuit; but the judges found the law clean against him; the huge reptile head was declared to be "treasure trove" [Pg 104] escheat to the lord of the manor; and Hoffmann, half broken-hearted, with but his labor and the lawyer's bills for his pains, saw it transferred by rude hands from its place in his museum, to the residence of the grasping churchman. The huge fossil head experienced the fate of Dr. Chalmer's two hundred churches. Hoffmann was a philosopher, however, and he continued to observe and collect as before; but he never found such another fossil; and at length, in the midst of his ingenious labors, the vital energies failed within him, and he broke down and died. The useless canon lived on. The French Revolution broke out; the republican army invested Maestricht; the batteries were opened; and shot and shell fell thick on the devoted city. But in one especial quarter there alighted neither shot nor shell. All was safe around the canon's house. Ordinary relics would have availed him nothing in the circumstances,—no, not "the three kings of Cologne," had he possessed the three kings entire, or the jaw-bones of the "eleven thousand virgins;" but there was virtue in the jaw-bones of the Mosasaurus, and safety in their neighborhood. The French *savans*, like all the other *savans* of Europe, had heard of Hoffmann's fossil, and the French artillery had been directed to play wide of the place where it lay. Maestricht surrendered; the fossil was found secreted in a vault, and sent away to the *Jardin des Plantes* at Paris, maugre the canon, to delight there the heart of Cuvier; and the French, generously addressing themselves to the heirs of Hoffmann as its legitimate owners, made over to them a considerable sum of money as its price. They reversed the finding of the Maestricht judges; and all save the monks of St. Peter's have acquiesced in the justice of the decision.

[Pg 105]

CHAPTER VI.

Something for Non-geologists—Man Destructive—A Better and Last Creation coming—A Rainy Sabbath—The Meeting House—The Congregation—The Sermon in Gaelic—The Old Wondrous Story—The Drunken Minister of Eigg—Presbyterianism without Life—Dr. Johnson's Account of the Conversion of the People of Rum—Romanism at Eigg—The Two Boys—The Freebooter of Eigg—Voyage Resumed—The Homeless Minister—Harbor of Isle Ornsay—Interesting Gneiss Deposit—A Norwegian Keep—Gneiss at Knock—Curious Chemistry—Sea-cliffs beyond Portsea—The Goblin Luidag—Scenery of Skye.

I reckon among my readers a class of non-geologists, who think my geological chapters would be less dull if I left out the geology; and another class of semi-geologists, who say there was decidedly too much geology in my last. With the present chapter, as there threatens to be an utter lack of science in the earlier half of it, and very little, if any, in the latter half, I trust both classes may be in some degree satisfied. It will bear reference to but the existing system of things,—assuredly not the last of the consecutive creations,—and to a species of animal that, save in the celebrated Guadaloupe specimens, has not yet been found locked up in stone. There have been much of violence and suffering in the old immature stages of being,—much, from the era of the Holoptychius, with its sharp murderous teeth and strong armor of bone, down to that of the cannibal Ichthyosaurus, that bears the broken remains of its own kind in its bowels,—much, again, from the times of the crocodile of the Oölite, down to the times of the fossil hyena and gigantic shark of the Tertiary. Nor, I fear, have matters greatly improved in that latest-born [Pg 106] creation in the series, that recognizes as its delegated lord the first tenant of earth accountable to his Maker. But there is a better and a last creation coming, in which man shall reappear, not to oppress and devour his fellow-men, and in which there shall be no such wrongs perpetrated as it is my present purpose to record,—"new heavens and a new earth, wherein dwelleth righteousness." Well sung the Ayrshire ploughman, when musing on the great truth that the present scene of being "is surely not the last,"—a truth corroborated since his day by the analogies of a new science,—

"The poor, oppressed, honest man,
Had never sure been born,
Had not there been some recompense
To comfort those that mourn."

It was Sabbath, but the morning rose like a hypochondriac wrapped up in his night-clothes,—gray in fog, and sad with rain. The higher grounds of the island lay hid in clouds, far below the level of the central hollow; and our whole prospect from the deck was limited to the nearer slopes, dank, brown, and uninhabited, and to the rough black crags that frown like sentinels over the beach. Now the rime thickened as the rain pattered more loudly on the deck; and even the nearer stacks and precipices showed as unsolid and spectral in the cloud as moonlight shadows thrown on a ground of vapor; anon it cleared up for a few hundred yards, as the shower lightened; and then there came in view, partially at least, two objects that spoke of man,—a deserted boat harbor, formed of loosely piled stone, at the upper extremity of a sandy bay; and a roofless dwelling beside it, with two ruinous gables rising over the broken walls. The entire scene suggested the idea [Pg 107] of a land with which man had done for ever;—the vapor-enveloped rocks,—the waste of ebb-uncovered sand,—the deserted harbor,—the ruinous house,—the melancholy rain-fretted tides eddying along the strip of brown tangle in the foreground,—and, dim over all, the thick, slant lines of the beating shower!—I know not that of themselves they would have furnished materials enough for a finished picture in the style of Hogarth's "End of all Things;" but right sure am I that in the hands of Bewick they would have been grouped into a tasteful and poetic vignette. We set out for church a little after eleven,—the minister encased in his ample-skirted storm-jacket of oiled canvas, and protected atop by a genuine *sou-wester*, of which the broad posterior rim eloped half a yard down his back; and I closely wrapped up in my gray maud, which proved, however, a rather indifferent protection against the penetrating powers of a true Hebridean drizzle. The building in which the congregation meets is a low dingy cottage of turf and stone, situated nearly opposite to the manse windows. It had been built by my friend, previous to the Disruption, at his own expense, for a Gaelic school, and it now serves as a place of worship for the people.

We found the congregation already gathered, and that the very bad morning had failed to lessen their numbers. There were a few of the male parishioners keeping watch at the door, looking wistfully out through the fog and rain for their minister; and at his approach nearly twenty more came issuing from the place,—like carder bees from their nest of dried grass and moss,—to gather round him, and shake him by the hand. The islanders of Eigg are an active, middle-sized race, with well-developed heads, acute intellects, and singularly warm feelings. And on this occasion at least there could be no possibility of mistake [Pg 108] respecting the feelings with which they regarded their minister. Rarely have I seen human countenances so eloquently vocal with veneration and love. The gospel message, which my friend had been the first effectually to bring home to their hearts,—the palpable fact of his sacrifice for the sake of the high principles which he has taught,—his own kindly disposition,—the many services which he has rendered them, for not only has he been the minister, but also the sole medical man, of the Small Isles, and the benefit of his practice they have enjoyed, in every instance, without fee or reward,—his new life of hardship and danger, maintained for their sakes amid sinking health and great privation,—their frequent fears for his safety when stormy nights close over the sea,—and they have seen his little vessel driven from her anchorage, just as the evening has fallen,—all these are circumstances that have concurred in giving him a strong hold on their affections.

The rude turf-building we found full from end to end, and all asteam with a particularly wet congregation, some of whom, neither very robust nor young, had travelled in the soaking drizzle from the farther extremities of the island. And, judging from the serious attention with which they listened to the discourse, they must have deemed it full value for all it cost them. I have never yet seen a congregation more deeply impressed, or that seemed to follow the preacher more intelligently; and I was quite sure, though ignorant of the language in which my friend addressed them, that he preached to them neither heresy nor nonsense. There was as little of the reverence of externals in the place as can well be imagined: an uneven earthen floor,—turf-walls on every side, and a turf-roof above,—two little windows of four panes a-piece, adown which the rain-drops were coursing thick and fast,—a pulpit grotesquely rude, that had never [Pg 109] employed the bred carpenter,—and a few ranges of seats of undressed deal, such were the mere materialisms of this lowly church of the people; and yet here, notwithstanding, was the living soul of a Christian community,—understandings convinced of the truth of the gospel, and hearts softened and impressed by its power.

My friend, at the conclusion of his discourse, gave a brief digest of its contents in English, for the benefit of his one Saxon auditor; and I found, as I had anticipated, that what had so moved the simple islanders was just the old wondrous story, which, though repeated and re-repeated times beyond number, from the days of the apostles till now, continues to be as full of novelty and interest as ever,—"God so loved the world, that he gave his only begotten Son, that whosoever believeth on Him should not perish, but have everlasting life." The great truths which had affected many of these poor people to tears, were exactly those which, during the last eighteen hundred years, have been active in effecting so many moral revolutions in the world, and which must ultimately triumph over all error and all oppression. On this occasion, as on many others, I had to regret my want of Gaelic. It was my misfortune to miss being born to this ancient language, by barely a mile of ferry. I first saw light on the southern shore of the Frith of Cromarty, where the strait is narrowest, among an old established Lowland community, marked by all the characteristics, physical and mental, of the Lowlanders of the southern districts; whereas, had I been born on the northern shore, I would have been brought up among a Celtic tribe, and Gaelic would have been my earliest language. Thus distinct was the line between the two races preserved, even after the commencement of the present century.

In returning to the Betsey during the mid-day interval [Pg 110] in the service, we passed the ruinous two-gabled house beside the boat-harbor. During the incumbency of my friend's predecessor, it had been the public-house of the island, and the parish minister was by far its best customer. He was in the practice of sitting in one of its dingy little rooms, day after day, imbibing whisky and peat-reek;

and his favorite boon companion on these occasions was a Roman Catholic tenant who lived on the opposite side of the island, and who, when drinking with the minister, used regularly to fasten his horse beside the door, till at length all the parish came to know that when the horse was standing outside the minister was drinking within. In course of time, through the natural gravitation operative in such cases, the poor incumbent became utterly scandalous, and was libelled for drunkenness before the General Assembly; but, as the island of Eigg lies remote from observation, evidence was difficult to procure; and had not the infatuated man got senselessly drunk one evening, when in Edinburgh on his trial, and staggered, of all places in the world, into the General Assembly, he would probably have died minister of Eigg. As the event happened, however, the testimony thus unwittingly furnished in the face of the Court that tried him was deemed conclusive;—he was summarily deposed from his office, and my friend succeeded him. Presbyterianism without the animating life is a poor shrunken thing: it never lies in state when it is dead; for it has no body of fine forms, or trapping of imposing ceremonies, to give it bulk or adornment: without the vitality of evangelism it is nothing; and in this low and abject state my friend found the Presbyterianism of Eigg. His predecessor had done it only mischief; nor had it been by any means vigorous before. Rum is one of the four islands of the parish; and all my readers must be familiar with Dr. Johnson's celebrated [Pg 111] account of the conversion to Protestantism of the people of Rum. "The inhabitants," says the Doctor, in his "Journey to the Western Islands," "are fifty-eight families, who continued Papists for some time after the laird became a Protestant. Their adherence to their old religion was strengthened by the countenance of the laird's sister, a zealous Romanist; till one Sunday, as they were going to mass under the conduct of their patroness, Maclean met them on the way, gave one of them a blow on the head with a yellow stick,—I suppose a cane, for which the Erse had no name, and drove them to the kirk, from which they have never departed. Since the use of this method of conversion, the inhabitants of Eigg and Canna who continue Papists call the Protestantism of Rum the religion of the yellow stick." Now, such was the kind of Protestantism that, since the days of Dr. Johnson, had also been introduced, I know not by what means, into Eigg. It had lived on the best possible terms with the

Popery of the island; the parish minister had soaked day after day in the public-house with a Roman Catholic boon companion; and when a Papist man married a Protestant woman, the woman, as a matter of course, became Papist also; whereas, when it was the man who was a Protestant, and the woman a Papist, the woman remained what she had been. Roman Catholicism was quite content with terms, actual though not implied, of a kind so decidedly advantageous; and the Roman Catholics used good-humoredly to urge on their neighbors the Protestants, that, as it was palpable they had no religion of any kind, they had better surely come over to them, and have some. In short, all was harmony between the two Churches. My friend labored hard, as a good and honest man ought, to impart to Protestantism in his parish the animating life of [Pg 112] the Reformation; and, through the blessing of God, after years of anxious toil, he at length fully succeeded.

I had got wet, and the day continued bad; and so, instead of returning to the evening sermon, which began at six, I remained alone aboard of the vessel. The rain ceased in little more than an hour after, and in somewhat more than two hours I got up on deck to see whether the congregation was not dispersing, and if it was not yet time to hang on the kettle for our evening tea. The unexpected apparition of some one aboard the Free Church yacht startled two ragged boys who were manœuvring a little boat a stone-cast away, under the rocky shores of *Eilean Chaisteil*, and who, on catching a glimpse of me, flung themselves below the thwarts for concealment. An oar dropped into the water; there was a hasty arm and half a head thrust over the gunwale to secure it; and then the urchin to whom they belonged again disappeared. Meanwhile the boat drifted slowly away: first one little head would appear for a moment over the gunwale, then another, as if reconnoitering the enemy; but I still kept my place on deck; and at length, tired out, the ragged little crew took to their oars, and rowed into a shallow bay at the lower extremity of the glebe, with a cottage, in size and appearance much resembling an ant-hill, peeping out at its inner extremity among some stunted bushes. I had marked the place before, and had been struck with the peculiarity of the choice that could have fixed on it as a site for a dwelling: it is at once the most inconvenient and picturesque on this side the island. A semi-circular line of columnar precipices, that somewhat resembles an amphitheatre turned outside in,—for the columns that overlook the area are quite as lofty as those which should form the amphitheatre's outer wall,—sweeps round a little bay, flat and sandy at half-tide, but bordered higher up by [Pg 113] a dingy, scarce passable beach of columnar fragments that have toppled from above. Between the beach and the line of columns there is a bosky talus, more thickly covered with brushwood than is at all common in the Hebrides, and scarce more passable than the rough beach at its feet. And at the bottom of this talus, with its one gable buried in the steep ascent,—for there is scarce a foot-breadth of platform between the slope and the beach,—and with the other gable projected to the tide-line on rugged columnar masses, stands the cottage. The story of the inmate,—the father of the two ragged boys,—is such a one as Crabbe would have delighted to tell, and as he could have told better than any one else.

He had been, after a sort, a freebooter in his time, but born an age or two rather late; and the law had proved over strong for him. On at least one occasion, perhaps oftener,—for his adventures are not all known in Eigg,—he had been in prison for sheep-stealing. He had the dangerous art of subsisting without the ostensible means, and came to be feared and avoided by his neighbors as a man who lived on them without asking their leave. With neither character nor a settled way of living, his wits, I am afraid, must have been often whetted by his necessities: he stole lest he should starve. For some time he had resided in the adjacent island of Muck; but, proving a bad tenant, he had been ejected by the agent of the landlord, I believe a very worthy man, who gave him half a boll of meal to get quietly rid of him, and pulled down his house, when he had left the island, to prevent his return. Betaking himself, with his boys, to a boat, he set out in quest of some new lodgment. He made his first attempt or two on the mainland, where he strove to drive a trade in begging, but he was always recognized as the convicted sheep-stealer, and driven back to the shore. At [Pg 114] length, after a miserable term of wandering, he landed in the winter season on Eigg, where he had a grown-up son, a miller; and, erecting a wretched shed with some spars and the old sail of a boat placed slantways against the side of a rock, he squatted on the beach, determined, whether he lived or died, to find a home on the island. The islanders were no strangers to the character of the poor forlorn creature, and kept aloof from him,—none of them, however, so much as his own son; and, for a time, my friend the minister, aware that he had been the pest of every community among which he had lived, stood aloof from him too, in the hope that at length, wearied out, he might seek for himself a lodgment elsewhere. There came on, however, a dreary night of sleet and rain, accompanied by a fierce storm from the sea; and intelligence reached the manse late in the evening, that the wretched sheep-stealer had been seized by sudden illness, and was dying on the beach. There could be no room for further hesitation in this case; and my friend the minister gave instant orders that the poor creature should be carried to the manse. The party, however, which he had sent to remove him found the task impracticable. The night was pitch dark; and the road, dangerous with precipices, and blocked up with rough masses of rock and stone, they found wholly impassable with so helpless a burden. And so, administering some cordials to the poor, hapless wretch, they had to leave him in the midst of the storm, with the old wet sail flapping about his ears, and the half-frozen rain pouring in upon him in torrents. He must have passed a miserable night, but it could not have been a whit more miserable than that passed by the minister in the manse. As the wild blast howled around his comfortable dwelling, and shook the casements as if some hand outside were assaying to open them, or as the rain [Pg 115] pattered sharp and thick on the panes, and the measured roar of the surf rose high over every other sound, he could think of only the wretched creature exposed to the fury of a tempest so terrible, as perchance wrestling in his death agony in the darkness beside the breaking wave, or as already stiffening on the shore. He was early astir next morning, and almost the first person he met was the poor sheep-stealer, looking more like a ghost than a living man. The miserable creature had mustered strength enough to crawl up from the beach. My friend has often met better men with less pleasure. He found a shelter for the poor outcast; he tended him, prescribed for him, and, on his recovery, gave him leave to build for himself the hovel at the foot of the crags. The islanders were aware they had got but an indifferent neighbor through the transaction, though none of them, with the exception of the poor creature's son, saw what else their minister could have done in the circumstances. But the miller could sustain no apology for the arrangement that had given him his vagabond father as a neighbor; and oftener than once the site of the rising hovel became a scene of noisy contention between parent and son. Some of the islanders informed me that they had seen the son engaged in pulling down the stones of the walls as fast as the father raised them up; and, save for the interference of the minister, the hut, notwithstanding the permission he gave, would scarce have been built.

On the morning of Monday we unloosed from our moorings, and set out with a light variable breeze for Isle Ornsay, in Skye, where the wife and family of Mr. Swanson resided, and from which he had now been absent for a full month. The island diminished, and assumed its tint of diluting blue, that waxed paler and paler hour after hour, as we left it slowly behind us; and the Scuir, projected boldly from its steep hill-top, resembled a sharp hatchet-edge [Pg 116] presented to the sky. "Nowhere," said my friend, "did I so thoroughly realize the Disruption of last year as at this spot. I had just taken my last leave of the manse; Mrs. Swanson had staid a day behind me in charge of a few remaining pieces of furniture, and I was bearing some of the rest, and my little boy Bill, scarce five years of age at the time, in the yacht with me to Skye. The little fellow had not much liked to part from his mother, and the previous unsettling of all sorts of things in the manse had bred in him thoughts he had not quite words to express. The further change to the yacht, too, he had deemed far from an agreeable one. But he had borne up, by way of being very manly; and he seemed rather amused that papa should now have to make his porridge for him, and to put him to bed, and that it was John Stewart, the sailor, who was to be the servant girl. The passage, however, was tedious and disagreeable; the wind blew a-head, and heart and spirits failing poor Bill, and somewhat seasick to boot, he lay down on the floor, and cried bitterly to be taken home. 'Alas, my boy!' I said, 'you have no home now: your father is like the poor sheep-stealer whom you saw on the shore of Eigg.' This view of matters proved in no way consolatory to poor Bill. He continued his sad wail, 'Home, home, home!' until at length he fairly sobbed himself asleep; and I never, on any other occasion, so felt the desolateness of my condition as when the cry of my boy,—'Home, home, home!'—was ringing in my ears."

We passed, on the one hand, Loch Nevis and Loch Hourn, two fine arms of the sea that run far into the mainland, and open up noble vistas among the mountains; and, on the other, the long undulating line of Sleat in Skye, with its intermingled patches of woodland and arable on the coast, and its mottled ranges of heath and rock above. [Pg 117] Towards evening we entered the harbor of Isle Ornsay, a quiet, well-sheltered bay, with a rocky islet for a breakwater on the one side, and the rudiments of a Highland village, containing a few good houses, on the other. Half a dozen small vessels were riding at anchor, curtained round, half-mast high, with herring nets; and a fleet of herring-boats lay moored beside them a little nearer the shore. There had been tolerable takes for a few nights in the neighboring sea, but the fish had again disappeared, and the fishermen, whose worn-out tackle gave such evidence of a long-continued run of ill-luck, as I had learned to interpret on the east coast, looked gloomy and spiritless, and reported a deficient fishery. I found Mrs. Swanson and her family located in one of the two best houses in the village, with a neat enclosure in front, and a good kitchen-garden behind. The following day I spent in exploring the rocks of the district,—a primary region with regard to organic existence, "without *form* and void." From Isle Ornsay to the Point of Sleat, a distance of thirteen miles, gneiss is the prevailing deposit; and in no place in the district are the strata more varied and interesting than in the neighborhood of Knockhouse, the residence of Mr. Elder, which I found pleasingly situated at the bottom of a little open bay, skirted with picturesque knolls partially wooded, that present to the surf precipitous fronts of rock. One insulated eminence, a gun-shot from the dwelling-house, that presents to the sea two mural fronts of precipice, and sinks in steep grassy slopes on two sides more, bears atop a fine old ruin. There is a blind-fronted massy keep, wrapped up in a mantle of ivy, perched at the one end, where the precipice sinks steepest; while a more ruinous though much more modern pile of building, perforated by a double row of windows, occupies the rest of the area. The square keep has lost its [Pg 118] genealogy in the mists of the past, but a vague tradition attributes its erection to the Norwegians. The more modern pile is said to have been built about three centuries ago by a younger son of M'Donald of the Isles; but it is added that, owing to the jealousy of his elder brother, he was not permitted to complete or inhabit it. I find it characteristic of most Highland traditions, that they contain speeches: they constitute true oral specimens of that earliest and rudest style of historic composition in which dialogue alternates with narrative. "My wise brother is building a fine house," is the speech preserved in this tradition as that of the elder son: "it is rather a pity for himself that he should be building it on another man's lands." The remark was repeated to the builder, says the story, and at once arrested the progress of the work. Mr. Elder's boys showed me several minute pieces of brass, somewhat resembling rust-eaten coin, that they had dug out of the walls of the old keep; but the pieces bore no impress of the dye, and seemed mere fragments of metal beaten thin by the hammer.

The gneiss at Knock is exceedingly various in its composition, and many of its strata the geologist would fail to recognize as gneiss at all. We find along the precipices its two unequivocal varieties, the schistose and the granitic, passing not unfrequently, the former into a true mica schist, the latter into a pale feldspathose rock, thickly pervaded by needle-like crystals of tremolite, that, from the style of the grouping, and the contrast existing between the dark green of the enclosed mineral, and the pale flesh-color of the ground, frequently furnishes specimens of great beauty. In some pieces the tremolite assumes the common fan-like form; in some, the crystals, lying at nearly right angles with each other, present the appearance of ancient characters inlaid in the rock; in some they [Pg 119] resemble the footprints of birds in a thin layer of snow; and in one curious specimen picked up by Mr. Swanson, in which a dark linear strip is covered transversely by crystals that project thickly from both its sides, the appearance presented is that of a minute stigmaria of the Coal Measures, with the leaves, still bearing their original green color, bristling thick around it. Mr. Elder showed me, intercalated among the gneiss strata of a little ravine in the neighborhood of Isle Ornsay, a thin band of a bluish-colored indurated clay, scarcely distinguishable, in the hand specimen, from a weathered clay-stone, but unequivocally a stratum of the rock. I have found the same stone existing, in a decomposed state, as a very tenacious clay, among the gneiss strata of the hill of Cromarty; and oftener than once had I amused myself in fashioning it, with tolerable success, into such rude pieces of pottery as are sometimes found in old sepulchral tumuli. Such are a few of the rocks included in the general gneiss deposit of Sleat. If we are to hold, with one of the most distinguished of living geologists, that the stratified primary rocks are aqueous deposits altered by heat, to how various a chemistry must they not have been subjected in this district! In one stratum, so softened that all its particles were disengaged to enter into new combinations, and yet not so softened but that it still maintained its lines of division from the strata above and below, the green tremolite was shooting its crystals into the pale homogeneous mass; while in another stratum the quartz drew its atoms apart in masses that assumed one especial form, the feldspar drew its atoms apart into masses that assumed another and different form, and the glittering mica built up its multitudinous layers between. Here the unctuous chlorite constructed its soft felt; there the micaceous schist arranged its undulating layers; yonder the [Pg 120] dull clay hardened amid the intense heat, but, when all else was changing, retained its structure unchanged. Surely a curious chemistry, and conducted on an enormous scale!

It had been an essential part of my plan to explore the splendid section of the Lower Oölite furnished by the line of sea-cliffs that, to the north of the Portree, rise full seven hundred feet over the beach; and on the morning of Wednesday I set out with this intention from Isle Ornsay, to join the mail gig at Broadford, and pass on to Portree,—a journey of rather more than thirty miles. I soon passed over the gneiss, and entered on a wide deposit, extending from side to side of the island, of what is generally laid down in our geological maps as Old Red Sandstone, but which, in most of its beds, quite as much resembles a quartz rock, and which, unlike any Old Red proper I have ever seen, passes, by insensible gradations, into the gneiss. [2] Wherever it has been laid bare in flat tables among the heath, we find it bearing those mysterious scratches on a polished surface which we so commonly find associated on the mainland with the boulder clay; but here, as in the Hebrides generally, the boulder clay is wanting. To the tract of Red Sandstone there succeeds a tract of Lias, which, also extending across the island, forms by far the most largely-developed deposit of this formation in Scotland. It occupies a flat dingy valley, about six miles in length, and that varies from two to four miles in breadth. The dreary interior is covered with mosses, and studded with inky pools, in which the botanist finds a few rare plants, and which were dimpled, as I passed them this [Pg 121] morning, with countless eddies, formed by myriads of small quick glancing trout, that seemed busily engaged in fly-catching. The rock appears but rarely,—all is moss, marsh, and pool; but in a few localities on the hill-sides, where some stream has cut into the slope, and disintegrated the softer shales, the shepherd finds shells of strange form strewed along the water-courses, or bleaching white among the heath. The valley,—evidently a dangerous one to the night traveller, from its bogs and its tarns,—is said to be haunted by a spirit peculiar to itself,—a mischievous, eccentric, grotesque creature, not unworthy, from the monstrosity of its form, of being associated with the old monsters of the Lias. Luidag—for so the goblin is called—has but one leg, terminating, like an ancient satyr's, in a cloven foot; but it is furnished with two arms, bearing hard fists at the end of them, with which it has been known to strike the benighted traveller in the face, or to tumble him over into some dark pool. The spectre may be seen at the close of evening hopping vigorously among the distant bogs, like a felt ball on its electric platform; and when the mist lies thick in the hollows, an occasional glimpse may be caught of it even by day. But when I passed the way there was no fog: the light, though softened by a thin film of cloud, fell equally over the heath, revealing hill and hollow; and I was unlucky enough not to see this goblin of the Liasic valley.

A deep indentation of the coast, which forms the bay of Broadford, corresponds with the hollow of the valley. It is simply a portion of the valley itself occupied by the sea; and we find the Lias, from its lower to its upper beds, exposed in unbroken series along the beach. In the middle of the opening lies the green level island of Pabba, altogether composed of this formation, and which, differing, in consequence, both in outline and color, from every [Pg 122] neighboring island and hill, seems a little bit of flat fertile England, laid down, as if for contrast's sake, amid the wild rough Hebrides. Of Pabba and its wonders, however, more anon. I explored a considerable range of shore along the bay; but as I made it the subject of two after explorations ere I mastered its deposits, I shall defer my description till a subsequent chapter. It was late this evening ere the post-gig arrived from the south, and the night and several hours of the following morning were spent in travelling to Portree. I know not, however, that I could have seen some of the wildest and most desolate tracts in Skye to greater advantage. There was light enough to show the bold outlines of the hills,—lofty, abrupt, pyramidal,—just such hills, both in form and grouping, as a profile in black showed best; a low blue vapor slept in the calm over the marshes at their feet; the sea, smooth as glass, reflected the dusk twilight gleam in the north, revealing the narrow sounds and deep mountain-girdled lochs along which we passed; gray crags gleamed dimly on the sight; birch-feathered acclivities presented against sea and sky their rough bristly edges; all was vast, dreamy, obscure, like one of Martin's darker pictures: the land of the seer and the spectre could not have been better seen. Morning broke dim and gray, while we were yet several miles from Portree; and I reached the inn in time to see from my bed-room windows the first rays of the rising sun gleaming on the hill-tops.

[Pg 123]

CHAPTER VII.

Exploration resumed—Geology of Rasay—An Illustration—Storr of Skye—From Portree to Holm—Discovery of Fossils—An Island Rain—Sir R. Murchison—Labor of drawing a Geological Line—Three Edinburgh Gentlemen—*Prosopolepsia*—Wrong surmises corrected—The Mail Gig—The Portree Postmaster—Isle Ornsay—An Old Acquaintance—Reminiscences—A Run for Rum—"Semi-fossil Madeira"—Idling on Deck—Prognostics of a Storm—Description of the Gale—Loch Scresort—The Minister's lost *Sou-wester*—The Free Church Gathering—The weary Minister.

I breakfasted in the travellers' room with three gentlemen from Edinburgh; and then, accompanied by a boy, whom I had engaged to carry my bag, set out to explore. The morning was ominously hot and breathless; and while the sea lay moveless in the calm, as a floor of polished marble, mountain and rock, and distant island, seemed tremulous all over, through a wavy medium of thick rising vapor. I judged from the first that my course of exploration for the day was destined to terminate abruptly; and as my arrangements with Mr.

Swanson left me, for this part of the country, no second day to calculate upon, I hurried over deposits which in other circumstances I would have examined more carefully,—content with a glance. Accustomed in most instances to take long aims, as Cuddy Headrig did, when he steadied his musket on a rest behind the hedge, and sent his ball through Laird Oliphant's forehead, I had on this occasion to shoot flying; and so, selecting a large object for a mark, that I might run the less risk of missing, I strove to acquaint myself rather with the general structure of the district than with the organisms of its various fossiliferous beds.

[Pg 124]

The long narrow island of Rasay lies parallel to the coast of Skye, like a vessel laid along a wharf, but drawn out from it as if to suffer another vessel of the same size to take her berth between; and on the eastern shores of both Skye and Rasay we find the same Oölitic deposits tilted up at nearly the same angle. The section presented on the eastern coast of the one is nearly a duplicate of the section presented on the eastern coast of the other. During one of the severer frosts of last winter I passed along a shallow pond, studded along the sides with boulder stones. It had been frozen over; and then, from the evaporation so common in protracted frosts, the water had shrunk, and the sheet of ice which had sunk down over the central portion of the pond exhibited what a geologist would term very considerable marks of disturbance among the boulders at the edges. Over one sharp-backed boulder there lay a sheet tilted up like the lid of a chest half-raised; and over another boulder immediately behind it there lay another uptilted sheet, like the lid of a second half-open chest; and in both sheets, the edges, lying in nearly parallel lines, presented a range of miniature cliffs to the shore. Now, in the two uptilted ice-sheets of this pond I recognized a model of the fundamental Oölitic deposits Rasay and Skye. The mainland of Scotland had its representative in the crisp snow-covered shore of the pond, with its belt of faded sedges; the place of Rasay was indicated by the inner, that of Skye by the outer boulder; while the ice-sheets, with their shoreward-turned line of cliffs, represented the Oölitic beds, that turn to the mainland their dizzy range of precipices, varying from six to eight hundred feet in height, and then, sloping outwards and downwards, disappear under mountain wildernesses of overlying trap. And it was along a portion of the range of cliff that forms the outermost of the two uptilted [Pg 125] lines, and which presents in this district of Skye a frontage of nearly twenty continuous miles to the long Sound of Rasay, that my to-day's course of exploration lay. From the top of the cliff the surface slopes downwards for about two miles into the interior, like the half-raised chest-lid of my illustration sloping towards the hinges, or the uptilted ice-table of the boulder sloping towards the centre of the pond; and the depression behind forms a flat moory valley, full fifteen miles in length, occupied by a chain of dark bogs and treeless lochans. A long line of trap-hills rises over it, in one of which, considerably in advance of the others, I recognized the Storr of Skye, famous among lovers of the picturesque for its strange group of mingled pinnacles and towers; while directly crossing into the valley from the Sound, and then running southwards for about two miles along its bottom, is the noble sea-arm, Loch Portree, in which, as indicated by the name (the King's Port) a Scottish king of the olden time, in his voyage round his dominions, cast anchor. The opening of the loch is singularly majestic;—the cliffs tower high on either side in graceful magnificence: but from the peculiar inward slope of the land, all within, as the loch reaches the line of the valley, becomes tame and low, and a black dreary moor stretches from the flat terminal basin into the interior. The opening of Loch Portree is a palace gateway, erected in front of some homely suburb, that occupies the place which the palace itself should have occupied.

There was, however, no such mixture of the homely and the magnificent in the route I had selected to explore. It lay under the escarpment of the cliff; and I purposed pursuing it from Portree to Holm, a distance of about six miles, and then returning by the flat interior valley. On the one hand rose a sloping rampart, full seven hundred [Pg 126] feet in height, striped longitudinally with alternating bands of white sandstone and dark shale, and capped atop by a continuous coping of trap, that lacked not massy tower, and overhanging turret, and projecting sentry-box; while, on the other hand, spreading outwards in the calm from the line of dark trap-rocks below, like a mirror from its carved frame of black oak, lay the Sound of Rasay, with its noble background of island and main rising bold on the east, and its long mountain vista opening to the south. The first fossiliferous deposit which gave me occasion this morning to use my hammer occurs near the opening of the loch, beside an old Celtic burying-ground, in the form of a thick bed of hard sandstone, charged with Belemnites,—a bed that must at one time have existed as a widely-spread accumulation of sand,—the bottom, mayhap, of some extensive bay of the Oölite, resembling the Loch Portree of the present day, in which eddy tides deposited the sand swept along by the tidal currents of some neighboring sound, and which swarmed as thickly with Cephalopoda as the loch swarmed this day with minute purple-tinged Medusæ. I found detached on the shore, immediately below this bed, a piece of calcareous fissile sandstone, abounding in small sulcated Terebratulæ, identical, apparently, with the Terebratula of a specimen in my collection from the inferior Oölite of Yorkshire. A colony of this delicate Brachiopod must have once lain moored near this spot, like a fleet of long-prowed galleys at anchor, each one with its cable of many strands extended earthwards from the single *dead-eye* in its umbone. For a full mile after rounding the northern boundary of the loch, we find the immense escarpment composed from top to bottom exclusively of trap; but then the Oölite again begins to appear, and about two miles further on the section becomes truly magnificent,—one [Pg 127] of the finest sections of this formation exhibited anywhere in Britain, perhaps in the world. In a ravine furrowed in the face of the declivity by the headlong descent of a small stream, we may trace all the beds of the system in succession, from the Cornbrash, an upper deposit of the Lower Oölite, down to the Lias, the formation on which the Oölite rests. The only modifying circumstance to the geologist is, that though the sandstone beds run continuously along the cliff for miles together, distinct as the white bands in a piece of onyx, the intervening beds of shale are swarded over, save where we here and there see them laid bare in some abrupter acclivity or deeper water-course. In the shale we find numerous minute Ammonites, sorely weathered; in the sandstone, Belemnites, some of them of great size; and dark carbonaceous markings, passing not unfrequently into a glossy cubical coal. At the foot of the cliff I picked up an ammonite of considerable size and well-marked character,—the *Ammonites Murchisonæ,* first discovered on this coast by Sir R. Murchison about fifteen years ago. It measures, when full grown, from six to seven inches in diameter; the inner whorls, which are broadly visible, are ribbed; whereas the two, and sometimes the three outer ones, are smooth,—a marked characteristic of the species. My specimen merely enabled me to examine the peculiarities of the shell just a little more minutely than I could have done in the pages of Sowerby; for such was its state of decay, that it fell to pieces in my hands. I had now come full in view of the rocky island of Holm, when the altered appearance of the heavens led me to deliberate, just as I was warming in the work of exploration, whether, after all, it might not be well to scale the cliffs, and strike directly on the inn. It was nearly three o'clock; the sky had been gradually darkening since noon, as if one thin covering of gauze [Pg 128] after another had been drawn over it; hill and island had first dimmed and then disappeared in the landscape; and now the sun stood up right over the fast-contracting vista of the Sound, round and lightless as the moon in a haze; and the downward cataract-like streaming of the gray vapor on the horizon showed that there the rain had already broken, and was descending in torrents. We had been thirsty in the hot sun, and had found the springs few and scanty; but the boy now assured me, in very broken English, that we were to get a great deal more water than would be good for us, and that it might be advisable to get out of its way. And so, climbing to the top of the cliffs, along a water-course, we reached the ridge, just as the fog came rolling downwards from the peaked brow of the Storr into the flat moory valley, and the melancholy lochans roughened and darkened in the rain. We were both particularly wet ere we reached Portree.

In exploring our Scotch formations, I have had frequent occasion, in Ross, Sutherland, Caithness, and now once more in Skye, to pass over ground described by Sir R. Murchison; and in every instance have I found myself immensely his debtor. His descriptions possess the merit of being true: they are simple outlines often, that leave much to be filled up by after discovery; but, like those outlines of the skilful geographer that fix the place of some island or strait, though they may not entirely define it, they always indicate the exact position in the scale of the formations to which they refer. They leave a good deal to be done in the way of mapping out the interior of a deposit, if I may so speak; but they leave nothing to be done in the way of ascertaining its place. The work accomplished is *bona fide* work,—actual, solid, not to be done over again,—work such as could be achieved in only the school of Dr. William Smith, the father of English Geology. I have found much to admire, too, in the sections [Pg 129] of Sir R. Murchison. His section of this part of the coast, for example, strikes from the extreme northern part of Skye to the island of Holm, thence to Scrapidale in Rasay, thence along part of the coast of Scalpa, thence direct through the middle of Pabba, and thence to the shore of the Bay of Laig. The line thus taken includes, in regular sequence in the descending order, the whole Oölitic deposits of the Hebrides, from the Cornbrash, with its overlying fresh-water outliers of mayhap the Weald, down to where the Lower Lias rests on the primary red sandstones of Sleat. It would have cost M'Culloch less exploration to have written a volume than it must have cost Sir R. Murchison to draw this single line; but the line once drawn, is work done to the hands of all after explorers. I have followed repeatedly in the track of another geologist, of, however, a very different school, who explored, at a comparatively recent period, the deposits of not a few of our Scotch counties. But his labors, in at least the fossiliferous formations, seem to have accomplished nothing for Geology,—I am afraid, even less than nothing. So far as they had influence at all, it must have been to throw back the science. A geologist who could have asserted only three years ago ("Geognostical Account of Banffshire," 1842), that the Old Red Sandstone of Scotland forms merely "a part of the great coal deposit," could have known marvellously little of the fossils of the one system, and nothing whatever of those of the other. Had he examined ere he decided, instead of deciding without any intention of examining, he would have found that, while both systems abound in organic remains, they do not possess, in Scotland at least, a single species in common, and that even their types of being, viewed in the group, are essentially distinct.

The three Edinburgh gentlemen whom I had met at breakfast were still in the inn. One of them I had seen before, as one of the guests at a Wesleyan soiree, though I saw he [Pg 130] failed to remember that I had been there as a guest too. The two other gentlemen were altogether strangers to me. One of them,—a man on the right side of forty, and a superb specimen of the powerful, six-feet two-inch Norman Celt,—I set down as a scion of some old Highland family, who, as the broadsword had gone out, carried on the internal wars of the country with the formidable artillery of Statute and Decision. The other, a gentleman more advanced in life, I predicated to be a Highland proprietor, the uncle of the younger of the two,—a man whose name, as he had an air of business about him, occurred, in all probability, in the Almanac, in the list of Scotch advocates. Both were of course high Tories,—I was quite sure of that,—zealous in behalf of the Establishment, though previous to the Disruption they had not cared for it a pin's point,—and prepared to justify the virtual suppression of the toleration laws in the case of the Free Church. I was thus decidedly guilty of what old Dr. More calls a *prosopolepsia,—i.e.* of the crime of judging men by their looks. At dinner, however, we gradually ate ourselves into conversation: we differed, and disputed, and agreed, and then differed, disputed and agreed again. I found first, that my chance companions were really not very high Tories; and then, that they were not Tories at all; and then, that the younger of the two was very much a Whig, and the more advanced in life,—strange as the fact might seem,—very considerably a *Presbyterian* Whig; and finally, that this latter gentleman, whom I had set down as an intolerant Highland proprietor, was a respected writer to the signet, a Free Church elder in Edinburgh; and that the other, his equally intolerant nephew, was an Edinburgh advocate, of vigorous talent, much an enemy of all oppression, and a brother contributor of my own to one of the Quarterlies. Of all my surmisings regarding the stranger gentlemen, only two points held [Pg 131] true,—they were both gentlemen of the law, and both had Celtic blood in their veins. The evening passed pleasantly; and I can now recommend from experience, to the hapless traveller who gets thoroughly wet thirty miles from a change of dress, that some of the best things he can resort to in the circumstances are, a warm room, a warm glass, and agreeable companions.

On the morrow I behooved to return to Isle Ornsay, to set out on the following day, with my friend the minister, for Rum, where he purposed preaching on the Sabbath. To have lost a day would have been to lose the opportunity of exploring the island, perhaps forever; and, to make all sure, I had taken a seat in the mail gig, from the postman who drives it, ere going to bed, on the morning of my arrival; and now, when it drove up, I went to take my place in it. The postmaster of the village, a lean, hungry-looking man, interfered to prevent me. I had secured my seat, I said, two days previous. Ah, but I had not secured it from him. "I know nothing of you," I replied; "but I secured it from one who deemed himself authorized to receive the fare; was he so?" "Yes." "Could you have received it?" "No." "Show me a copy of your regulations." "I have no copy of regulations; but I have given the place in the gig to another." "Just so; and what say you, postman?" "That you took the place from me, and that *he* has no right to give a place to any one: I carry the Portree letters to him, but he has nothing to do with the passengers." A person present, the proprietor or stabler of the horse, I believe, also interfered on the same side; but what Carlyle terms the "gigmanity" of the postmaster was all at stake,—his whole influence in the mail-gig of Portree; and so he argued, and threatened withal, and, what was the more serious part of the business, the person he had given the seat to had taken possession of the gig; and so we had to [Pg 132] compound the matter by carrying a passenger additional. The incident is scarce worth relating; but the postmaster was so vehement and terrible, so defiant of us all,—post, stabler, and simple passenger,—and so justly impressed with the importance of being postmaster of Portree, that, as I am in the way of describing rare specimens at any rate, I must refer to him among the rest, as if he had been one of the minor carnivoræ of a Skye deposit,—a cuttle-fish, that preyed on the weaker molluscs, or a hungry polypus, terrible among the animalculæ.

We drove heavily, and had to dismount and walk afoot over every steeper acclivity; but I carried my hammer, and only grieved that in some one or two localities the road should have been so level. I regretted it in especial on the southern and eastern side of Loch Sligachan, where I could see from my seat, as we drove past, the dark blue rocks in the water-courses on each side the road, studded over with that characteristic shell of the Lias, the *Gryphæa incurva*, and that the dry-stone fences in the moor above exhibit fossils that might figure in a museum. But we rattled by. At Broadford, twenty-five miles from Portree, and nine miles from Isle Ornsay, I partook of a hospitable meal in the house of an acquaintance; and in little more than two hours after was with my friend the minister at Isle Ornsay. The night wore pleasantly by. Mrs. Swanson, a niece of the late Dr. Smith of Campbelton, so well known for his Celtic researches and his exquisite translations of ancient Celtic poetry, I found deeply versed in the legendary lore of the Highlands. The minister showed me a fine specimen of Pterichthys which I had disinterred for him, out of my first discovered fossiliferous deposit of the Old Red Sandstone, exactly thirteen years before, and full seven years ere I had introduced the creature to the notice of Agassiz. And the minister's [Pg 133] daughter, a little chubby girl of three summers, taking part in the general entertainment, strove to make her Gaelic sound as like English as she could, in my especial behalf. I remembered, as I listened to the unintelligible prattle of the little thing, unprovided with a word of English, that just eighteen years before, her father had had no Gaelic; and wondered what he would have thought, could he have been told, when he first sat down to study it, the story of his island charge in Eigg, and his Free Church yacht the Betsey. Nineteen years before, we had been engaged in beating over the Eathie Lias together, collecting Belemnites, Ammonites, and fossil wood, and striving in friendly emulation the one to surpass the other in the variety and excellence of our specimens. Our leisure hours were snatched, at the time, from college studies by the one, from the mallet by the other: there were few of them that we did not spend together, and that we were not mutually the better for so spending. I at least, owe much to these hours,—among other things, views of theologic truth, that determined the side I have taken in our ecclesiastical controversy. Our courses at an after period lay diverse; the young minister had greatly more important business to pursue than any which the geologic field furnishes; and so our amicable rivalry ceased early. In the words in which an English poet addresses his brother,—the clergyman who sat for the picture in the "Deserted Village,"—my friend "entered on a sacred office, where the harvest is great and the laborers are few, and left to me a field in which the laborers are many, and the harvest scarce worth carrying away."

Next day at noon we weighed anchor, and stood out for Rum, a run of about twenty-five miles. A kind friend had, we found, sent aboard in our behalf two pieces of [Pg 134] rare antiquity,—rare anywhere, but especially rare in the lockers of the Betsey,—in the agreeable form of two bottles of semi-fossil Madeira,—Madeira that had actually existed in the grape exactly half a century before, at the time when Robespierre was startling Paris from its propriety, by mutilating at the neck the busts of other people, and multiplying casts and medals of his own; and we found it, explored in moderation, no bad study for geologists, especially in coarse weather, when they had got wet and somewhat fatigued. It was like Landlord Boniface's ale, mild as milk, had exchanged its distinctive flavor as Madeira for a better one, and filled the cabin with fragrance every time the cork was drawn. Old observant Homer must have smelt some such liquor somewhere, or he could never have described so well the still more ancient and venerable wine with which wily Ulysses beguiled one-eyed Polypheme:—

> "Unmingled wine,
> Mellifluous, undecaying, and divine,
> Which now, some ages from his race concealed,
> The hoary sire in gratitude revealed....
> Scarce twenty measures from the living stream
> To cool one cup sufficed: the goblet crowned,
> Breathed aromatic fragrances around."

Winds were light and variable. As we reached the middle of the sound opposite Armadale, there fell a dead calm; and the Betsey, more actively idle than the ship manned by the Ancient Mariner, dropped sternwards along the tide, to the dull music of the flapping sail. The minister spent the day in the cabin, engaged with his discourse for the morrow; and I, that he might suffer as little from interruption as possible, *mis*-spent it upon the deck. I tried fishing with the yacht's set of lines, but there were no fish to bite,—got into the boat, but there were no [Pg 135] neighboring islands to visit,—and sent half a dozen pistol-bullets after a shoal of porpoises, which, coming from the Free Church yacht, must have astonished the fat sleek fellows pretty considerably, but did them, I am afraid, no serious damage. As the evening began to close gloomy and gray, a tumbling swell came heaving in right ahead from the west; and a bank of cloud, which had been gradually rising higher and darker over the horizon in the same direction, first changed its abrupt edge atop for a diffused and broken line, and then spread itself over the central heavens. The calm was evidently not to be a calm long; and the minister issued orders that the gaff-topsail should be taken down, and the storm-jib bent; and that we should lower our top-mast, and have all tight and ready for a smart gale ahead. At half past ten, however, the Betsey was still pitching to the swell, with not a breath of wind to act on the diminished canvas, and with the solitary circumstance in her favor, that the tide ran no longer against her, as before. The cabin was full of all manner of creakings; the close lamp swung to and fro over the head of my friend; and a refractory Concordance, after having twice travelled from him along the entire length of the table, flung itself pettishly upon the floor. I got into my snug bed about eleven; and at twelve, the minister, after poring sufficiently over his notes, and drawing the final score, turned into his. In a brief hour after, on came the gale, in a style worthy of its previous hours of preparation; and my friend,—his Saturday's work in his ministerial capacity well over when he had completed his two discourses,—had to begin the Sabbath morning early as the morning itself began, by taking his stand at the helm, in his capacity of skipper of the Betsey. With the prospect of the services of the Sabbath before him, and after working all Saturday to boot, it was rather hard to set him down to a [Pg 136] midnight spell at the helm, but he could not be wanted at such a time, as we had no other such helmsman aboard. The gale, thickened with rain, came down, shrieking like a maniac, from off the peaked hills of Rum, striking away the tops of the long ridgy billows that had risen in the calm to indicate its approach, and then carrying them in sheets of spray aslant the furrowed surface, like snow-drift hurried across a frozen field. But the Betsey, with her storm-jib set, and her mainsail reefed to the cross, kept her weather bow bravely to the blast, and gained on it with every tack. She had been the pleasure yacht, in her day, of a man of fortune, who had used, in running south with her at times as far as Lisbon, to encounter, on not worse terms than the stateliest of her neighbors in the voyage, the swell of the Bay of Biscay; and she still kept true to her old character, with but this drawback, that she had now got somewhat crazy in her fastenings, and made rather more water in a heavy sea than her one little pump could conveniently keep under. As the fitful gust struck her headlong, as if it had been some invisible missile hurled at us from off the hill-tops, she stooped her head lower and lower, like old stately Hardyknute under the blow of the "King of Norse," till at length the lee chain-plate rustled sharp through the foam; but, like a staunch Free Churchwoman, the lowlier she bent, the more steadfastly did she hold her head to the storm. The strength of the opposition served but to speed her on all the more surely to the desired haven. At five o'clock in the morning we cast anchor in Loch Scresort,—the only harbor of Rum in which a vessel can moor,—within two hundred yards of the shore, having, with the exception of the minister, gained no loss in the gale. He, luckless man, had parted from his excellent *sou-wester*; a sudden gust had seized it by the flap, and hurried it away far to the lee. He had yielded it to the winds, as he had done the temporalities, [Pg 137] but much more unwillingly, and less as a free agent. Should any conscientious mariner pick up any where in the Atlantic a serviceable ochre-colored *sou-wester*, not at all the worse for the wear, I give him to wit that he holds Free Church property, and that he is heartily welcome to hold it, leaving it to himself to consider whether a benefaction to its full value, deducting salvage, is not owing, in honor, to the Sustenation Fund.

It was ten o'clock ere the more fatigued aboard could muster resolution enough to quit their beds a second time; and then it behooved the minister to prepare for his Sabbath labors ashore. The gale still blew in fierce gusts from the hills, and the rain pattered like small shot on the deck. Loch Scresort, by no means one of our finer island lochs, viewed under any circumstances, looked particularly dismal this morning. It forms the opening of a dreary moorland valley, bounded on one of its sides, to the mouth of the loch, by a homely ridge of Old Red Sandstone, and on the other by a line of dark augitic hills, that attain, at the distance of about a mile from the sea, an elevation of two thousand feet. Along the slopes of the sandstone ridge I could discern, through the haze, numerous green patches, that had once supported a dense population, long since "cleared off" to the backwoods of America, but not one inhabited dwelling; while along a black moory acclivity under the hills on the other side I could see several groups of turf cottages, with here and there a minute speck of raw-looking corn beside them, that, judging from its color, seemed to have but a slight chance of ripening. The hill-tops were lost in cloud and storm; and ever and anon, as a heavier shower came sweeping down on the wind, the intervening hollows closed up their gloomy vistas, and all was fog and rime to the water's edge. Bad as the morning was, however, we could see the people wending their way, in threes and fours, [Pg 138] through the dark moor, to the place of worship,—a black turf hovel, like the meeting-house in Eigg. The appearance of the Betsey in the loch had been the gathering signal; and the Free Church islanders,—three-fourths of the entire population—had all come out to meet their minister.

On going ashore, we found the place nearly filled. My friend preached two long energetic discourses, and then returned to the yacht, "a worn and weary man." The studies of the previous day, and the fatigues of the previous night, added to his pulpit duties, had so fairly prostrated his strength, that the sternest teetotaller in the kingdom would scarce have forbidden him a glass of our fifty-year-old Madeira. But even the fifty-year-old Madeira proved no specific in the case. He was suffering under excruciating headache, and had to stretch himself in his bed, with eyes shut but sleepless, waiting till the fit should pass,—every pulse that beat in his temples a throb of pain.

[Pg 139]

CHAPTER VIII.

Geology of Rum—Its curious Character illustrated—Rum famous for Bloodstones—Red Sandstones—"Scratchings" in the Rocks—A Geological Inscription without a Key—The Lizard—Vitality broken into two—Illustrations—Speculation—Scuir More—Ascent of the Scuir—The Bloodstones—An Illustrative Set of the Gem—M'Culloch's Pebble—A Chemical Problem—The solitary Shepherd's House—Sheep *versus* Men—The Depopulation of Rum—A Haul of Trout—Rum Mode of catching Trout—At Anchor in the Bay of Glenelg.

The geology of the island of Rum is simple, but curious. Let the reader take, if he can, from twelve to fifteen trap-hills, varying from one thousand to two thousand three hundred feet in height; let him pack them closely and squarely together, like rum-bottles in a case-basket; let him surround them with a frame of Old Red Sandstone, measuring rather more than seven miles on the side, in the way the basket surrounds the bottles; then let him set them down in the sea a dozen miles off the land,—and he shall have produced a second island of Rum, similar in structure to the existing one. In the actual island, however, there is a defect in the inclosing basket of sandstone: the basket, complete on three of its sides, wants the fourth: and the side opposite to the gap which the fourth should have occupied is thicker than the two other sides put together. Where I now write there is an old dark-colored picture on the wall before me. I take off one of the four bars of which the frame is composed,—the end-bar,—and stick it on to the end-bar opposite, and then the picture is fully framed on two of its sides, and doubly framed on a third, but the fourth side lacks framing altogether. And such is [Pg 140] the geology of the island of Rum. We find the one loch of the island,—that in which the Betsey lies at anchor,—and the long withdrawing valley, of which the loch is merely a prolongation, occurring in the double sandstone bar: it seems to mark—to return to my illustration—the line in which the superadded piece of frame has been stuck on to the frame proper. The origin of the island is illustrated by its structure: it has left its story legibly written, and we have but to run our eye over the characters and read. An extended sea-bottom, composed of Old Red Sandstone, already tilted up by previous convulsions, so that the strata presented their edges, tier beyond tier, like roofing slate laid aslant on a floor, became a centre of Plutonic activity. The molten trap broke through at various times, and presenting various appearances, but in nearly the same centre; here existing as an augitic rock, there as a syenite, yonder as a basalt or amygdaloid. At one place it uptilted the sandstone; at another it overflowed it; the dark central masses raised their heads above the surface, higher and higher with every earthquake throe from beneath; till at length the gigantic Ben More attained to its present altitude of two thousand three hundred feet over the sea-level, and the sandstone, borne up from beneath like floating sea-wrack on the back of a porpoise, reached in long outside bands its elevation of from six to eight hundred. And such is the piece of history, composed in silent but expressive language, and inscribed in the old geological character, on the rocks of Rum.

The wind lowered and the rain ceased during the night, and the morning of Monday was clear, bracing, and breezy. The island of Rum is chiefly famous among mineralogists for its heliotropes or bloodstones; and we proposed devoting the greater part of the day to an examination of the hill of Scuir More, in which they occur, and [Pg 141] which lies on the opposite side of the island, about eight miles from the mooring ground of the Betsey. Ere setting out, however, I found time enough, by rising some two or three hours before breakfast, to explore the Red Sandstones on the southern side of the loch. They lie in this bar of the frame,—to return once more to my old illustration,—as if it had been cut out of a piece of cross-grained deal, in which the annular bands, instead of ranging lengthwise, ran diagonally from side to side; stratum leans over stratum, dipping towards the west at an angle of about thirty degrees; and as in a continuous line of more than seven miles there seem no breaks or repetitions in the strata, the thickness of the deposit must be enormous,—not less, I should suppose, than from six to eight thousand feet. Like the Lower Old Red Sandstones of Cromarty and Moray, the red arenaceous strata occur in thick beds, separated from each other by bands of a grayish-colored stratified clay, on the planes of which I could trace with great distinctness ripple markings; but in vain did I explore their numerous folds for the plates, scales, and fucoid impressions which abound in the gray argillaceous beds of the shores of the Moray and Cromarty Friths. It would, however, be rash to pronounce them non-fossiliferous, after the hasty search of a single morning,—unpardonably so in one who had spent very many mornings in putting to the question the gray stratified beds of Ross and Cromarty, ere he succeeded in extorting from them the secret of their organic riches.

We set out about half-past ten for Scuir More, through the Red Sandstone valley in which Loch Scresort terminates, with one of Mr. Swanson's people, a young active lad of twenty, for our guide. In passing upwards for nearly a mile along the stream that falls into the upper part of the loch, and lays bare the strata, we saw no [Pg 142] change in the character of the sandstone. Red arenaceous beds of great thickness alternate with grayish-colored bands, composed of a ripple-marked micaceous slate and a stratified clay. For a depth of full three thousand feet, and I know not how much more,—for I lacked time to trace it further,—the deposit presents no other variety: the thick red bed of at least a hundred yards succeeds the thin gray band of from three to six feet, and is succeeded by a similar gray band in turn. The ripple-marks I found as sharply relieved in some of the folds as if the wavy undulations to which they owed their origin had passed over them within the hour. The comparatively small size of their alternating ridges and furrows give evidence that the waters beneath which they had formed had been of no very profound depth. In the upper part of the valley, which is bare, trackless, and solitary, with a high monotonous sandstone ridge bounding it on the one side, and a line of gloomy trap-hills rising over it on the other, the edges of the strata, where they protrude through the mingled heath and moss, exhibit the mysterious scratchings and polishings now so generally connected with the glacial theory of Agassiz. The scratchings run in nearly the line of the valley, which exhibits no trace of moraines; and they seem to have been produced rather by the operation of those extensively developed causes, whatever their nature, that have at once left their mark on the sides and summits of some of our highest hills, and the rocks and boulders of some of our most extended plains, than by the agency of forces limited to the locality. They testify, Agassiz would perhaps say, not regarding the existence of some local glacier that descended from the higher grounds into the valley, but respecting the existence of the great polar glacier. I felt, however, in this bleak and solitary hollow, with the grooved and polished platforms at my [Pg 143] feet, stretching away amid the heath, like flat tombstones in a graveyard, that I had arrived at one geologic inscription to which I still wanted the key. The vesicular structure of the traps on the one hand, identical with that of so many of our modern lavas,—the ripple-markings of the arenaceous beds on the other, indistinguishable from those of the sea-banks on our coasts,—the upturned strata and the overlying trap,—told all their several stories of fire, or wave, or terrible convulsion, and told them simply and clearly; but here was a story not clearly told. It summoned up doubtful, ever-shifting visions,—now of a vast ice continent, abutting on this far isle of the Hebrides from the Pole, and trampling heavily over it,—now of the wild rush of a turbid, mountain-high flood breaking in from the west, and hurling athwart the torn surface, rocks, and stones, and clay,—now of a dreary ocean rising high along the hills, and bearing onward with its winds and currents, huge icebergs, that now brushed the mountain-sides, and now grated along the bottom of the submerged valleys. The inscription on the polished surfaces, with its careless mixture of groove and scratch, is an inscription of very various readings.

We passed along a transverse hollow, and then began to ascend a hill-side, from the ridge of which the water sheds to the opposite shore of the island, and on which we catch our first glimpse of Scuir More, standing up over the sea, like a pyramid shorn of its top. A brown lizard, nearly five inches in length, startled by our approach, ran hurriedly across the path; and our guide, possessed by the general Highland belief that the creature is poisonous, and injures cattle, struck at it with a switch, and cut it in two immediately behind the hinder legs. The upper half, containing all that anatomists regard as the vitals, heart, brain, and viscera, all the main nerves, and all the larger [Pg 144] arteries, lay stunned by the blow, as if dead; nor did it manifest any signs of vitality so long as we remained beside it; whereas the lower half, as if the whole life of the animal had retired into *it*, continued dancing upon the moss for a full minute after, like a young eel scooped out of some stream, and thrown upon the bank; and then lay wriggling and palpitating for about half a minute more. There are few things more inexplicable in the province of the naturalist than the phenomenon of what may be termed divided life,—vitality broken into two, and yet continuing to exist as vitality in both the dissevered pieces. We see in the nobler animals mere glimpses of the phenomenon,—mere indications of it, doubtfully apparent for at most a few minutes. The blood drawn from the human arm by the lancet continues to live in the cup until it has cooled and begun to coagulate; and when head and body have parted company under the guillotine, both exhibit for a brief space such unequivocal signs of life, that the question arose in France during the horrors of the Revolution, whether there might not be some glimmering of consciousness attendant at the same time on the fearfully opening and shutting eyes and mouth of the one, and the beating heart and jerking neck of the other. The lower we descend in the scale of being, the more striking the instances which we receive of this divisibility of the vital principle. I have seen the two halves of the heart of a ray pulsating for a full quarter of an hour after they had been separated from the body and from each other. The blood circulates in the hind leg of a frog for many minutes after the removal of the heart, which meanwhile keeps up an independent motion of its own. Vitality can be so divided in the earthworm, that, as demonstrated by the experiments of Spalanzani, each of the severed parts carries life enough away to set it up as an independent animal; [Pg 145] while the polypus, a creature of still more imperfect organization, and with the vivacious principle more equally diffused over it, may be multiplied by its pieces nearly as readily as a gooseberry bush by its slips. It was sufficiently curious, however, to see, in the case of this brown lizard, the least vital half of the creature so much more vivacious, apparently, than the half which contained the heart and brain. It is not improbable, however, that the presence of these organs had only the effect of rendering the upper portion which contained them more capable of being thrown into a state of insensibility. A blow dealt one of the vertebrata on the head at once renders it insensible. It is after this mode the fisherman kills the salmon captured in his wear, and a single blow, when well directed, is always sufficient; but no single blow has the same effect on the earthworm; and here it was vitality in the inferior portion of the reptile,—the earthworm portion of it, if I may so speak,—that refused to participate in the state of syncope into which the vitality of the superior portion had been thrown. The nice and delicate vitality of the brain seems to impart to the whole system in connection with it an aptitude for dying suddenly,—a susceptibility of instant death, which would be wanting without it. The heart of the rabbit continues to beat regularly long after the brain has been removed by careful excision, if respiration be artificially kept up; but if, instead of amputating the head, the brain be crushed in its place by a sudden blow of a hammer, the heart ceases its motion at once. And such seemed to be the principle illustrated here. But why the agonized dancing on the sward of the inferior part of the reptile?—why its after painful writhing and wriggling? The young eel scooped from the stream, whose motions it resembled, is impressed by terror, and can feel pain; was *it* also impressed by terror, [Pg 146] or susceptible of suffering? We see in the case of both exactly the same signs,—the dancing, the writhing, the wriggling; but are we to interpret them after the same manner? In the small red-headed earthworm divided by Spalanzani, that in three months got upper extremities to its lower part, and lower extremities, in as many weeks, to its upper part, the dividing blow must have dealt duplicate feelings,—pain and terror to the portion below, and pain and terror to the portion above,—so far, at least, as a creature so low in the scale was susceptible of these feelings; but are we to hold that the leaping, wriggling tail of the reptile possessed in any degree a similar susceptibility? *I* can propound the riddle, but who shall resolve it? It may be added, that this brown lizard was the only recent saurian I chanced to see in the Hebrides, and that, though large for its kind, its whole bulk did not nearly equal that of a single vertebral joint of the fossil saurians of Eigg. The reptile, since his deposition from the first place in the scale of creation, has sunk sadly in those parts: the ex-monarch has become a low plebeian.

We came down upon the coast through a swampy valley, terminating in the interior in a frowning wall of basalt, and bounded on the south, where it opens to the sea, by the Scuir More. The Scuir is a precipitous mountain, that rises from twelve to fifteen hundred feet direct over the beach. M'Culloch describes it as inaccessible, and states that it is only among the debris at its base that its heliotropes can be procured; but the distinguished mineralogist must have had considerably less skill in climbing rocks than in describing them, as, indeed, some of his descriptions, though generally very admirable, abundantly testify. I am inclined to infer from his book, after having passed over much of the ground which he describes, that he must have been a man of the type so [Pg 147] well hit off by Burns in his portrait of Captain Grose,—round, rosy, short-legged, quick of eye but slow of foot, quite as indifferent a climber as Bailie Nicol Jarvie, and disposed at times, like the elderly gentleman drawn by Crabbe, to prefer the view at the hill-foot to the prospect from its summit. I found little difficulty in scaling the sides of Scuir More for a thousand feet upwards,—in one part by a route rarely attempted before,—and in ensconcing myself among the bloodstones. They occur in the amygdaloidal trap of which the upper part of the hill is mainly composed, in great numbers, and occasionally in bulky masses; but it is rare to find other than small specimens that would be recognized as of value by the lapidary. The inclosing rock must have been as thickly vesicular in its original state as the scoria of a glass-house; and all the vesicles, large and small, like the retorts and receivers of a laboratory, have been vessels in which some curious chemical process has been carried on. Many of them we find filled with a white semi-translucent or opaque chalcedony; many more with a pure green earth, which, where exposed to the bleaching influences of the weather, exhibits a fine verdigris hue, but which in the fresh fracture is generally of an olive green, or of a brownish or reddish color. I have never yet seen a rock in which this earth was so abundant as in the amygdaloid of Scuir More. For yards together in some places we see it projecting from the surface in round globules, that very much resemble green peas, and that occur as thickly in the inclosing mass as pebbles in an Old Red Sandstone conglomerate. The heliotrope has formed among it in centres, to which the chalcedony seems to have been drawn, as if by molecular attraction. We find a mass, varying from the size of a walnut to that of a man's head, occupying some larger vesicle or crevice of the amygdaloid, [Pg 148] and all the smaller vesicles around it, for an inch or two, filled with what we may venture to term satellite heliotropes, some of them as minute as grains of wild mustard, and all of them more or less earthy, generally in proportion to their distance from the first formed heliotrope in the middle. No one can see them in their place in the rock, with the abundant green earth all around, and the chalcedony, in its uncolored state, filling up so many of the larger cavities, without acquiescing in the conclusion respecting the origin of the gem first suggested by Werner, and afterwards adopted and illustrated by M'Culloch. The heliotrope is merely a chalcedony, stained in the forming with an infusion of green earth, as the colored waters in the apothecary's window are stained by the infusions, vegetable and mineral, from which they derive their ornamental character. The red mottlings which so heighten the beauty of the stone occur in comparatively few of the specimens of Scuir More. They are minute jasperous formations, independent of the inclosing mass; and, from their resemblance to streaks and spots of blood, suggest the name by which the heliotrope is popularly known. I succeeded in making up, among the crags, a set of specimens curiously illustrative of the origin of the gem. One specimen consists of white, uncolored chalcedony; a second, of a rich verdigris-hued green earth; a third, of chalcedony barely tinged with green; a fourth, of chalcedony tinged just a shade more deeply; a fifth, tinged more deeply still; a sixth, of a deep green on one side, and scarce at all colored on the other; and a seventh, dark and richly toned,—a true bloodstone,—thickly streaked and mottled with red jasper. In the chemical process that rendered the Scuir More a mountain of gems there were two deteriorating circumstances, which operated to the disadvantage of its larger heliotropes: the [Pg 149] green earth, as if insufficiently stirred in the mixing, has gathered, in many of them, into minute soft globules, like air-bubbles in glass, that render them valueless for the purposes of the lapidary, by filling them all over with little cavities; and in not a few of the others, an infiltration of lime, that refused to incorporate with the chalcedonic mass, exists in thin glassy films and veins, that, from their comparative softness, have a nearly similar effect with the impalpable green earth in roughing the surface under the burnisher.

We find figured by M'Culloch, in his "Western Islands," the internal cavity of a pebble of Scuir More, which he picked up on the beach below, and which had been formed evidently within one of the larger vesicles of the amygdaloid. He describes it as curiously illustrative of a various chemistry; the outer crust is composed of a pale-zoned agate, inclosing a cavity, from the upper side of which there depends a group of chalcedonic stalactites, some of them, as in ancient spar caves, reaching to the floor; and bearing on its under side a large crystal of carbonate of lime, that the longer stalactites pass through. In the vesicle in which this hollow pebble was formed three consecutive processes must have gone on. First, a process of infiltration coated the interior all around with layer after layer, now of one mineral substance, now of another, as a plasterer coats over the sides and ceiling of a room with successive layers of lime, putty, and stucco; and had this process gone on, the whole cell would have been filled with a pale-zoned agate. But it ceased, and a new process began. A chalcedonic infiltration gradually entered from above; and, instead of coating over the walls, roof, and floor, it hardened into a group of spear-like stalactites, that lengthened by slow degrees, till some of them had traversed the entire cavity from top to bottom. And then this second process ceased like the first, and a [Pg 150] third commenced. An infiltration of lime took place; and the minute calcareous molecules, under the influence of the law of crystallization, built themselves up on the floor into a large smooth-sided rhomb, resembling a closed sarcophagus resting in the middle of some Egyptian cemetery. And then, the limestone crystal completed, there ensued no after change. As shown by some other specimens, however, there was a yet farther process: a pure quartzose deposition took place, that coated not a few of the calcareous rhombs with sprigs of rock-crystal. I found in the Scuir More several cellular agates in which similar processes had gone on,—none of them quite so fine, however, as the one figured by M'Culloch; but there seemed no lack of evidence regarding the strange and multifarious chemistry that had been carried on in the vesicular cavities of this mountain, as in the retorts of some vast laboratory. Here was a vesicle filled with green earth,—there a vesicle filled with calcareous spar,—yonder a vesicle crusted round on a thin chalcedonic shell with rock-crystal,—in one cavity an agate had been elaborated, in another a heliotrope, in a third a milk-white chalcedony, in a fourth a jasper. On what principle, and under what direction, have results so various taken place in vesicles of the same rock, that in many instances occur scarce half an inch apart? Why, for instance, should that vesicle have elaborated only green earth, and the vesicle separated from it by a partition barely a line in thickness, have elaborated only chalcedony? Why should this chamber contain only a quartzose compound of oxygen and silica, and that second chamber beside it contain only a calcareous compound of lime and carbonic acid? What law directed infiltrations so diverse to seek out for themselves vesicles in such close neighborhood, and to keep, in so many instances, each to his own vesicle? I can but state the problem,—not solve it. The groups of heliotropes clustered each around [Pg 151] its bulky centrical mass seem to show that the principle of molecular attraction may be operative in very dense mediæ,—in a hard amygdaloidal trap even; and it seems not improbable, that to this law, which draws atom to its kindred atom, as clansmen of old used to speed at the mustering signal to their gathering place, the various chemistry of the vesicles may owe its variety.

I shall attempt stating the chemical problem furnished by the vesicles here in a mechanical form. Let us suppose that every vesicle was a chamber furnished with a door, and that beside every door there watched, as in the draught doors of our coal-pits, some one to open and shut it, as circumstances might require. Let us suppose further, that for a certain time an infusion of green earth pervaded the surrounding mass, and percolated through it, and that every door was opened to receive a portion of the infusion. We find that no vesicle wants its coating of this earthy mineral. The coating received, however, one-half the doors shut, while the other half remained agap, and filled with green earth entirely. Next followed a series of alternate infusions of chalcedony, jasper, and quartz; many doors opened and received some two or three coatings, that form around the vesicles skull-like shells of agate, and then shut; a few remained open, and became as entirely occupied with agate as many of the previous ones had become filled with green earth. Then an ample infusion of chalcedony pervaded the mass. Numerous doors again opened; some took in a portion of the chalcedony, and then shut; some remained open, and became filled with it; and many more that had been previously filled by the green earth opened their doors again, and the chalcedony pervading the green porous mass, converted it into heliotrope. Then an infusion of lime took place. Doors opened, many of which had been hitherto shut, save for a short time, when the green earth infusion [Pg 152] obtained, and became filled with lime; other doors opened for a brief space, and received lime enough to form a few crystals. Last of all, there was a pure quartzose infusion, and doors opened, some for a longer time, some for a shorter, just as on previous occasions. Now, by mechanical means of this character,—by such an arrangement of successive infusions, and such a device of shutting and opening of doors,—the phenomena exhibited by the vesicles could be produced. There is no difficulty in working the problem mechanically, if we be allowed to assume in our data successive infusions, well-fitted doors, and watchful door-keepers; and if any one can work it chemically,—certainly without door-keepers, but with such doors and such infusions as he can show to have existed,—he shall have cleared up the mystery of the Scuir More. I have given their various cargoes to all its many vesicles by mechanical means, at no expense of ingenuity whatever. Are there any of my readers prepared to give it to them by means purely chemical?

There is a solitary house in the opening of the valley, over which the Scuir More stands sentinel,—a house so solitary, that the entire breadth of the island intervenes between it and the nearest human dwelling. It is inhabited by a shepherd and his wife,—the sole representatives in the valley of a numerous population, long since expatriated to make way for a few flocks of sheep, but whose ranges of little fields may still be seen green amid the heath on both sides, for nearly a mile upwards from the opening. After descending along the precipices of the Scuir, we struck across the valley, and, on scaling the opposite slope sat down on the summit to rest us, about a hundred yards over the house of the shepherd. He had seen us from below, when engaged among the bloodstones, and had seen, withal, that we were not coming his way; and, "on hospitable thoughts intent," [Pg 153] he climbed to where we sat, accompanied by his wife, she bearing a vast bowl of milk, and he a basket of bread and cheese. And we found the refreshment most seasonable, after our long hours of toil, and with a rough journey still before us. It is an excellent circumstance, that hospitality grows best where it is most needed. In the thick of men it dwindles and disappears, like fruits in the thick of a wood; but where man is planted sparsely, it blossoms and matures, like apples on a standard or espalier. It flourishes where the inn and the lodging-house cannot exist, and dies out where they thrive and multiply.

We reached the cross valley in the interior of the island about half an hour before sunset. The evening was clear, calm, golden-tinted; even wild heaths and rude rocks had assumed a flush of transient beauty; and the emerald-green patches on the hill-sides, barred by the plough lengthwise, diagonally, and transverse, had borrowed an aspect of soft and velvety richness, from the mellowed light and the broadening shadows. All was solitary. We could see among the deserted fields the grass-grown foundations of cottages razed to the ground; but the valley, more desolate than that which we had left, had not even its single inhabited dwelling: it seemed as if man had done with it forever. The island, eighteen years before, had been divested of its inhabitants, amounting at the time to rather more than four hundred souls, to make way for one sheep-farmer and eight thousand sheep. All the aborigines of Rum crossed the Atlantic; and at the close of 1828, the entire population consisted of but the sheep-farmer, and a few shepherds, his servants; the island of Rum reckoned up scarce a single family at this period for every five square miles of area which it contained. But depopulation on so extreme a scale was found inconvenient; the place had been rendered too thoroughly a desert for the comfort of the occupant; [Pg 154] and on the occasion of a clearing which took place shortly after in Skye, he accommodated some ten or twelve of the ejected families with sites for cottages, and pasturage for a few cows, on the bit of morass beside Loch Scresort, on which I had seen their humble dwellings. But the whole of the once-peopled interior remains a wilderness, without inhabitant,—all the more lonely in its aspect from the circumstance that the solitary valleys, with their plough-furrowed patches, and their ruined heaps of stone, open upon shores every whit as solitary as themselves, and that the wide untrodden sea stretches drearily around. The armies of the insect world were sporting in the light this evening by millions; a brown stream that runs through the valley yielded an incessant poppling sound, from the myriads of fish that were ceaselessly leaping in the pools, beguiled by the quick glancing wings of green and gold that fluttered over them; along a distant hill-side there ran what seemed the ruins of a gray-stone fence, erected, says tradition, in a remote age, to facilitate the hunting of the deer; there were fields on which the heath and moss of the surrounding moorlands were fast encroaching, that had borne many a successive harvest; and prostrate cottages, that had been the scenes of christenings, and bridals, and blythe new-year's days;—all seemed to bespeak the place a fitting habitation for man, in which not only the necessaries, but also a few of the luxuries of life, might be procured; but in the entire prospect not a man nor a man's dwelling could the eye command. The landscape was one without figures. I do not much like extermination carried out so thoroughly and on system;—it seems bad policy; and I have not succeeded in thinking any the better of it though assured by economists that there are more than people enough in Scotland still. There are, I believe, more than enough in our workhouses,—more than enough on our [Pg 155] pauper-rolls,—more than enough huddled up, disreputable, useless, and unhappy, in the miasmatic alleys and typhoid courts of our large towns; but I have yet to learn how arguments for local depopulation are to be drawn from facts such as these. A brave and hardy people, favorably placed for the development of all that is excellent in human nature, form the glory and strength of a country;—a people sunk into an abyss of degradation and misery, and in which it is the whole tendency of external circumstances to sink them yet deeper, constitute its weakness and its shame; and I cannot quite see on what principle the ominous increase which is taking place among us in the worse class, is to form our solace or apology for the wholesale expatriation of the better. It did not seem as if the depopulation of Rum had tended much to any one's advantage. The single sheep-farmer who had occupied the holdings of so many had been unfortunate in his speculations, and had left the island: the proprietor, his landlord, seemed to have been as little fortunate as the tenant, for the island itself was in the market; and a report went current at the time, that it was on the eve of being purchased by some wealthy Englishman, who purposed converting it into a deer-forest. How strange a cycle! Uninhabited originally save by wild animals, it became at an early period a home of men, who, as the gray wall on the hill-side testified, derived, in part at least, their sustenance from the chase. They broke in from the waste the furrowed patches on the slopes of the valleys,—they reared herds of cattle and flocks of sheep,—their number increased to nearly five hundred souls,—they enjoyed the average happiness of human creatures in the present imperfect state of being,—they contributed their portion of hardy and vigorous manhood to the armies of the country,—and a few of their more adventurous spirits, impatient of the narrow bounds which confined them, and [Pg 156] a course of life little varied by incident, emigrated to America. Then came the change of system so general in the Highlands; and the island lost all its original inhabitants, on a wool and mutton speculation,—inhabitants, the descendants of men who had chased the deer on its hills five hundred years before, and who, though they recognized some wild island lord as their superior, and did him service, had regarded the place as indisputably their own. And now yet another change was on the eve of ensuing, and the island was to return to its original state, as a home of wild animals, where a few hunters from the mainland might enjoy the chase for a month or two every twelvemonth, but which could form no permanent place of human abode. Once more, a strange and surely most melancholy cycle!

There was light enough left, as we reached the upper part of Loch Scresort, to show us a shoal of small silver-coated trout, leaping by scores at the effluence of the little stream along which we had set out in the morning on our expedition. There was a net stretched across where the play was thickest; and we learned that the haul of the previous tide had amounted to several hundreds. On reaching the Betsey, we found a pail and basket laid against the companion-head,—the basket containing about two dozen small trout,—the minister's unsolicited teind of the morning draught; the pail filled with razor-fish of great size. The people of my friend are far from wealthy; there is scarce any circulating medium in Rum; and the cottars in Eigg contrive barely enough to earn at the harvest in the Lowlands money sufficient to clear with their landlord at rent-day. Their contributions for ecclesiastical purposes make no great figure, therefore, in the lists of the Sustentation Fund. But of what they have they give willingly and in a kindly spirit; and if baskets of small trout, or pailfuls of spout-fish, went current in the Free Church, [Pg 157] there would, I am certain, be a per centage of both the fish and the mollusc, derived from the Small Isles, in the half-yearly sustentation dividends. We found the supply of both,—especially as provisions were beginning to run short in the lockers of the Betsey,—quite deserving of our gratitude. The razor-fish had been brought us by the worthy catechist of the island. He had gone to the ebb in our special behalf, and had spent a tide in laboriously filling the pail with these "treasures hid in the sand;" thoroughly aware, like the old exiled puritan, who eked out his meals in a time of scarcity with the oysters of New England, that even the razor-fish, under this head, is included in the promises. There is a peculiarity in the razor-fish of Rum that I have not marked in the razor-fish of our eastern coasts. The gills of the animal, instead of bearing the general color of its other parts, like those of the oyster, are of a deep green color, resembling, when examined by the microscope, the fringe of a green curtain.

We were told by John Stewart, that the expatriated inhabitants of Rum used to catch trout by a simple device of ancient standing, which preceded the introduction of nets into the island, and which, it is possible, may in other localities have not only preceded the use of the net, but may have also suggested it: it had at least the appearance of being a first beginning of invention in this direction. The islanders gathered large quantities of heath, and then tying it loosely into bundles, and stripping it of its softer leafage, they laid the bundles across the stream on a little mound held down by stones, with the tops of the heath turned upwards to the current. The water rose against the mound for a foot or eighteen inches, and then murmured over and through, occasioning an expansion among the hard elastic sprays. Next a party of the islanders came down the stream, beating the banks and pools, and sending a still thickening [Pg 158] shoal of trout before them, that, on reaching the miniature dam formed by the bundles, darted forward for shelter, as if to a hollow bank, and stuck among the slim hard branches, as they would in the meshes of a net. The stones were then hastily thrown off,—the bundles pitched ashore,—the better fish, to the amount not unfrequently of several scores, secured,—and the young fry returned to the stream, to take care of themselves, and grow bigger. We fared richly this evening, after our hard day's labor, on tea and trout; and as the minister had to attend a meeting of the Presbytery of Skye on the following Wednesday, we sailed next morning for Glenelg, whence he purposed taking the steamer for Portree. Winds were light and baffling, and the currents, like capricious friends, neutralized at one time the assistance which they lent us at another. It was dark night ere we had passed Isle Ornsay, and morning broke as we cast anchor in the Bay of Glenelg. At ten o'clock the steamer heaved-to in the bay to land a few passengers, and the minister went on board, leaving me in charge of the Betsey, to follow him, when the tide set in, through the Kyles of Skye.

[Pg 159]

CHAPTER IX.

Kyles of Skye—A Gneiss District—Kyle Rhea—A Boiling Tide—A "Take" of Sillocks—The Betsey's "Paces"—In the Bay at Broadford—Rain—Island of Pabba—Description of the Island—Its Geological Structure—Astrea—Polypifers—*Gryphæa incurva*—Three groups of Fossils in the Lias of Skye—Abundance of the Petrifactions of Pabba—Scenery—Pabba a "piece of smooth, level England"—Fossil Shells of Pabba—Voyage resumed—Kyle Akin—Ruins of Castle Maoil—A "Thornback" Dinner—The Bunch of Deep Sea Tangle—The Caileach Stone—Kelp Furnaces—Escape of the Betsey from sinking.

No sailing vessel attempts threading the Kyles of Skye from the south in the face of an adverse tide. The currents of Kyle Rhea care little for the wind-filled sail, and battle at times, on scarce unequal terms, with the steam-propelled paddle. The Toward Castle this morning had such a struggle to force her way inwards, as may be seen maintained at the door of some place of public meeting during the heat of some agitating controversy, when seat and passage within can hold no more, and a disappointed crowd press eagerly for admission from without. Viewed from the anchoring place at Glenelg, the opening of the Kyle presents the appearance of the bottom of a landlocked bay;—the hills of Skye seem leaning against those of the mainland: and the tide-buffeted steamer looked this morning as if boring her way into the earth, like a disinterred mole, only at a rate vastly slower. First, however, with a progress resembling that of the minute-hand of a clock, the bows disappeared amid the heath, then the midships, then the quarter-deck and stern, and then, last of all, the red tip of the sun-brightened union-jack that streamed gaudily behind. [Pg 160] I had at least two hours before me ere the Betsey might attempt weighing anchor; and, that they might leave some mark, I went and spent them ashore in the opening of Glenelg,—a gneiss district, nearly identical in structure with the district of Knock and Isle Ornsay. The upper part of the valley is bare and treeless, but not such its character where it opens to the sea; the hills are richly wooded; and cottages, and cornfields, with here and there a reach of the lively little river, peep out from among the trees. A group of tall roofless buildings, with a strong wall in front, form the central point in the landscape; these are the dismantled Berera Barracks, built, like the line of forts in the great Caledonian Valley,—Fort George, Fort Augustus, and Fort William,—to overawe the Highlands at a time when the loyalty of the Highlander pointed to a king beyond the water; but all use for them has long gone by, and they now lie in dreary ruin,—mere sheltering places for the toad and the bat. I found in a loose silt on the banks of the river, at some little distance below tide-mark, a bed of shells and coral, which might belong, I at first supposed, to some secondary formation, but which I ascertained, on examination, to be a mere recent deposit, not so old by many centuries as our last raised sea-beaches. There occurs in various localities on these western coasts, especially on the shores of the island of Pabba, a sprig coral, considerably larger in size than any I have elsewhere seen in Scotland; and it was from its great abundance in this bed of silt that I was at first led to deem the deposit an ancient one.

We weighed anchor about noon, and entered the opening of Kyle Rhea. Vessel after vessel, to the number of eight or ten in all, had been arriving in the course of the morning, and dropping anchor, nearer the opening or farther away, each according to its sailing ability, to await the turn of the tide; and we now found ourselves one of [Pg 161] the components of a little fleet, with some five or six vessels sweeping up the Kyle before us, and some three or four driving on behind. Never, except perhaps in a Highland river big in flood, have I seen such a tide. It danced and wheeled, and came boiling in huge masses from the bottom; and now our bows heaved abruptly round in one direction, and now they jerked as suddenly round in another; and, though there blew a moderate breeze at the time, the helm failed to keep the sails steadily full. But whether our sheets bellied out, or flapped right in the wind's eye, on we swept in the tideway, like a cork caught during a thunder shower in one of the rapids of the High Street. At one point the Kyle is little more than a quarter of a mile in breadth; and here, in the powerful eddy which ran along the shore, we saw a group of small fishing-boats pursuing a shoal of sillocks in a style that blent all the liveliness of the chase with the specific interest of the angle. The shoal, restless as the tides among which it disported, now rose in the boilings of one eddy, now beat the water into foam amid the stiller dimplings of another. The boats hurried from spot to spot wherever the quick glittering scales appeared. For a few seconds, rods would be cast thick and fast, as if employed in beating the water, and captured fish glanced bright to the sun; and then the take would cease, and the play rise elsewhere, and oars would flash out amain, as the little fleet again dashed into the heart of the shoal. As the Kyle widened, the force of the current diminished, and sail and helm again became things of positive importance. The wind blew a-head, steady though not strong; and the Betsey, with companions in the voyage against which to measure herself, began to show her paces. First she passed one bulky vessel, then another: she lay closer to the wind than any of her fellows, glided more quickly through the water, turned in [Pg 162] her stays like Lady Betty in a minuet; and, ere we had reached

Kyle Akin, the fleet in the middle of which we had started were toiling far behind us, all save one vessel, a stately brig; and just as we were going to pass her too, she cast anchor, to await the change of the tide, which runs from the west during flood at Kyle Akin, as it runs from the east through Kyle Rhea. The wind had freshened; and as it was now within two hours of full sea, the force of the current had somewhat abated; and so we kept on our course, tacking in scant room, however, and making but little way. A few vessels attempted following us, but, after an inefficient tack or two, they fell back on the anchoring ground, leaving the Betsey to buffet the currents alone. Tack followed tack sharp and quick in the narrows, with an iron-bound coast on either hand. We had frequent and delicate turning: now we lost fifty yards, now we gained a hundred. John Stewart held the helm; and as none of us had ever sailed the way before, I had the vessel's chart spread out on the companion-head before me, and told him when to wear and when to hold on his way,—at what places we might run up almost to the rock edge, and at what places it was safest to give the land a good offing. Hurrah for the Free Church yacht Betsey! and hurrah once more! We cleared the Kyle, leaving a whole fleet tide-bound behind us; and, stretching out at one long tack into the open sea, bore, at the next, right into the bay at Broadford, where we cast anchor for the night, within two hundred yards of the shore. Provisions were running short; and so I had to make a late dinner this evening on some of the razor-fish of Rum, topped by a dish of tea. But there is always rather more appetite than food in the country;—such, at least, is the common result under the present mode of distribution: the hunger overlaps and outstretches the provision; [Pg 163] and there was comfort in the reflection, that with the razor-fish on which to fall back, it overlapped it but by a very little on this occasion in the cabin of the Betsey. The steam-boat passed southwards next morning, and I was joined by my friend the minister a little before breakfast.

The day was miserably bad: the rain continued pattering on the skylight, now lighter, now heavier, till within an hour of sunset, when it ceased, and a light breeze began to unroll the thick fogs from off the landscape, volume after volume, like coverings from off a mummy,—leaving exposed in the valley of the Lias a brown and cheerless prospect of dark bogs and of debris-covered hills, streaked this evening with downward lines of foam. The seaward view is more pleasing. The deep russet of the interior we find bordered for miles along the edge of the bay with a many-shaded fringe of green; and the smooth grassy island of Pabba lies in the midst, a polished gem, all the more advantageously displayed from the roughness of the surrounding setting. We took boat, and explored the Lias in our immediate neighborhood till dusk. I had spent several hours among its deposits when on my way to Portree, and several hours more when on my journey across the country to the east coast; but it may be well, for the sake of maintaining some continuity of description, to throw together my various observations on the formation, as if made at one time, and to connect them with my exploration of Pabba, which took place on the following morning. The rocks of Pabba belong to the upper part of the Lias; while the lower part may be found leaning to the south, towards the Red Sandstones of the Bay of Lucy. Taking what seems to be the natural order, I shall begin with the base of the formation first.

In the general indentation of the coast, in the opening [Pg 164] of which the island of Pabba lies somewhat like a long green steamboat at anchor, there is included a smaller indentation, known as the Bay or Cove of Lucy. The central space in the cove is soft and gravelly; but on both its sides it is flanked by low rocks, that stretch out into the sea in long rectilinear lines, like the foundations of drystone fences. On the south side the rocks are red; on the north they are of a bluish-gray color; their hues are as distinct as those of the colored patches in a map; and they represent geological periods that lie widely apart. The red rocks we find laid down in most of our maps as Old Red, though I am disposed to regard them as of a much higher antiquity than even that ancient system; while the bluish-gray rocks are decidedly Liasic. [3] The cove between represents a deep ditch-like hollow, which occurs in Skye, both in the interior and on the sea-shore, in the line of boundary betwixt the Red Sandstone and the Lias; and it "seems to have originated," says M'Culloch, "in the decomposition of the exposed parts of the formations at their junction." "Hence," he adds, "from the wearing of the materials at the surface, a cavity has been produced, which becoming subsequently filled with rubbish, and generally covered over with a vegetable soil of unusual depth, effectually prevents a view of the contiguous parts." The first strata exposed on the northern side are the oldest Liasic rocks anywhere seen in Scotland. They are composed chiefly of greenish-colored fissile sandstones and calciferous grits, in which we meet a few fossils, very imperfectly preserved. But the organisms increase as we go on. We see in passing, near a picturesque little cottage,—the only one on the shores of the bay,—a crag [Pg 165] of a singularly rough appearance, that projects mole-like from the sward upon the beach, and then descending abruptly to the level of the other strata, runs out in a long ragged line into the sea. The stratum, from two to three feet in thickness, of which it is formed, seems wholly built up of irregularly-formed rubbly concretions, just as some of the garden-walls in the neighborhood of Edinburgh are built of the rough scoria of our glass-houses; and we find, on examination, that every seeming concretion in the bed is a perfectly formed coral of the genus Astrea. We have arrived at an entire bed of corals, all of one species. Their surfaces, wherever they have been washed by the sea, are of great beauty: nothing can be more irregular than the outline of each mass, and yet scarce anything more regular than the sculpturings on every part of it. We find them fretted over with polygons, like those of a honeycomb, only somewhat less mathematically exact, and the centre of every polygon contains its many-rayed star. It is difficult to distinguish between species in some of the divisions of corals: one Astrea, recent or extinct, is sometimes found so exceedingly like another of some very different formation or period, that the more modern might almost be deemed a lineal descendant of the more ancient species. With an eye to the fact, I brought with me some characteristic specimens of this Astrea [4] of the Lower Lias, which I have ranged side by side with the Astreæ of the Oölite I had found so abundant a twelvemonth before in the neighborhood of Helmsdale. In some of the hand specimens, that present merely a piece of polygonal surface, bounded by fractured sides, the difference is not easily distinguishable: the polygonal depressions are generally smaller in the Oölitic species, and shallower in the Liasic one; but [Pg 166] not unfrequently these differences disappear, and it is only when compared in the entire unbroken coral that their specific peculiarities acquire the necessary prominence. The Oölitic Astrea is of much greater size than the Liasic one: it occurs not unfrequently in masses of from two to three feet in diameter; and as its polygons are tubes that converge to the footstalk on which it originally formed, it presents in the average outline a fungous-like appearance; whereas in the smaller Liasic coral, which rarely exceeds a foot in diameter, there is no such general convergency of the tubes; and the form in one piece, save that there is a certain degree of flatness common to all, bears no resemblance to the form in another. Some of the recent Astreæ are of great beauty when inhabited by the living zoöphites whose skeleton framework they compose. Every polygonal star in the mass is the house of a separate animal, that, when withdrawn into its cell, presents the appearance of a minute flower, somewhat like a daisy stuck flat to the surface, and that, when stretched out, resembles a small round tower, with a garland of leaves bound round it atop for a cornice. The *Astrea viridis*, a coral of the tropics, presents on a ground of velvety brown myriads of deep green florets, that ever and anon start up from the level in their tower-like shape, contract and expand their petals, and then, shrinking back into their cells, straightway became florets again. The Lower Lias presented in one of its opening scenes, in this part of the world, appearances of similar beauty widely spread. For miles together,—we know not how many,—the bottom of a clear shallow sea was paved with living Astreæ: every irregular rock-like coral formed a separate colony of polypora, that, when in motion, presented the appearance of continuous masses of many-colored life, and when at rest, the places they occupied were more thickly studded with the living [Pg 167] florets than the richest and most flowery piece of pasture the reader ever saw, with its violets or its daisies. And mile beyond mile this scene of beauty stretched on through the shallow depths of the Liasic sea. The calcareous framework of most of the recent Astreæ are white; but in the species referred to,—the *Astrea viridis*,—it is of a dark-brown color. It is not unworthy of remark, in connection with these facts, that the Oölitic Astrea of Helmsdale occurs as a white, or, when darkest, as a cream-colored petrifaction; whereas the Liasic Astrea of Skye is invariably of a deep earthy hue. The one was probably a white, the other a dingy-colored coral.

The Liasic bed of Astreæ existed long enough here to attain a thickness of from two to three feet. Mass rose over mass,—the living upon the dead,—till at length, by a deposit of mingled mud and sand,—the effect, mayhap, of some change of currents, induced we know not how,—the innumerable polypedes of the living surface were buried up and killed, and then, for many yards, layer after layer of a calciferous grit was piled over them. The fossils of the grit are few and ill preserved; but we occasionally find in it a coral similar to the Astrea of the bed below, and, a little higher up, in an impure limestone, specimens, in rather indifferent keeping, of a genus of polypifer which somewhat resembles the Turbinolia of the Mountain Limestone. It presents in the cross section the same radiated structure as the *Turbinolia fungites,* and nearly the same furrowed appearance in the longitudinal one; but, seen in the larger specimens, we find that it was a branched coral, with obtuse forky boughs, in each of which, it is probable, from their general structure, there lived a single polype. It may have been the resemblance which these bear, when seen in detached branches, to the elder Caryophyllia, taken in connection with the fact that the deposit in which they occur [Pg 168] rests on the ancient Red Sandstone of the district, that led M'Culloch to question whether this fossiliferous formation had not nearly as clear a claim to be regarded as an analogue of the Carboniferous Limestone of England as of its Lias; and hence he contented himself with terming it simply the Gryphite Limestone. Sir R. Murchison, whose much more close and extensive acquaintance with fossils enabled him to assign to the deposit its true place, was struck, however, with the general resemblance of its polypifers to "those of the Madreporite Limestone of the Carboniferous series." These polypifers occur in only the lower Lias of Skye. [5] I found no corals in its higher beds, though these are charged with other fossils, more characteristic of the formation, in vast abundance. In not a few of the middle strata, composed of a mud-colored fissile sandstone, the gryphites lie as thickly as currants in a Christmas cake; and as they weather white, while the stone in which they are embedded retains its dingy hue, they somewhat remind one of the white-lead tears of the undertaker mottling a hatchment of sable. In a fragment of the dark sandstone, six inches by seven, which I brought with me, I reckon no fewer than twenty-two gryphites; and it forms but an average specimen of the bed from which I detached it. By far the most abundant species is that not inelegant shell so characteristic of the formation, the *Gryphæa incurva*. We find detached specimens scattered over the beach by hundreds, mixed up with the remains of recent shells, as if the

Gryphæa incurva were a recent shell too. They lie, bleached white by the weather, among the valves of defunct oysters and dead buccinidæ; and, from their resemblance to lamps cast in the classic model, remind one, in the corners where they have accumulated [Pg 169] most thickly, of the old magician's stock in trade, who wiled away the lamp of Aladdin from Aladdin's simple wife. The *Gryphæa obliquita* and *Gryphæa M'Cullochii* also occur among these middle strata of the Lias, though much less frequently than the other. We, besides, found in them at least two species of Pecten, with two species of Terebratula,—the one smooth, the other sulcated; a bivalve resembling a Donax; another bivalve, evidently a Gervillia, though apparently of a species not yet described; and the ill-preserved rings of large Ammonites, from ten inches to a foot in diameter. Towards the bottom of the bay the fossils again become more rare, though they re-appear once more in considerable abundance as we pass along its northern side; but in order to acquaint ourselves with the upper organisms of the formation, we have to take boat and explore the northern shores of Pabba. The Lias of Skye has its three distinct groups of fossils: its lower coraline group, in which the Astrea described is most abundant; its middle group, in which the *Gryphæa incurva* occurs by millions; and its upper group, abounding in Ammonites, Nautili, Pinnæ, and Serpulæ.

Friday made amends for the rains and fogs of its disagreeable predecessor: the morning rose bright and beautiful, with just wind enough to fill, and barely fill, the sail, hoisted high, with miser economy, that not a breath might be lost; and, weighing anchor, and shaking out all our canvass, we bore down on Pabba, to explore. This island, so soft in outline and color, is formidably fenced round by dangerous reefs; and, leaving the Betsey in charge of John Stewart and his companion, to dodge on in the offing, I set out with the minister in our little boat, and landed on the north-eastern side of the island, beside a trap-dyke that served us as a pier. He would be a happy geologist who, with a few thousands to spare, could call Pabba his own. It contains less [Pg 170] than a square mile of surface; and a walk of little more than three miles and a half along the line where the waves break at high water brings the traveller back to his starting point; and yet, though thus limited in area, the petrifactions of its shores might of themselves fill a museum. They rise by thousands and tens of thousands on the exposed planes of its sea-washed strata, standing out in bold relief, like sculpturings on ancient tombstones, at once mummies and monuments,—the dead and the carved memorials of the dead. Every rock is a tablet of hieroglyphics, with an ascertained alphabet; every rolled pebble a casket with old pictorial records locked up within. Trap-dykes, beyond comparison finer than those of the Water of Leith, which first suggested to Hutton his theory, stand up like fences over the sedimentary strata, or run out like moles far into the sea. The entire island, too, so green, rich, and level, is itself a specimen illustrative of the effect of geologic formation on scenery. We find its nearest neighbor,—the steep, brown, barren island of Longa, which is composed of the ancient Red Sandstone of the district,—differing as thoroughly from it in aspect as a bit of granite differs from a bit of clay-slate; and the whole prospect around, save the green Liasic strip that lies along the bottom of the Bay of Broadford, exhibits, true to its various components, Plutonic or sedimentary, a character of picturesque roughness or bold sublimity. The only piece of smooth, level England, contained in the entire landscape, is the fossil-mottled island of Pabba. We were first struck, on landing this morning, by the great number of Pinnæ embedded in the strata,—shells varying from five to ten inches in length,—one species of the common flat type, exemplified in the existing *Pinna sulcata*, and another nearly quadrangular, in the cross section, like the *Pinna lanceolata* of the Scarborough limestone. The quadrangular species is more deeply [Pg 171] crisped outside than the flat one. Both species bear the longitudinal groove in the centre, and when broken across, are found to contain numerous smaller shells,—Terebratulæ of both the smooth and sulcated kinds, and a species of minute smooth Pecten resembling the *Pecten demissus*, but smaller. The Pinnæ, ere they became embedded in the original sea-bottom, long since hardened into rock around them, were, we find, dead shells, into which, as into the dead open shells of our existing beaches, smaller shells were washed by the waves. Our recent Pinnæ are all sedentary shells, some of them full two feet in length, fastened to their places on their deep-sea floors by flowing silky byssi,—cables of many strands,—of which beautiful pieces of dress, such as gloves and hose, have been manufactured. An old French naturalist, the Abbe Le Pluche, tells us that "the Pinna with its fleshy tongue" (foot),—a rude inefficient looking implement for work so nice,—"spins such threads as are more valuable than silk itself, and with which the most beautiful stuffs that ever were seen have been made by Sicilian weavers." Gloves made of the byssus of recent Pinnæ may be seen in the British Museum. Associated with the numerous Pinnæ of Pabba we found a delicately-formed Modiola, a small Ostrya, Plagiostoma, Terebratula, several species of Pectens, a triangular univalve resembling a Trochus, innumerable groups of Serpulæ, and the star-like joints of Pentacrinites. The Gryphæ are also abundant, occurring in extensive beds; and Belemnites of various species lie as thickly scattered over the rock as if they had been the spindles of a whole kingdom thrown aside in consequence of some such edict framed to put them down as that passed by the father of the Sleeping Beauty. We find, among the detached masses of the beach, specimens of Nautilus, which, though rarely perfect, are sufficiently so to show the peculiarities [Pg 172] of the shell; and numerous Ammonites project in relief from almost every weathered plane of the strata. These last shells, in the tract of shore which we examined, are chiefly of one species,—the *Ammonites spinatus*,—one of which, considerably broken, the reader may find figured in Sowerby's "Mineral Conchology," from a specimen brought from Pabba sixteen years ago by Sir R. Murchison. It is difficult to procure specimens tolerably complete. We find bits of outer rings existing as limestone, with every rib sharply preserved, but the rest of the fossil lost in the shale. I succeeded in finding but two specimens that show the inner whorls. They are thickly ribbed; and the chief peculiarity which they exhibit, not so directly indicated by Mr. Sowerby's figure, is, that while the ribs of the outer whorl are broad and deep, as in the *Ammonites obtusus*, they suddenly change their character, and become numerous and narrow in the inner whorls, as in the *Ammonites communis.*

The tide began to flow, and we had to quit our explorations, and return to the Betsey. The little wind had become less, and all the canvas we could hang out enabled us to draw but a sluggish furrow. The stern of the Betsey "wrought no buttons" on this occasion; but she had a good tide under her keel; and ere the dinner-hour we had passed through the narrows of Kyle Akin. The village of this name was designed by the late Lord M'Donald for a great seaport town; but it refused to grow; and it has since become a gentleman in a small way, and does nothing. It forms, however, a handsome group of houses, pleasantly situated on a flat green tongue of land, on the Skye side, just within the opening of the Kyle; and there rises on an eminence beyond it a fine old tower, rent open, as if by an earthquake, from top to bottom, which forms one of the most picturesque objects I have almost ever seen in a landscape. [Pg 173] There are bold hills all around, and rocky islands, with the ceaseless rush of tides in front; while the cloven tower, rising high over the shore, is seen, in threading the Kyles, whether from the south or north, relieved dark against the sky, as the central object in the vista. We find it thus described by the Messrs. Anderson of Inverness, in their excellent "Guide Book,"—by far the best companion of the kind with which the traveller who sets himself to explore our Scottish Highlands can be provided. "Close to the village of Kyle Akin are the ruins of an old square keep, called Castle Muel or Maoil, the walls of which are of a remarkable thickness. It is said to have been built by the daughter of a Norwegian king, married to a Mackinnon or Macdonald, for the purpose of levying an impost on all vessels passing the Kyles, excepting, says the tradition, those of her own country. For the more certain exaction of this duty, she is reported to have caused a strong chain to be stretched across from shore to shore; and the spot in the rocks to which the terminal links were attached is still pointed out." It was high time for us to be home. The dinner hour came; but, in meet illustration of the profound remark of Trotty-Veck, not the dinner. We had been in a cold Moderate district, whence there came no half-dozens of eggs, or whole dozens of trout, or pailfuls of razor-fish, and in which hard cabin-biscuit cost us sixpence per pound. And now our stores were exhausted, and we had to dine as best we could, on our last half-ounce of tea, sweetened by our last quarter of a pound of sugar. I had marked, however, a dried thornback hanging among the rigging. It had been there nearly three weeks before, when I came first aboard, and no one seemed to know for how many weeks previous; for as it had come to be a sort of fixture in the vessel, it could be looked at without being seen. But necessity sharpens the discerning faculty, and [Pg 174] on this pressing occasion I was fortunate enough to see it. It was straightway taken down, skinned, roasted, and eaten; and, though rather rich in ammonia,—a substance better suited to form the food of the organisms that do not unite sensation to vitality, than organisms so high in the scale as the minister and his friend,—we came deliberately to the opinion, that on the whole, we could scarce have dined so well on one of Major Bellenden's jack-boots,—"so thick in the soles," according to Jenny Dennison, "forby being tough in the upper leather." The tide failed us opposite the opening of Loch Alsh; the wind, long dying, at length died out into a dead calm; and we cast anchor in ten fathoms water, to wait the ebbing current that was to carry us through Kyle Rhea.

The ebb-tide set in about half an hour after sunset; and in weighing anchor to float down the Kyle,—for we still lacked wind to sail down it,—we brought up from below, on one of the anchor-flukes, an immense bunch of deep-sea tangle, with huge soft fronds and long slender stems, that had lain flat on the rocky bottom, and had here and there thrown out roots along its length of stalk, to attach itself to the rock, in the way the ivy attaches itself to the wall. Among the intricacies of the true roots of the bunch, if one may speak of the true roots of an alga, I reckoned from eighteen to twenty different forms of animal life,—Flustræ, Sertulariæ, Serpulæ, Anomiæ, Modiolæ, Astarte, Annelida, Crustacea, and Radiata. Among the Crustaceans I found a female crab of a reddish-brown color, considerably smaller than the nail of my small finger, but fully grown apparently, for the abdominal flap was loaded with spawn; and among the Echinoderms, a brownish-yellow sea-urchin about the size of a pistol-bullet, furnished with comparatively large but thinly-set spines. There is a dangerous rock in the Kyle Rhea, the Caileach stone, on which the Commissioners [Pg 175] for the Northern Lighthouses have stuck a bit of board about the size of a pot-lid, which, as it is known to be there, and as no one ever sees it after sunset, is really very effective, considering how little it must have cost the country, in wrecking vessels. I saw one of its victims, the sloop of an honest Methodist, in whose bottom the Caileach had knocked out a hole, repairing at Isle Ornsay; and I was told, that if I wished to see more, I had only just to wait a little. The honest Methodist, after looking out in vain for the bit of board, was just stepping into the shrouds, to try whether he could not see the rock on which the bit of board is placed, when all at once his vessel found out both board and rock for herself. We also had anxious looking out this evening for the bit of board: one of us thought he saw it right a-head; and when some of the others were trying to see it too, John Stewart succeeded in discovering it half a pistol-shot astern. The evening was one of the loveliest. The moon rose in cloudy majesty over the mountains of Glenelg, brightening as it rose, till the boiling eddies around us curled on the darker surface in pale circlets of light, and the shadow of the Betsey lay as sharply defined on the brown patch of calm to the larboard as if it were her portrait taken in black. Immediately at the water-edge, under a tall dark hill, there were two smouldering fires, that now shot up a sudden tongue of bright flame, and now dimmed into blood-red specks, and sent thick strongly-scented trails of smoke athwart the surface of the Kyle. We could hear, in the calm, voices from beside them, apparently those of children; and learned that they indicated the places of two kelp-furnaces,—things which have now become comparatively rare along the coasts of the Hebrides. There was the low rush of tides all around, and the distant voices from the shore, but no other sounds; and, dim in the moonshine, we could see [Pg 176] behind us several spectral-looking sails threading their silent way through the narrows, like twilight ghosts traversing some haunted corridor.

It was late ere we reached the opening of Isle Ornsay; and as it was still a dead calm we had to tug in the Betsey to the anchoring ground with a pair of long sweeps. The minister pointed to a low-lying rock on the left-hand side of the opening,—a favorite haunt of the seal. "I took farewell of the Betsey there last winter," he said. "The night had worn late, and was pitch dark; we could see before us scarce the length of our bowsprit; not a single light twinkled from the shore; and, in taking the bay, we ran bump on the skerry, and stuck fast. The water came rushing in, and covered over the cabin-floor. I had Mrs. Swanson and my little daughter aboard with me, with one of our servant-maids who had become attached to the family, and insisted on following us from Eigg; and, of course, our first care was to get them ashore. We had to land them on the bare uninhabited island yonder, and a dreary enough place it was at midnight, in winter, with its rocks, bogs, and heath, and with a rude sea tumbling over the skerries in front; but it had at least the recommendation of being safe, and the sky, though black and wild, was not stormy. I had brought two lanterns ashore: the servant girl, with the child in her lap, sat beside one of them, in the shelter of a rock; while my wife, with the other, went walking up and down along a piece of level sward yonder, waving the light, to attract notice from the opposite side of the bay. But though it was seen from the windows of my own house by an attached relative, it was deemed merely a singularly-distinct apparition of Will o' the Wisp, and so brought us no assistance. Meanwhile we had carried out a kedge astern of the Betsey, as the sea was flowing at the time, to keep her from beating in over the rocks; [Pg 177] and then, taking our few movables ashore, we hung on till the tide rose, and, with our boat alongside ready for escape, succeeded in warping her into deep water, with the intention of letting her sink somewhere beyond the influence of the surf, which, without fail, would have broken her up on the skerry in a few hours, had we suffered her to remain there. But though, when on the rock, the tide had risen as freely over the cabin sole inside as over the crags without, in the deep water the Betsey gave no sign of sinking. I went down to the cabin; the water was knee-high on the floor, dashing against bed and locker, but it rose no higher;—the enormous leak had stopped, we knew not how; and, setting ourselves to the pump, we had in an hour or two a clear ship. The Betsey is clinker-built below. The elastic oak planks had yielded inwards to the pressure of the rock, tearing out the fastenings, and admitted the tide at wide yawning seams; but no sooner was the pressure removed, than out they sprung again into their places, like bows when the strings are slackened; and when the carpenter came to overhaul, he found he had little else to do than to remove a split plank, and to supply a few dozens of drawn nails."

[Pg 178]

CHAPTER X.

Isle Ornsay—The Sabbath—A Sailor-minister's Sermon for Sailors—The Scuir Sermon—Loch Carron—Groups of Moraines—A sheep District—The Editor of the *Witness* and the Establishment Clergyman—Dingwall—Conon-side revisited—The Pond and its Changes—New Faces—The Stonemason's Mark—The Burying Ground of Urquhart—An old acquaintance—Property Qualification for Voting in Scotland—Montgerald Sandstone Quarries—

Geological Science in Cromarty—The Danes at Cromarty—The Danish Professor and the "Old Red Sandstone"—Harmonizing tendencies of Science.

The anchoring ground at Isle Ornsay was crowded with coasting vessels and fishing boats; and when the Sabbath came round, no inconsiderable portion of my friend's congregation was composed of sailors and fishermen. His text was appropriate,—"He bringeth them into their desired haven;" and as his sea-craft and his theology were alike excellent, there were no incongruities in his allegory, and no defects in his mode of applying it, and the seamen were hugely delighted. John Stewart, though less a master of English than of many other things, told me he was able to follow the minister from beginning to end,—a thing he had never done before at an English preaching. The sea portion of the sermon, he said, was very plain: it was about the helm, and the sails, and the anchor, and the chart, and the pilot,—about rocks, winds, currents, and safe harborage; and by attending to this simpler part of it, he was led into the parts that were less simple, and so succeeded in comprehending the whole. I would fain see this unique discourse, preached by a sailor minister to a sailor congregation, preserved in some permanent form, with at least one [Pg 179] other discourse,—of which I found trace in the island of Eigg, after the lapse of more than a twelvemonth,—that had been preached about the time of the Disruption, full in sight of the Scuir, with its impregnable hill-fort, and in the immediate neighborhood of the cave of Frances, with its heaps of dead men's bones. One note stuck fast to the islanders. In times of peril and alarm, said the minister, the ancient inhabitants of the island had two essentially different kinds of places in which they sought security; they had the deep, unwholesome cave, shut up from the light and the breath of heaven, and the tall rock summit, with its impregnable fort, on which the sun shone and the wind blew. Much hardship might no doubt be encountered on the one, when the sky was black with tempest, and rains beat, or snows descended; but it was found associated with no story of real loss or disaster,—it had kept safe all who had committed themselves to it; whereas, in the close atmosphere of the other there was warmth, and, after a sort, comfort; and on one memorable day of trouble the islanders had deemed it the preferable sheltering place of the two.

And there survived mouldering skeletons and a frightful tradition, to tell the history of their choice. Places of refuge of these very opposite kinds, said the minister, continuing his allegory, are not peculiar to your island; never was there a day or a place of trial in which they did not advance their opposite claims: they are advancing them even now all over the world. The one kind you find described by one great prophet as low-lying "refuges of lies," over which the desolating "scourge must pass," and which the destroying "waters must overflow;" while the true character of the other may be learned from another great prophet, who was never weary of celebrating his "rock and his fortress." "Wit succeeds more from being happily addressed," says Goldsmith, "than even from its native [Pg 180] poignancy." If my friend's allegory does not please quite as well in print and in English as it did when delivered *viva voce* in Gaelic, it should be remembered that it was addressed to an outdoor congregation, whose minds were filled with the consequences of the Disruption,—that the bones of *Uamh Fraingh* lay within a few hundred yards of them,—and that the Scuir, with the sun shining bright on its summit, rose tall in the background, scarce a mile away.

On Monday I spent several hours in reëxploring the Lias of Lucy Bay and its neighborhood, and then walked on to Kyle-Akin, where I parted from my friend Mr. Swanson, and took boat for Loch Carron. The greater part of the following day was spent in crossing the country to the east coast in the mail-gig, through long dreary glens, and a fierce storm of wind and rain. In the lower portion of the valley occupied by the river Carron, I saw at least two fine groups of moraines. One of these, about a mile and a half above the parish manse, marks the place where a glacier, that had once descended from a hollow amid the northern range of hills, had furrowed up the gravel and earth before it in long ridges, which we find running nearly parallel to the road; the other group, which lies higher up the valley, and seems of considerably greater extent, indicates where one of those river-like glaciers that fill up long hollows, and impel their irresistible flood downwards, slow as the hour-hand of a timepiece, had terminated towards the sea. I could but glance at the appearances as the gig drove past, and point them out to a fellow passenger, the Establishment minister of——, remarking, at the same time, how much more dreary the prospect must have seemed than even it did to-day, though the fog was thick and the drizzle disagreeable, when the lateral hollows on each side were blocked up with ice, and overhanging glaciers, that ploughed the rock bare in their descent, glistened on the bleak hill-sides. [Pg 181] I wore a gray maud over a coat of rough russet, with waist-coat and trowsers of plaid; and the minister, who must have taken me, I suppose, for a southland shepherd looking out for a farm, gave me much information of a kind I might have found valuable had such been my condition and business, regarding the various districts through which we passed. On one high-lying farm, the grass, he said, was short and thin, but sweet and wholesome, and the flocks throve steadily, and were never thinned by disease; whereas on another farm, that lay along the dank bottom of a valley, the herbage was rank and rich, and the sheep fed and got heavy, but braxy at the close of autumn fell upon them like a pestilence, and more than neutralized to the farmer every advantage of the superior fertility of the soil. It was not uninteresting, even for one not a sheep-farmer, to learn that the life of the sheep is worth fewer years' purchase in one little track of country than in another adjacent one; and that those differences in the salubrity of particular spots which obtain in other parts of the world in regard to our own species, and which make it death to linger on the luxuriant river-side, while on the arid plain or elevated hill-top there is health and safety, should exist in contiguous walks in the Highlands of Scotland in reference to some of the inferior animals. The minister and I became wonderfully good friends for the time. All the seats in the gig, both back and front, had been occupied ere he had taken his passage, and the postman had assigned him a miserable place on the narrow elevated platform in the middle, where he had to coil himself up like a hedgehog in its hole, sadly to the discomfort of limbs still stout and strong, but stiffened by the long service of full seventy years. And, as in the case made famous by Cowper, of the "softer sex" and the old-fashioned iron-cushioned arm-chairs, the old man had, as became his years, "'gan murmur." I contrived, by sitting [Pg 182] on the edge of the gig on the one side, and by getting the postman to take a similar seat on the other, to find room for him in front; and there, feeling he had not to do with savages, he became kindly and conversible. We beat together over a wide range of topics;—the Scotch banks, and Sir

Robert Peel's intentions regarding them,—the periodical press of Scotland,—the Edinburgh literati,—the Free Church even: he had been a consistent Moderate all his days, and disliked renegades, he said; and I, of course, disliked renegades too. We both remembered that, though civilized nations give quarter to an enemy overpowered in open fight, they are still in the habit of shooting deserters. In short, we agreed on a great many different matters; and, by comparing notes, we made the best we could of a tedious journey and a very bad day. At the inn at Garve, a long stage from Dingwall, we alighted, and took the road together, to straighten our stiffened limbs, while the post man was engaged in changing horses. The minister stopped short in the middle of a discussion. We are not on equal terms, he said: you know who I am, and I don't know you: we did not start fair at the beginning, but let us start fair now. Ah, we have agreed hitherto, I replied; but I know not how we are to agree when you know who I am: are you sure you will not be frightened? Frightened! said the minister sturdily; no, by no man. Then, I am the Editor of the *Witness*. There was a momentary pause. "Well," said the minister, "it's all the same: I'm glad we should have met. Give me, man, a shake of your hand." And so the conversation went on as before till we parted at Dingwall,—the Establishment clergyman wet to the skin, the Free Church editor in no better condition; but both, mayhap, rather less out of conceit with the ride than if it had been ridden alone.

I had intended passing at least two days in the neighborhood [Pg 183] of Dingwall, where I proposed renewing an acquaintance, broken off for three-and-twenty years, with those bituminous shales of Strathpeffer in which the celebrated mineral waters of the valley take their rise,—the Old Red Conglomerate of Brahan, the vitrified fort of Knockferrel, the ancient tower of Fairburn, above all, the pleasure-grounds of Conon-side. I had spent the greater portion of my eighteenth and nineteenth years in this part of the country; and I was curious to ascertain to what extent the man in middle life would verify the observations of the lad,—to recall early incidents, revisit remembered scenes, return on old feelings, and see who were dead and who were alive among the casual acquaintances of nearly a quarter of a century ago. The morning of Wednesday rose dark with fog and rain, but the wind had fallen; and as I could not afford to miss seeing Conon-side, I sallied out under cover of an umbrella. I crossed the bridge, and reached the pleasure-grounds of Conon-house. The river was big in flood: it was exactly such a river Conon as I had lost sight of in the winter of 1821; and I had to give up all hope of wading into its fords, as I used to do early in the autumn of that year, and pick up the pearl muscles that lie so thickly among the stones at the bottom. I saw, however, amid a thicket of bushes by the river-side, a heap of broken shells, where some herd-boy had been carrying on such a pearl fishery as I had sometimes used to carry on in my own behalf so long before; and I felt it was just something to see it. The flood eddied past, dark and heavy, sweeping over bulwark and bank. The low-stemmed alders that rose on islet and mound seemed shorn of half their trunks in the tide; here and there an elastic branch bent to the current, and rose and bent again; and now a tuft of withered heath came floating down, and now a soiled wreath of foam. How [Pg 184] vividly the past rose up before me!—boyish day-dreams forgotten for twenty years,—the fossils of an early formation of mind, produced at a period when the atmosphere of feeling was warmer than now, and the immaturities of the mental kingdom grew rank and large, like the ancient Cryptogamiæ, and bore no specific resemblance to the productions of a present time. I had passed in the neighborhood the first season I anywhere spent among strangers, at an age when home is not a country, nor a province even, but simply a little spot of earth inhabited by friends and relatives; and the rude verses, long forgotten, in which my joy had found vent when on the eve of returning to that home,—a home little more than twenty miles away,—came chiming as freshly into my memory as if scarce a month had passed since I had composed them beside the Conon. [6]

Three-and-twenty years form a large portion of the short life of man,—one-third, as nearly as can be expressed in unbroken numbers, of the entire term fixed by the psalmist, and full one-half, if we strike off the twilight periods of childhood and immature youth, and of senectitude weary of its toils. I found curious indications among the grounds of Conon-side, of the time that had elapsed since I had last seen them. There was a rectangular pond in a corner of a moor, near the public road, inhabited by about a dozen voracious, frog-eating pike, that I used frequently to visit. The water in the pond was exceedingly limpid; and I could watch from the banks every motion of the hungry, energetic inmates. And now I struck off from the river-side by a narrow tangled pathway, to visit it once more. I could have [Pg 185] found out the place blindfold: there was a piece of flat brown heath that stretched round its edges, and a mossy slope that rose at its upper side, at the foot of which the taste of the proprietor had placed a rustic chair. The spot, though itself bare and moory, was nearly surrounded by wood, and looked like a clearing in an American forest. There were lines of graceful larches on two of its sides, and a grove of vigorous beeches that directly fronted the setting sun on a third; and I had often found it a place of delightful resort, in which to saunter alone in the calm summer evenings, after the work of the day was over. Such was the scene as it existed in my recollection. I came up to it this day through dripping trees, along a neglected pathway; and found, for the open space and the rectangular pond, a gloomy patch of water in the middle of a tangled thicket, that rose some ten or twelve feet over my head. What had been bare heath a quarter of a century before had become a thick wood; and I remembered, that when I had been last there, the open space had just been planted with forest-trees, and that some of the taller plants rose half-way to my knee. Human lifetimes, as now measured, are not intended to witness both the seed-times and the harvests of forests,—both the planting of the sapling, and the felling of the huge tree into which it has grown; and so the incident impressed me strongly. It reminded me of the sage Shalum in Addison's antediluvian tale, who became wealthy by the sale of his great trees, two centuries after he had planted them. I pursued my walk, to revisit another little patch of water which I had found so very entertaining a volume three-and-twenty years previous, that I could still recall many of its lessons; but the hand of improvement had been busy among the fields of Cononside; and when I came up to the spot which it had occupied, I found but a [Pg 186] piece of level arable land, bearing a rank swathe of grass and clover. [7]

Not a single individual did I find on the farm who had been there twenty years before. I entered into conversation with one of the ploughmen, apparently a man of some intelligence; but he had come to the place only a summer or two previous, and the names of most of his predecessors sounded unfamiliar in his ears: he knew scarce anything of the old laird or his times, and but little of the general history of the district. The frequent change of servants incident to the large-farm system has done scarce less to wear out the oral antiquities of the country than has been done by its busy ploughs in obliterating antiquities of a more material cast. The mythologic legend and traditionary story have shared the same fate, through the influence of the one cause, which has been experienced by the sepulchral tumulus and the ancient encampment under the operations of the other. I saw in the pillars and archways of the farm-steading some of the hewn stones bearing my own mark,—an anchor, to which I used to attach a certain symbolical meaning; and I pointed them out to the ploughman. I had hewn these stones, I said, in the days of the old laird, the grandfather of the present proprietor. The ploughman wondered how a man still in middle life could have such a story to tell. I must surely have begun work early in the day, he remarked, which was perhaps the best way for getting it soon over. I remembered having seen similar markings on the hewn-work of ancient castles, and of indulging in, I daresay, idle enough speculations regarding what was doing at court and in the field, in Scotland and elsewhere, when the old long-departed mechanics had been engaged in their work. When this mark was affixed, [Pg 187] I have said, all Scotland was in mourning for the disaster at Flodden, and the folk in the work-shed would have been, mayhap, engaged in discussing the supposed treachery of Home, and in arguing whether the hapless James had fallen in battle, or gone on a pilgrimage to merit absolution for the death of his father. And when this other more modern mark was affixed, the Gowrie conspiracy must have been the topic of the day, and the mechanics were probably speculating,—at worst not more doubtfully than the historians have done after them,—on the guilt or innocence of the Ruthvens. It now rose curiously enough in memory, that I was employed in fashioning one of the stones marked by the anchor,—a corner stone in a gate-pillar,—when one of my brother apprentices entered the work-shed, laden with a bundle of newly sharpened irons from the smithy, and said he had just been told by the smith that the great Napoleon Bonaparte was dead. I returned to the village of Conon Bridge, through the woods of Conon House. The day was still very bad: the rain pattered thick on the leaves, and fell incessantly in large drops on the pathways. There is a solitary, picturesque burying-ground on a wooded hillock beside the river, with thick dark woods all around it,—one of the two burying-grounds of the parish of Urquhart,—which I would fain have visited, but the swollen stream had risen high around, converting the hillock into an island, and forbade access. I had spent many an hour among the tombs. They are few and scattered, and of the true antique cast,—roughened with death's heads, and cross-bones, and rudely sculptured armorial bearings; and on a broken wall, that marked where the ancient chapel once had stood, there might be seen, in the year 1821, a small, badly-cut sun-dial, with its iron gnomon wasted to a saw-edged film, that contained more oxide than metal. The only fossils described in my present chapter are fossils of mind; and the reader [Pg 188] will, I trust, bear with me should I produce one fossil more of this somewhat equivocal class. It has no merit to recommend it,—it is simply an organism of an immature intellectual formation, in which, however, as in the Carboniferous period, there was provision made for the necessities of an after time. [8] If a young man born on the wrong side of the Tweed for *speaking* English, is desirous to acquire the ability of *writing* it, he should by all means begin by trying to write it in verse.

I passed, on my return to Dingwall, through the village of Conon Bridge; and remembering that one of the masons who had hewn beside me in the work-shed so many years before lived in the village at the time, I went direct to the house he had inhabited, to see whether he might not be there still. It was a low-roofed domicile beside the river, but in the days of my old acquaintance it had presented an appearance of great comfort and neatness; and as there now hung an air of neglect about it, I inferred that it had found some other tenant. I inquired, however, at the door, and was informed that Mr. —— now lived higher up the street. I would find him, it was added, in the best house on the right-hand side,—the house with a hewn front, and a shop in it. He kept the shop, and was the owner of the house, and had another house besides, and was one of the elders of the Free Church in Urquhart. Such was the standing of my old acquaintance the journeyman mason of twenty-three years ago. He had been, when I knew him, a steady, industrious, religious man,—with but one exception the only contributor to missionary and Bible societies among a numerous party of workmen; and [Pg 189] he was now occupying a respectable place in his village, and was one of the voters of the county. Let Chartism assert what it pleases on the one hand, and Toryism what it may on the other, the property-qualification of the Reform Bill is essentially a good one for such a country as Scotland. In our cities it no doubt extends the political franchise to a fluctuating class, ill hafted in society, who possess it one year and want it another; but in our villages and smaller towns it hits very nearly the right medium for forming a premium on steady industry and character, and for securing that at least the mass of those who possess it should be sober-minded men, with a stake in the general welfare. In running over the histories of the various voters in one of our smaller towns, I found that nearly one-half of the whole had, like my old comrade at Conon Bridge, acquired for themselves, through steady and industrious habits, the qualification from which they derive their vote. My companion failed to recognize in the man turned of forty the smooth-cheeked stripling of eighteen, with whom he had wrought so long before. I soon succeeded, however, in making good my claim to his acquaintance. He had previously established the identity of the editor of his newspaper with his quondam fellow-workman, and a single link more was all the chain wanted. We talked over old matters for half an hour. His wife, a staid respectable matron, who, when I had been last in the district, was exactly such a person as her eldest daughter, showed me an Encyclopædia, with colored prints, which she wished to send, if she knew but how, to the Free Church library. I walked with him through his garden, and saw trees loaded with yellow-cheeked pippins, where I had once seen only unproductive heath, that scantily covered a barren soil of ferruginous sand, and unwillingly declining an invitation to wait tea,—for a previous engagement interfered,—I took leave of [Pg 190] the family, and returned to Dingwall. The following morning was gloomy, and threatened rain; and giving up my intention of exploring Strathpeffer, I took the morning coach for Invergordon, and then walked to Cromarty, where I arrived just in time for breakfast.

I marked, from the top of the coach, about two miles to the north-east of Dingwall, beds of a deep gray sandstone, identical in color and appearance with some of the gray sandstones of the Middle Old Red of Forfarshire, and learned that quarries had lately been opened in these beds near Montgerald. The Old Red Sandstone lies in immense development on the flanks of Ben-Wevis; and it is just possible that the analogue of the gray flagstones of Forfar may be found among its upper beds. If so, the quarriers should be instructed to look hard for organic remains,—the broad-headed Cephalaspis, so characteristic of the formation, and the huge Crustacean, its contemporary, that disported in plates large as those of the steel mail of the later ages of chivalry. The geologists of Dingwall,—if Dingwall has yet got its geologists,—might do well to attempt determining the point. I found the science much in advance in Cromarty, especially among the ladies,—its great patronizers and illustrators everywhere,—and, in not a few localities, extensive contributors to its hoards of fact. Just as I arrived, there was a pic-nic party of young people setting out for the Lias of Shandwick. They spent the day among its richly fossiliferous shales and limestones, and brought back with them in the evening, Ammonites and Gryphites enough to store a museum. Cromarty had been visited during the summer by geologists speaking a foreign tongue, but thoroughly conversant with the occult yet common language of the rocks, and deeply interested in the stories which the rocks told. The vessels in which the Crown Prince of Denmark voyaged to the Faroe Isles [Pg 191] had been for some time in the bay; and the Danes, his companions, votaries of the stony science, zealously plied chisel and hammer among the Old Red Sandstones of the coast. A townsman informed me that he had seen a Danish Professor hammering like the tutelary Thor of his country among the nodules in which I had found the first Pterichthys and first Diplacanthus ever disinterred; and that the Professor, ever and anon as he laid open a specimen, brought it to a huge smooth boulder, on which there lay a copy of the "Old Red Sandstone," to ascertain from the descriptions and prints its family and name. Shall I confess that the circumstance gratified me exceedingly? There are many elements of Discord among mankind in the present time, both at home and abroad,—so many, that I am afraid we need entertain no hope of seeing an end, in at least our day, to controversy and war. And we should be all the better pleased, therefore, to witness the increase of those links of union,—such as the harmonizing bonds of a scientific sympathy,—

the tendency of which is to draw men together in a kindly spirit, and the formation of which involves no sacrifice of principle, moral or religious. I do not think that the foreigner, after geologizing in my company, would have had any very vehement desire, in the event of a war, to cut me down, or to knock me on the head. I am afraid this chapter would require a long apology, and for a long apology space is wanting. But there will be no egotism, and much geology, in my next.

[Pg 192]

CHAPTER XI

Ichthyolite Beds—An interesting Discovery—Two Storeys of Organic Remains in the Old Red Sandstone—Ancient Ocean of Lower Old Red—Two great Catastrophes—Ancient Fish Scales—Their skilful Mechanism displayed by examples—Bone Lips—Arts of the Slater and Tiler as old as Old Red Sandstone—Jet Trinkets—Flint Arrow-heads—Vitrified Forts of Scotland—Style of grouping Lower Old Red Fossils—Illustration from Cromarty Fishing Phenomena—Singular Remains of Holoptychius—Ramble with Mr. Robert Dick—Color of the Planet Mars—Tombs never dreamed of by Hervey—Skeleton of the Bruce—Gigantic Holoptychius—"Coal money Currency"—Upper Boundary of Lower Old Red—Every one may add to the Store of Geological Facts—Discoveries of Messrs. Dick and Peach.

I spent one long day in exploring the ichthyolite beds on both sides the Cromarty Frith, and another long day in renewing my acquaintance with the Liasic deposit at Shandwick. In beating over the Lias, though I picked up a few good specimens, I acquired no new facts; but in re-examining the Old Red Sandstone and its organisms I was rather more successful. I succeeded in eliciting some curious points not yet recorded, which, with the details of an interesting discovery made in the far north in this formation, I may be perhaps able to weave into a chapter somewhat more geological than my last.

Some of the readers of my little work on the Old Red Sandstone will perhaps remember that I described the organisms of that ancient system as occurring in the neighborhood of Cromarty mainly on one platform, raised rather more than a hundred feet over the great Conglomerate; and that on this platform, as if suddenly overtaken by some wide-spread catastrophe, the ichthyolites lie by [Pg 193] thousands and tens of thousands, in every attitude of distortion and terror. We see the spiked wings of the Pterichthys elevated to the full, as they had been erected in the fatal moment of anger and alarm, and the bodies of the Cheirolepis and Cheiracanthus bent head to tail, in the stiff posture into which they had curled when the last pang was over. In various places in the neighborhood the ichthyolites are found *in situ* in their coffin-like nodules, where it is impossible to trace the relation of the beds in which they occur to the rocks above and below; and I had suspected for years that in at least some of the localities, they could not have belonged to the lower platform of death, but to some posterior catastrophe that had strewed with carcasses some upper platform. I had thought over the matter many a time and oft when I should have been asleep,—for it is marvellous how questions of the kind grow upon a man; and now, selecting as a hopeful scene of inquiry the splendid section under the Northern Sutor, I set myself doggedly to determine whether the Old Red Sandstone in this part of the country has not at least its two storeys of organic remains, each of which had been equally a scene of sudden mortality. I was entirely successful. The lower ichthyolite bed occurs exactly one hundred and fourteen feet over the great Conglomerate; and three hundred and eighteen feet higher up I found a second ichthyolite bed, as rich in fossils as the first, with its thorny Acanthodians twisted half round, as if still in the agony of dissolution, and its Pterichthyes still extending their spear-like arms in the attitude of defence. The discovery enabled me to assign to their true places the various ichthyolite beds of the district. Those in the immediate neighborhood of the town, and a bed which abuts on the Lias at Eathie, belong to the upper platform; while those which appear in Eathie Burn, and along the shores [Pg 194] at Navity, belong to the lower. The chief interest of the discovery, however, arises from the light which it throws on the condition of the ancient ocean of the Lower Old Red, and on the extreme precariousness of the tenure on which the existence of its numerous denizens was held. In a section of little more than a hundred yards there occur at least two platforms of violent death,—platforms inscribed with unequivocal evidence of two great catastrophes which over wide areas depopulated the seas. In the Old Red Sandstone of Caithness there are many such platforms: storey rises over storey; and the floor of each bears its closely-written record of disaster and sudden extinction. Pompeii in this northern locality lies over Herculaneum, and Anglano over both. We cease to wonder why the higher order of animals should not have been introduced into a scene of being that had so recently arisen out of chaos, and over which the reign of death so frequently returned. In a somewhat different sense from that indicated by the poet of the "Seasons,"

> "As yet the trembling *year* was unconfirmed.
> And *winter* oft at eve resumed the gale."

Lying detached in the stratified clay of the fish-beds, there occur in abundance single plates and scales of ichthyolites, which, as they can be removed entire, and viewed on both sides, illustrate points in the mechanism of the creatures to which they belonged that cannot be so clearly traced in the same remains when locked up in stone. There is a vast deal of skilful carpentry exhibited—if carpentry I may term it—in the coverings of these ancient ichthyolites. In the commoner fish of our existing seas the scales are so thin and flexible,—mere films of horn,—that there is no particularly nice fitting required in their arrangement. The condition, [Pg 195] too, through which portions of unprotected skin may be presented to the water, as over and between the rays of the fins, and on the snout and lips, obviates many a mechanical difficulty of the earlier period, when it was a condition, as the remains demonstrate, that no bit of naked skin, should be exposed, and when the scales and plates were formed, not of thin horny films, but of solid pieces of bone. Thin slates lie on the roof of a modern dwelling, without any nice fitting;—they are scales of the modern construction: but it required much nice fitting to make thick flagstones lie on the roof of an ancient cathedral;—*they*, on the other hand, were scales of the ancient type. Again, it requires no ingenuity whatever, to suffer the hands and face to go naked,—and such is the condition of our existing fish, with their soft skinny snouts and membranous fins; but to cover the hands with flexible steel gauntlets, and the face with such an iron mask as that worn by the mysterious prisoner of Louis XIV., would require a very large amount of ingenuity indeed; and the ancient ichthyolites of the Old Red were all masked and gauntleted. Now the detached plates and scales of the stratified clay exhibit not a few of the mechanical contrivances through which the bony coverings of these fish were made to unite—as in coats of old armor—great strength with great flexibility. The scales of the Osteolepis and Diplopterus I found nicely bevelled atop and at one of the sides; so that where they overlapped each other,—for at the joints not a needle-point could be insinuated,—the thickness of the two scales equalled but the thickness of one scale in the centre, and thus an equable covering was formed. I brought with me some of these detached scales, and they now lie fitted together on the table before me, like pieces of complicated hewn work carefully arranged on the ground ere the workman transfers them to their place on the wall. In the smaller-scaled fish, [Pg 196] such as the Cheiracanthus and Cheirolepis, a different principle obtained. The minute glittering rhombs of bone were set thick on the skin, like those small scales of metal sewed on leather, that formed an inferior kind of armor still in use in eastern nations, and which was partially used in our own country just ere the buff coat altogether superseded the coat of mail. I found a beautiful piece of jaw in the clay, with the enamelled tusks bristling on its brightly enamelled edge, like iron teeth in an iron rake. Mr. Parkinson expresses some wonder, in his work on fossils, that in a fine ichthyolite in the British Museum, not only the teeth should have been preserved, but also the lips; but we now know enough of the construction of the more ancient fish, to cease wondering. The lips were formed of as solid bone as the teeth themselves, and had as fair a chance of being preserved entire; just as the metallic rim of a toothed wheel has as fair a chance of being preserved as the metallic teeth that project from it. I was interested in marking the various modes of attachment to the body of the animal which the detached scales exhibit. The slater fastens on his slates with nails driven into the wood: the tiler secures his tiles by means of a raised bar on the under side of each, that locks into a corresponding bar of deal in the framework of the roof. Now in some of the scales I found the art of the tiler anticipated; there were bars raised on their inner sides, to lay hold of the skin beneath; while in others it was the art of the slater that had been anticipated,—the scales had been slates fastened down by long nails driven in slant-wise, which were, however, mere prolongations of the scale itself. Great truths may be repeated until they become truisms, and we fail to note what they in reality convey. The great truth that all knowledge dwelt without beginning in the adorable Creator must, I am afraid, have been thus common-placed in my mind; for at [Pg 197] first it struck me as wonderful that the humble arts of the tiler and slater should have existed in perfection in the times of the Old Red Sandstone.

I had often remarked amid the fossiliferous limestones of the Lower Old Red, minute specks and slender veins of a glossy bituminous substance somewhat resembling jet, sufficiently hard to admit of a tolerable polish, and which emitted in the fire a bright flame, I had remarked, further, its apparent identity with a substance used by the ancient inhabitants of the northern part of the country in the manufacture of their rude ornaments, as occasionally found in sepulchral urns, such as beads of an elliptical form, and flat parallelograms, perforated edge-wise by some four or five holes a-piece; but I had failed hitherto in detecting in the stone, portions of sufficient bulk for the formation of either the beads or the parallelograms. On this visit to the ichthyolite beds, however, I picked up a nodule that inclosed a mass of the jet large enough to admit of being fashioned into trinkets of as great bulk as any of the ancient ones I have yet seen, and a portion of which I succeeded in actually forming into a parallelogram, that could not have been distinguished from those of our old sepulchral urns. It is interesting enough to think, that these fossiliferous beds, altogether unknown to the people of the country for many centuries, and which, when I first discovered them, some twelve or fourteen years ago, were equally unknown to geologists, should have been resorted to for this substance, perhaps thousands of years ago, by the savage aborigines of the district. But our antiquities of the remoter class furnish us with several such facts. It is comparatively of late years that we have become acquainted with the yellow chalk-flints of Banffshire and Aberdeen; though before the introduction of iron into the country they seem to have [Pg 198] been well known all over the north of Scotland. I have never yet seen a stone arrow-head found in any of the northern localities, that had not been fashioned out of this hard and splintery substance,—a sufficient proof that our ancestors, ere they had formed their first acquaintance with the metals, were intimately acquainted with at least the mechanical properties of the chalk-flint, and knew where in Scotland it was to be found. They were mineralogists enough, too, as their stone battle-axes testify, to know that the best tool-making rock is the axe-stone of Werner; and in some localities they must have brought their supply of this rather rare mineral from great distances. A history of those arts of savage life, as shown in the relics of our earlier antiquities, which the course of discovery sereved thoroughly to supplant, but which could not have been carried on without a knowledge of substances and qualities afterwards lost, until re-discovered by scientific curiosity, would form of itself an exceedingly curious chapter. The art of the gun-flint maker (and it, too, promises soon to pass into extinction) is unquestionably a curious one, but not a whit more curious or more ingenious than the art possessed by the rude inhabitants of our country eighteen hundred years ago, of chipping arrow-heads with an astonishing degree of neatness out of the same stubborn material. They found, however, that though flint made a serviceable arrow-head, it was by much too brittle for an adze or battle-axe; and sought elsewhere than among the Banffshire gravels for the rock out of which these were to be wrought. Where they found it in our northern provinces I have not yet ascertained. It is but a short time since I came to know that they were beforehand with me in the discovery of the bituminous jet of the Lower Old Red Sandstone, and were excavators among its fossiliferous beds. The vitrified forts of the [Pg 199] north of Scotland give evidence of yet another of the obsolete arts. Before the savage inhabitants of the country were ingenious enough to know the uses of mortar, or were furnished with tools sufficiently hard and solid to dress a bit of sandstone, they must have been acquainted with the *chemical* fact, that with the assistance of fluxes, a pile of stones could be fused into a solid wall, and with the *mineralogical* fact, that there are certain kinds of stones which yield much more readily to the heat than others. The art of making vitrified forts was the art of making ramparts of rock through a knowledge of the less obstinate earths and the more powerful fluxes. I have been informed by Mr. Patrick Duff of Elgin, that he found, in breaking open a vitrified fragment detached from an ancient hill-fort, distinct impressions of the serrated kelp-weed of our shores,—the identical flux which, in its character as the kelp of commerce, was so extensively used in our glass-houses only a few years ago.

I was struck, during my explorations at this time, as I had been often before, by the style of grouping, if I may so speak, which obtains among the Lower Old Red fossils. In no deposit with which I am acquainted, however rich in remains, have all its ichthyolites been found lying together. The collector finds some one or two species very numerous; some two or three considerably less so, but not unfrequent; some one or two more, perhaps, exceedingly rare; and a few, though abundant in other localities, that never occur at all. In the Cromarty beds, for instance, I never found a Holoptychius, and a Dipterus only once; the Diplopterus is rare; the Glyptolepis not common; the Cheirolepis and Pterichthys more so, but not very abundant; the Cheiracanthus and Diplacanthus, on the other hand, are numerous; and the Osteolepis and Coccosteus more numerous still. But in other deposits of [Pg 200] the same formation, though a similar style of grouping obtains, the proportions are reversed with regard to species and genera: the fish rare in one locality abound in another. In Banniskirk, for instance, the Dipterus is exceedingly common, while the Osteolepis and Coccosteus are rare, and the Cheiracanthus and Cheirolepis seem altogether awanting. Again, in the Morayshire deposits, the Glyptolepis is abundant, and noble specimens of the Lower Old Red Holoptychius—of which more anon—are to be found in the neighborhood of Thurso, associated with remains of the Diplopterus, Coccosteus, Dipterus, and Osteolepis. The fact may be deemed of some little interest by the geologist, and may serve to inculcate caution, by showing that it is not always safe to determine regarding the place or age of subordinate formations from the per centage of certain fossils which they may be found to contain, or from the fact that they should want some certain organisms of the system to which they belong, and possess others. These differences may and do exist in contemporary deposits; and I had a striking example, on this occasion, of their dependence on a simple law of instinct, which is as active in producing the same kind of phenomena now as it seems to have been in the earlier days of the Old Red Sandstone. The Cromarty and Moray Friths, mottled with fishing boats (for the bustle of the herring fishers had just begun), stretched out before me. A few hundred yards from the shore there was a yawl lying at anchor, with an old fisherman and a few boys angling from the stern for sillocks (the young of the coal-fish) and for small rock-cod. A few miles higher up, where the Cromarty Frith expands into a wide landlocked basin, with shallow sandy shores, there was a second yawl engaged in fishing for flounders and small skate,—for such are the kinds of fish that frequent the [Pg 201] flat shallows of the basin. A turbot-net lay drying in the sun: it served to remind me that some six or eight miles away, in an opposite direction, there is a deep-sea bank, on which turbot, halibut, and large skate are found. Numerous boats were stretching down the Moray Frith, bound for the banks of a more distant locality, frequented at this early stage of the herring fishing by shoals of herrings, with their attendant dog-fish and cod; and I knew that in yet another deep-sea range there lie haddock and whiting banks. Almost every variety of existing fish in the two friths has its own peculiar habitat; and were they to be destroyed by some sudden catastrophe, and preserved by some geologic process, on the banks and shoals which they frequent, there would occur exactly the same phenomena of grouping in the fossiliferous contemporaneous deposits which they would thus constitute, as we find exhibited by the deposits of the Lower Old Red Sandstone.

The remains of Holoptychius occur, I have said, in the neighborhood of Thurso. I must now add, that very singular remains they are,—full of interest to the naturalist, and, in great part at least, new to Geology. My readers, votaries of the stony science, must be acquainted with the masterly paper of Mr. Sedgwick and Sir R. Murchison "On the Old Red Sandstone of Caithness and the North of Scotland generally," which forms part of the second volume (second series) of the "Transactions of the Geological Society," and with the description which it furnishes, among many others, of the rocks in the neighborhood of Thurso. Calcareo-bituminous flags, grits, and shales, of which the paving flagstones of Caithness may be regarded as the general type, occur on the shores, in reefs, crags, and precipices; here stretching along the coast in the form of flat, uneven bulwarks: there rising over it in steep [Pg 202] walls; yonder leaning to the surf, stratum against stratum, like flights of stairs thrown down from their slant position to the level; in some places severed by faults; in others cast about in every possible direction, as if broken and contorted by a thousand antagonist movements; but in their general bearing rising towards the east, until the whole calcareo-bituminous schists of which this important member of the system is composed disappear under the red sandstones of Dunnet Head. Such, in effect, is the general description of Mr. Sedgwick and Sir R. Murchison, of the rocks in the neighborhood of Thurso. It indicates further, that in at least three localities in the range there occur in the grits and shales, scales and impressions of fish. And such was the ascertained geology of the deposit when taken up last year by an ingenious tradesman of Thurso, Mr. Robert Dick, whose patient explorations, concentrated mainly on the fossil remains of this deposit, bid fair to add to our knowledge of the ichthyology of the Old Red Sandstone. Let us accompany Mr. Dick in one of his exploratory rambles. The various organisms which he disinterred I shall describe from specimens before me, which I owe to his kindness,—the localities in which he found them, from a minute and interesting description, for which I am indebted to his pen.

Leaving behind us the town at the bottom of its deep bay, we set out to explore a bluff-headed parallelogramical promontory, bounded by Thurso Bay on the one hand, and Murkle Bay on the other, and which presents to the open sea, in the space that stretches between, an undulating line of iron-bound coast, exposed to the roll of the northern ocean. We pass two stations in which the hard Caithness flagstones so well known in commerce are jointed by saws wrought by machinery. As is common in the Old Red Sandstone, in which scarce any stratum [Pg 203] solid enough to be of value to the workmen, whether for building or paving, contains good specimens, we find but little to detain us in the dark coherent beds from which the flags are quarried. Here and there a few glittering scales occur; here and there a few coprolitic patches; here and there the faint impression of a fucoid; but no organism sufficiently entire to be transferred to the bag. As we proceed outwards, however, and the fitful breeze comes laden with the keen freshness of the open sea, we find among the hard dark strata in the immediate neighborhood of Thurso Castle, a paler-colored bed of fine-grained semi-calcareous stone, charged with remains in a state of coherency and keeping better fitted to repay the labor of the specimen-collector. The inclosing matrix is comparatively soft: when employed in the neighboring fences as a building stone, we see it resolved by the skyey influences into well-nigh its original mud; whereas the organisms which it contains are composed of a hard, scarce destructible substance,—bone steeped in bitumen; and the enamel on their outer surfaces is still as glossy and bright as the japan on a *papier-maché* tray fresh from the hands of the workman. Their deep black, too, contrasts strongly with the pale hue of the stone. They consist chiefly of scales, spines, dermal plates, snouts, skull-caps, and vegetable impressions. A little farther on, in a thick bed interposed between two faults, the same kind of remains occur in the same abundance, largely mingled with scales and teeth of Holoptychius, tuberculated plates, and coprolitic blotches; and further on still, in a rubbly flagstone, near where a little stream comes trotting merrily from the uplands to the sea, there occur skull-plates,—at least one of which has been disinterred entire,—large and massy as the helmets of ancient warriors. We have now reached the outer point of the promontory, [Pg 204] where the seaward wave, as it comes rolling unbroken from the Pole, crosses, in nearing the shore, the eastward sweep of the great Gulf-stream, and then casts itself headlong on the rocks. The view has been extending with almost every step we have taken, and it has now expanded into a wide and noble prospect of ocean and bay, island and main, bold surf-skirted headlands, and green retiring hollows. Yonder, on the one hand, are the Orkneys, rising dim and blue over the foam-mottled currents of the Pentland Frith; and yonder, on the other, the far-stretching promontory of Holborn Head, with the line of coast that sweeps along the opposite side of the bay; here sinking in abrupt flagstone precipices direct into the tide; there receding in grassy banks formed of a dark blue diluvium. The fields and dwellings of living men mingle in the landscape with old episcopal ruins and ancient burying-grounds; and yonder, well-nigh in the opening of the Frith, gleams ruddy to the sun,—a true blood-colored blush, when all around is azure or pale,—the tall Red Sandstone precipices of Dunnet Head. It has been suggested that the planet Mars may owe its red color to the extensive development of some such formation as the Old Red Sandstone of our own planet: the existing formation in Mars may, at the present time, it is said, be a Red Sandstone formation. It seems much more probable, however, that the red flush which characterizes the whole of that planet,—its oceans as certainly as its continents,—should be rather owing to some widely-diffused peculiarity of the surrounding atmosphere, than to aught peculiar in the varied surface of land and water which that atmosphere surrounds; but certainly the extensive existence of such a red system might produce the effect. If the rocks and soils of Dunnet Head formed average specimens of those of our globe generally, we could look across [Pg 205] the heavens at Mars with a disk vastly more rubicund and fiery than his own. The earth, as seen from the moon, would seem such a planet bathed in blood as the moon at its rising frequently appears from the earth.

We have rounded the promontory. The beds exposed along the coast to the lashings of the surf are of various texture and character,—here tough, bituminous, and dark; there of a pale hue, and so hard that they ring to the hammer like plates of cast iron; yonder soft, unctuous, and green,—a kind of chloritic sandstone. And these very various powers of resistance and degrees of hardness we find indicated by the rough irregularities of the surface. The softer parts retire in long trench-like hollows,—the harder stand out in sharp irregular ridges. Fossils abound: the bituminous beds glitter bright with glossy quadrangular scales, that look like sheets of black mica inclosed in granite. We find jaws, teeth, tubercled plates, skull-caps, spines, and fucoids,—"tombs among which to contemplate," says Mr. Dick, "of which Hervey never dreamed." The condition of complete keeping in which we discover some of these remains, even when exposed to the incessant dash of the surf, seems truly wonderful. We see scales of Holoptychius standing up in bold relief from the hard cherty rock that has worn from around them, with all the tubercles and wavy ridges of their sculpture entire. This state of keeping seems to be wholly owing to the curious chemical change that has taken place in their substance. Ere the skeleton of the Bruce, disinterred entire after the lapse of five centuries, was recommitted to the tomb, there were such measures taken to secure its preservation, that were it to be again disinterred even after as many centuries more had passed, it might be found retaining unbroken its gigantic proportions. There was molten pitch poured over the bones in a state of sufficient fluidity to permeate all their [Pg 206] pores, and fill up the central hollows, and which, soon hardening around them, formed a bituminous matrix, in which they may lie unchanged for more than a thousand years. Now, exactly such was the process of keeping to which nature resorted with these skeletons of the Old Red Sandstone. The animal matter with which they were charged had been converted into a hard black bitumen. Like the bones of the Bruce, they are bones steeped in pitch; and so thoroughly is every pore and hollow still occupied, that, when cast into the fire, they flamed like torches. In one of the beds at which we have now arrived Mr. Dick found the occipital plates of a Holoptychius of gigantic proportions. The frontal plates measured full sixteen inches across, and from the nape of the neck to a little above the place of the eyes, full eighteen; while a single plate belonging to the lower part of the head measures thirteen and a half inches by seven and a half. I have remarked, in my little work on the Old Red Sandstone,—founding on a large amount of negative evidence, that a mediocrity of size and bulk seems to have obtained among the fish of the Lower Old Red, though in at least the Upper formation, a considerable increase in both took place. A single piece of positive evidence, however, outweighs whole volumes of a merely negative kind. From the entire plate now in my possession, which is identical with one figured in Mr. Noble of St. Madoes' specimen, and from the huge fragments of the upper plates now before me, some of which are full five-eighth parts of an inch in thickness, I am prepared to demonstrate that this Holoptychius of the Lower Old Red must have been at least thrice the size of the *Holoptychius Nobilissimus* of Clashbennie.

Still we pass on, though with no difficulty, over the rough contorted crags, worn by the surf into deep ruts and uneven ridges, gnarled protuberances, and crater-like hollows. [Pg 207] The fossiliferous beds are still very numerous, and largely charged with remains. We see dermal bones, spines, scales, and jaws, projecting in high relief from the sea-worn surface of the ledges below, and from the weatherworn faces of the precipices above; for an uneven wall of crags some thirty or forty feet high, now runs along the shore. We have reached what seems a large mole, that sloping downwards athwart the beach from the precipices, like a huge boat-pier, runs far into the surf. We find it composed of a siliceous bed, so intensely compact and hard, that it has preserved its proportions entire, while every other rock has worn from around it. For century after century have the storms of the fierce north-west sent their long ocean-nursed waves to dash against it in foam; for century after century have the never-ceasing currents of the Pentland chafed against its steep sides, or eddied over its rough crest; and yet still does it remain unwasted and unworn,—its abrupt wall retaining all its former steepness, and every angular jutting all the original sharpness of edge. As we advance the scenery becomes wilder and more broken: here an irregular wall of rock projects from the crags towards the sea; there a dock-like hollow, in which the water gleams green, intrudes from the sea upon the crags; we pass a deep lime-encrusted cave, with which tradition associates some wild legends, and which, from the supposed resemblance of the hanging stalactites to the entrails of a large animal wounded in the chase, bears the name of Pudding-Gno; and then, turning an angle of the coast, we enter a solitary bay, that presents at its upper extremity a flat expanse of sand. Our walk is still over sepulchres charged with the remains of the long-departed. Scales of Holoptychius abound, scattered like coin over the surface of the ledges. It would seem—to borrow from Mr. Dick—as if some old lord of the treasury, who flourished [Pg 208] in the days of the coal-money currency, had taken a squandering fit at Sanday Bay, and tossed the dingy contents of his treasure-chest by shovelfuls upon the rocks. Mr. Dick found in this locality some of his finest specimens, one of which—the inner side of the skull-cap of a Holoptychius, with every plate occupying its proper place, and the large angular holes through which the eyes looked out still entire—I trust to be able by and by to present to the public in a good engraving. There occur jaws, plates, scales spines,—the remains of fucoids, too, of great size and in vast abundance. Mr. Dick has disinterred from among the rocks of Sanday Bay flattened carbonaceous stems four inches in diameter. We are still within an hour's walk of Thurso; but in that brief hour how many marvels have we witnessed!—how vast an amount of the vital mechanisms of a perished creation have we not passed over! Our walk has been along ranges of sepulchres, greatly more wonderful than those of Thebes or Petræa, and mayhap a thousand times more ancient. There is no lack of life along the shores of the solitary little bay. The shriek of the sparrow-hawk mingles from the cliffs with the hoarse deep croak of the raven; the cormorant on some wave-encircled ledge, hangs out his dark wing to the breeze; the spotted diver, plying his vocation on the shallows beyond, dives and then appears, and dives and appears again, and we see the silver glitter of scales from his beak; and far away in the offing the sunlight falls on a scull of seagulls, that flutter upwards, downwards, and athwart, now in the air, thick as midges over some forest-brook in an evening of midsummer.

But we again pass onwards, amid a wild ruinous scene of abrupt faults, detached fragments of rocks, and reversed strata: again the ledges assume their ordinary position and aspect, and we rise from lower to higher and still higher beds in the formation,—for such, as I have already remarked, [Pg 209] is the general arrangement from west to east, along the northern coast of Caithness, of the Old Red Sandstone. The great Conglomerate base of the formation we find largely developed at Port Skerry, just where the western boundary line of the county divides it from the county of Sutherland; its thick upper coping of sandstone we see forming the tall cliffs of Dunnet Head; and the greater part of the space between, nearly twenty miles as the crow flies, is occupied chiefly by the shales, grits, and flagstones, which we have found charged so abundantly with the strangely-organized ichthyolites of the second stage of vertebrate existence. In the twenty intervening miles there are many breaks and faults, and so there may be, of course, recurrences of the same strata, and re-appearances of the same beds; but, after making large allowance for partial foldings and repetitions, we must regard the development of this formation, with which the twenty miles are occupied, as truly enormous. And yet it is but one of three that occur in a single system. We reach the long flat bay of Dunnet, and cross its waste of sands. The incoherent coils of the sand-worm lie thick on the surface; and here a swarm of buzzing flies, disturbed by the foot, rises in a cloud from some tuft of tangled sea-weed; and here myriads of gray crustaceous sand-hoppers dart sidelong in the little pools, or vault from the drier ridges a few inches into the air. Were the trilobites of the Silurian system,—at one period, as their remains testify, more than equally abundant,—creatures of similar habits? We have at length arrived at the tall sandstone precipices of Dunnet, with their broad decaying fronts of red and yellow; but in vain may we ply hammer and chisel among them: not a scale, not a plate, not even the stain of an imperfect fucoid appears. We have reached the upper boundary of the Lower Old Red formation, and find it bordered by a desert devoid of all trace of life. Some of [Pg 210] the characteristic types of the formation re-appear in the upper deposits; but though there is a reproduction of the original works in their more characteristic passages, if I may so speak, many of the readings are diverse, and the editions are all new.

It is one of the circumstances of peculiar interest with which Geology at its present stage is invested, that there is no man of energy and observation who may not rationally indulge in the hope of extending its limits by adding to its facts. Mr. Dick, an intelligent tradesman of Thurso, agreeably occupies his hours of leisure, for a few months, in detaching from the rocks in his neighborhood their organic remains; and thus succeeds in adding to the existing knowledge of palæozoic life, by disinterring ichthyolites which even Agassiz himself would delight to figure and describe. Several of the specimens in my possession, which I owe to the kindness of Mr. Dick, are so decidedly unique, that they would be regarded as strangers in the completest geological museums extant. It is a not uncurious fact, that when the Thurso tradesman was pursuing his labors of exploration among rocks beside the Pentland Frith, a man of similar character was pursuing exactly similar labors, with nearly similar results, among rocks of nearly the same era, that bound, on the coast of Cornwall, the British Channel. When the one was hammering in "Ready-money Cove," the other, at the opposite end of the island, was disturbing the echoes of "Pudding-Gno;" and scales, plates, spines, and occipital fragments of palæozoic fishes rewarded the labors of both. In an article on the scientific meeting at York, which appeared in "Chambers' Journal" in the November of last year, the reading public were introduced to a singularly meritorious naturalist, Mr. Charles Peach, [9] a [Pg 211] private in the mounted guard (preventive service), stationed on the southern coast of Cornwall, who has made several interesting discoveries on the outer confines of the animal kingdom, that have added considerably to the list of our British zoöphites and echinodermata. The article, a finely-toned one, redolent of that pleasing sympathy which Mr. Robert Chambers has ever evinced with struggling merit, referred chiefly to Mr. Peach's labors as a naturalist; but he is also well known in the geological field.

[Pg 212]

CHAPTER XII.

Ichthyolite Beds of Clune and Lethenbarn—Limestone Quarry—Destruction of Urns and Sarcophagi in the Lime-kiln—Nodules opened—Beautiful coloring of the Remains—Patrick Duff's Description—New Genus of Morayshire Ichthyolite described—Form and size of the Nodules or Stone Coffins—Illustration from Mrs. Marshall's Cements—Forest of Darnaway—The Hill of Berries—Sluie—Elgin—Outliers of the Weald and the Oölite—Description of the Weald at Linksfield—Mr. Duff's *Lepidotus minor*—Eccentric Types of Fish Scales—Visit to the Sandstones of Scat-Craig—Fine suit of Fossils at Scat-Craig—True graveyard Bones, not mere Impressions—Varieties of pattern—The Diker's "Carved Flowers"—*Stagonolepis*, a new genus—Termination of the Ramble.

My term of furlough was fast drawing to a close. It was now Wednesday the 14th August, and on Monday the 19th it behooved me to be seated at my desk in Edinburgh. I took boat, and crossed the Moray Frith from Cromarty to Nairn, and then walked on, in a very hot sun, over Shakspeare's Moor to Boghole, with the intention of examining the ichthyolite beds of Clune and Lethenbarn, and afterwards striking across the country to Forres, through the forest of Darnaway, where the forest abuts on the Findhorn, at the picturesque village of Sluie. When I had last crossed the moor, exactly ten years before, it was in a tremendous storm of rain and wind; and the dark platform of heath and bog, with its old ruinous castle standing sentry over it, seemed greatly more worthy of the genius of the dramatist, as cloud after cloud dashed over it, like ocean waves breaking on some low volcanic island, than it did on this clear, breathless afternoon, in the unclouded sunshine. But the sublimity of the moor on which Macbeth met the witches depends in no degree on that of the "heath near Forres," [Pg 213] whether seen in foul weather or fair; its topography bears relation to but the mind of Shakspeare; and neither tile-draining nor the plough will ever lessen an inch of its area.

The limestone quarry of Clune has been opened on the edge of an extensive moor, about three miles from the public road, where the province of Moray sweeps upwards from the broad fertile belt of corn-land that borders on the sea, to the brown and shaggy interior. There is an old-fashioned bare-looking farm-house on the one side, surrounded by a few uninclosed patches of corn; and the moorland, here dark with heath, there gray with lichens, stretches away on the other. The quarry itself is merely a piece of moor that has been trenched to the depth of some five or six feet from the surface, and that presents, at the line where the broken ground leans against the ground still unbroken, a low uneven frontage, somewhat resembling that of a ruinous stone-fence. It has been opened in the outcrop of an ichthyolite bed of the Lower Old Red Sandstone, on which in this locality the thin moory soil immediately rests, without the intervention of the common boulder clay of the country; and the fish-enveloping nodules, which are composed in this bed of a rich limestone, have been burnt, for a considerable number of years, for the purposes of the agriculturist and builder. There was a kiln smoking this evening beside the quarry; and a few laborers were engaged with shovel and pickaxe in cutting into the stratified clay of the unbroken ground, and throwing up its spindle-shaped nodules on the bank, as materials for their next burning. Antiquaries have often regretted that the sculptured marble of Greece and Egypt,—classic urns, to whose keeping the ashes of the dead had been consigned, and antique sarcophagi, roughened with hieroglyphics,—should have been so often condemned to the lime-kiln by the illiterate Copt or tasteless [Pg 214] Mohammedan; and I could not help experiencing a somewhat similar feeling here. The urns and sarcophagi, many times more ancient than those of Greece and Egypt, and that told still more wondrous stories, lay thickly ranged in this strange catacomb,—so thickly, that there were quite enough for the lime-kiln and the geologists too; but I found the kiln got all, and this at a time when the collector finds scarce any fossils more difficult to procure than those of the Lower Old Red Sandstone. I asked one of the laborers whether he did not preserve some of the better specimens, in the hope of finding an occasional purchaser. Not now, he said: he used to preserve them in the days of Lady Cumming of Altyre; but since her ladyship's death, no one in the neighborhood seemed to care for them, and strangers rarely came the way.

The first nodule I laid open contained a tolerably well-preserved Cheiracanthus; the second, an indifferent specimen of Glyptolepis; and three others, in succession, remains of Coccosteus. Almost every nodule of one especial layer near the top incloses its organism. The coloring is frequently of great beauty. In the Cromarty, as in the Caithness, Orkney, and Gamrie specimens, the animal matter with which the bones were originally charged has been converted into a dark glossy bitumen, and the plates and scales glitter from a ground of opaque gray, like pieces of japan-work suspended against a rough-cast wall. But here, as in the other Morayshire deposits, the plates and scales exist in nearly their original condition, as bone that retains its white color in the centre of the specimens, where its bulk is greatest, and is often beautifully tinged at its thinner edges by the iron with which the stone is impregnated. It is not rare to find some of the better preserved fossils colored in a style that reminds one of the more gaudy fishes of the tropics. We see the body of the ichthyolite, [Pg 215] with its finely arranged scales, of a pure snow-white. Along the edges, where the original substance of the bone, combining with the oxide of the matrix, has formed a phosphate of iron, there runs a delicately shaded band of plum-blue; while the out-spread fins, charged still more largely with the oxide, are of a deep red. The description of Mr. Patrick Duff, in his "Geology of Moray," so redolent of the quiet enthusiasm of the true fossil-hunter, especially applies to the ichthyolites of this quarry, and to those of a neighboring opening in the same bed,—the quarry of Lethenbarn. "The nodules," says Mr. Duff, "which in their external shape resemble the stones used in the game of curling, but are elliptical bodies instead of round, lie in the shale on their flat sides, in a line with the dip. When taken out, they remind one of water-worn pebbles, or rather boulders of a shore. A smart blow on the edge splits them along on the major axis, and exposes the interesting inclosure. The practised geologist knows well the thrilling interest attending the breaking up of the nodule: the uninitiated cannot sympathize with it. There is no time when a fossil looks so well as when first exposed. There is a clammy moisture on the surface of the scales or plates, which brings out the beautiful coloring, and adds brilliancy to the enamel. Exposure to the weather soon dims the lustre; and even in a cabinet an old specimen is easily known by its tarnished aspect."

I found at Clune no ichthyolite to which the geologists have not been already introduced, or with which I had not been acquainted previously in the Cromarty beds. The Lower Old Red of Morayshire furnishes, however, at least one genus not yet figured nor described, and of which, so far as I am aware, only a single specimen has yet been found. It seems to have been a small delicately-formed fish; its head covered with plates; its body with round [Pg 216] scales of a size intermediate between those of the Osteolepis and Cheiracanthus; its anterior dorsal fin placed, as in the Dipterus, Diplopterus, and Glyptolepis, directly opposite to its ventral fins; the enamelled surfaces of the minute scales were fretted with microscopic undulating ridges, that radiated from the centre to the circumference; similar furrows traversed the occipital plates; and the fins, unfurnished with spines, were formed, as in the Dipterus and Diplopterus, of thick-set, enamelled rays. The posterior fins and tail of the creature were not preserved. I may mention, for the satisfaction of the geologist, that I saw this unique fossil in the possession of the late Lady Cumming of Altyre, a few weeks previous to the lamented death of her ladyship; and that, on assuring her it was as new in relation to the Cromarty and Caithness fish-beds as to those of Moray, she intimated an intention of forthwith sending a drawing of it to Agassiz; but her untimely decease in all probability interfered with the design, and I have not since heard of this new genus of ichthyolite, or of her ladyship's interesting specimen, hitherto apparently its only representative and memorial. In the Morayshire, as in the Cromarty beds, the limestone nodules take very generally the form of the fish which they inclose: they are stone coffins, carefully moulded to express the outline of the corpses within. Is the fish entire?—the nodule is of a spindle form, broader at the head and narrower at the tail. Is it slightly curved, in the attitude of violent death?—the nodule has also its slight curve. Is it bent round, so that the extremities of the creature meet?—the nodule, in conformity with the outline, is circular. Is it disjointed and broken?—the nodule is correspondingly irregular. In nine cases out of ten, the inclosing coffin, like that of an old mummy, conforms to the outline of the organism which it incloses. It is further worthy of remark, too, that a large fish forms [Pg 217] generally a large nodule, and a small fish a small one. Here, for instance, is a nodule fifteen inches in length, here a nodule of only three inches, and here a nodule of intermediate size, that measures eight inches. We find that the large nodule contains a Cheirolepis thirteen inches in length, the small one a Diplacanthus of but two and a half inches in length, and the intermediate one a Cheiracanthus of seven inches. The size of the fish evidently regulated that of the nodule. The coffin is generally as good a fit in size as in form; and the bulk of the nodule bears almost always a definite proportion to the amount of animal matter round which it had formed. I was a good deal struck, a few weeks ago, in glancing over a series of experiments conducted for a different purpose by a lady of singular ingenuity,—Mrs. Marshall, the inventor and patentee of the beautiful marble-looking plaster, *Intonacco,*—to find what seemed a similar principle illustrated in the compositions of her various cements. These are all formed of a basis of lime, mixed in certain proportions with organic matter. The reader must be familiar with cements of this kind long known among the people, and much used in the repairing of broken pottery, such as a cement compounded of quicklime made of oyster shells, mixed up with a glue made of skim-milk cheese, and another cement made also of quicklime mixed up with the whites of eggs. In Mrs. Marshall's cements, the organic matter is variously compounded of both animal and vegetable substances, while the earth generally employed is sulphate of lime; and the result is a close-grained marble-like composition, considerably harder than the sulphate in its original crystalline state. She had deposited, in one set of her experiments, the calcareous earth, mixed up with sand, clay, and other extraneous matters, on some of the commoner molluscs of our shores; and universally found that the mass, incoherent everywhere else, [Pg 218] had acquired solidity wherever it had been permeated by the animal matter of the molluscs. Each animal, in proportion to its size, is found to retain, as in the fossiliferous spindles of the Old Red Sandstone, its coherent nodule around it. One point in the natural phenomenon, however, still remains unillustrated by the experiments of Mrs. Marshall. We see in them the animal matter giving solidity to the lime in immediate contact with it; but we do not see it possessing any such affinity for it as to form, in an argillaceous compound, like that of the ichthyolite beds, a centre of attraction powerful enough to draw together the lime diffused throughout the mass. It still remains for the geologic chemist to discover on what principle masses of animal matter should form the attracting nuclei of limestone nodules.

The declining sun warned me that I had lingered rather longer than was prudent among the ichthyolites of Clune; and so, striking in an eastern direction across a flat moor, through which I found the schistose gneiss of the district protruding in masses resembling half-buried boulders, I entered the forest of Darnaway. There was no path, and much underwood, and I enjoyed the luxury of steering my course, out of sight of road and landmark, by the sun, and of being not sure at times whether I had skill enough to play the part of the bush-ranger under his guidance. A sultry day had clarified and cooled down into a clear, balmy evening; the slant beam was falling red on a thousand tall trunks,—here gleaming along some bosky vista, to which the white silky wood-moths, fluttering by scores, and the midge and the mosquito dancing by myriads, imparted a motty gold-dust atmosphere; there penetrating in straggling rays far into some gloomy recess, and resting in patches of flame, amid the darkness, on gnarled stem, or moss-cushioned stump, or gray beard-like lichen. I dislodged, in [Pg 219] passing through the underwood, many a tiny tenant of the forest, that had a better right to harbor among its wild raspberries and junipers than I had to disturb them,—velvety night-moths, that had sat with folded wings under the leaves, awaiting the twilight, and that now took short blind flights of some two or three yards, to get out of my way,—and robust, well-conditioned spiders, whose elastic, well-tightened lines snapped sharp before me as I pressed through, and then curled up on the scarce perceptible breeze, like broken strands of wool. But every man, however Whiggish in his inclinations, entertains a secret respect for the powerful; and though I passed within a few feet of a large wasps' nest, suspended to a jutting bough of furze, the wasps I took especial care *not* to disturb. I pressed on, first through a broad belt of the forest, occupied mainly by melancholy Scotch firs; next through an opening, in which I found an American-looking village of mingled cottages, gardens, fields and wood; and then through another broad forest-belt, in which the ground is more varied with height and hollow than in the first, and in which I found only forest trees, mostly oaks and beeches. I heard the roar of the Findhorn before me, and premised I was soon to reach the river; but whether I should pursue it upwards or downwards, in order to find the ferry at Sluie, was more than I knew. There lay in my track a beautiful hillock, that reclines on the one side to the setting sun, and sinks sheer on the other, in a mural sandstone precipice, into the Findhorn. The trees open over it, giving full access to the free air and the sunshine; and I found it as thickly studded over with berries as if it had been the special care of half a dozen gardeners. The red light fell yet redder on the thickly inlaid cranberries and stone-brambles of the slope, and here and there, though so late in the season, on a patch of wild strawberries; while over all, dark, delicate [Pg 220] blueberries, with their flour-bedusted coats, were studded as profusely as if they had been peppered over it by a hailstone cloud. I have seldom seen such a school-boy's paradise, and I was just thinking what a rare discovery I would have deemed it had I made it thirty years sooner, when I heard a whooping in the wood, and four little girls, the eldest scarcely eleven, came bounding up to the hillock, their lips and fingers already dyed purple, and dropped themselves down among the berries with a shout. They were sadly startled to find they had got a companion in so solitary a recess; but I succeeded in convincing them that they were in no manner of danger from him; and on asking whether there was any of them skilful enough to show me the way to Sluie, they told me they all lived there, and were on their way home from school, which they attended at the village in the forest. Hours had elapsed since the master had *let them go,* but in so fine an evening the berries wouldn't, and so they were still in the wood. I accompanied them to Sluie, and was ferried over the river in a salmon coble. There is no point where the Findhorn, celebrated among our Scotch streams for the beauty of its scenery, is so generally interesting as in the neighborhood of this village; forest and river,—each a paragon in its kind,—uniting for several miles together what is most choice and characteristic in the peculiar features of both. In no locality is the surface of the great forest of Darnaway more undulated, or its trees nobler; and nowhere does the river present a livelier succession of eddying pools and rippling shallows, or fret itself in sweeping on its zig-zag course, now to the one bank, now to the other, against a more picturesque and imposing series of cliffs. But to the geologist the locality possesses an interest peculiar to itself. The precipices on both sides are charged with fossils of the Upper Old Red Sandstone: [Pg 221] they form part of a vast indurated graveyard, excavated to the depth of an hundred feet by the ceaseless wear of the stream; and when the waters are low, the teeth-plates and scales of ichthyolites, all of them specifically different from those of Clune and Lethenbarn, and most of them generically so, may be disinterred from the strata in handfuls. But the closing evening left me neither light nor time for the work of exploration. I heard the curfew in the woods from the yet distant town, and dark night had set in long ere I reached Forres. On the following morning I took a seat in one of the south coaches, and got on to Elgin an hour before noon.

Elgin, one of the finest of our northern towns, occupies the centre of a richly fossiliferous district, which wants only better sections to rank it among the most interesting in the kingdom. An undulating platform of Old Red Sandstone, in which we see, largely developed in one locality, the lower formation of the Coccosteus, and in another, still more largely, the upper formation of the *Holoptychius Nobilissimus*, forms, if I may so speak, the foundation deposit of the district,—the true geologic plane of the country; and, thickly scattered over this plane, we find numerous detached knolls and patches of the Weald and the Oölite, deposited like heaps of travelled soil, or of lime shot down by the agriculturist on the surface of a field. The Old Red platform is mottled by the outliers of a comparatively modern time: the sepulchral mounds of later races, that lived and died during the reptile age of the world, repose on the surface of an ancient burying-ground, charged with remains of the long anterior age of the fish; and over all, as a general covering, rest the red boulder-clay and the vegetable mould. Mr. Duff, in his valuable "Sketch of the Geology of Moray," enumerates five several localities in the neighborhood [Pg 222] of Elgin in which there occur outliers of the Weald; though, of course, in a country so flat, and in which the diluvium lies deep, we cannot hold that all have been discovered. And though the outliers of the Oölite have not yet been ascertained to be equally numerous, they seem of greater extent; the isolated masses detached from them by the denuding agencies lie thick over extensive areas; and in working out the course of improvement which has already rendered Elginshire the garden of the north, the ditcher at one time touches on some bed of shale charged with the characteristic Ammonites and Belemnites of the system, and at another on some calcareous sandstone bed, abounding with its Pectens, its Plagiostoma, and its Pinnæ. Some of these outliers, whether Wealden or Oölitic, are externally of great beauty. They occur in the parish of Lhanbryde, about three miles to the east of Elgin, in the form of green pyramidal hillocks, mottled with trees, and at Linksfield, as a confluent group of swelling grassy mounds. And from their insulated character, and the abundance of organisms which they inclose, they serve to remind one of those green pyramids of Central America in which the traveller finds deposited the skeleton remains of extinct races. It has been suggested by Mr. Duff, in his "Sketch,"—a suggestion which the late Sutherlandshire discoveries of Mr. Robertson of Inverugie have tended to confirm,—that the Oölite and Weald of Moray do not, in all probability, represent consecutive formations: they seem to bear the same sort of relation to each other as that mutually borne by the Mountain Limestone and the Coal Measures. The one, of lacustrine or of estuary origin, exhibits chiefly the productions of the land and its fresh waters; the other, as decidedly of marine origin, is charged with the remains of animals whose proper home was the sea. But the productions, though dissimilar, were [Pg 223] in all probability contemporary, just as the crabs and periwinkles of the Frith of Forth are contemporary with the frogs and lymnea of Flanders moss.

I had little time for exploration in the neighborhood of Elgin; but that little, through the kindness of my friend Mr. Duff, I was enabled to economize. We first visited together the outlier of the Weald at Linksfield. It may be found rising in the landscape, a short mile below the town, in the form of a green undulating hillock, half cut through by a limestone quarry; and the section thus furnished is of great beauty. The basis on which the hillock rests is formed of the well-marked calcareous band in the Upper Old Red, known as the Cornstone, which we find occurring here, as elsewhere, as a pale concretionary limestone of considerable richness, though in some patches largely mixed with a green argillaceous earth, and in others passing into a siliceous chert. Over the pale-colored base, the section of the hillock is ribbed like an onyx: for about forty feet, bands of gray, green, and blue clays alternate with bands of cream-colored, light-green, and dark-blue limestones; and over all there rests a band of the red boulder-clay, capped by a thin layer of vegetable mould. It is a curious circumstance, well fitted to impress on the geologist the necessity of cautious induction, that the boulder-clay not only *overlies*, but also *underlies*, this fresh-water deposit; a bed of unequivocally the same origin and character with that at the top lying intercalated, as if filling up two low flat vaults, between the upper surface of the Cornstone and the lower band of the Weald. It would, however, be as unsafe to infer that this intervening bed is older than the overlying ones, as to infer that the rubbish which chokes up the vaulted dungeon of an old castle is more ancient than the arch that stretches over it. However introduced into the cavity which it [Pg 224] occupies,—whether by land-springs or otherwise,—we find it containing fragments of the green and pale limestones that lie above, just as the rubbish of the castle dungeon might be found to contain fragments of the castle itself. When the bed of red boulder-clay was intercalated, the rocks of the overlying Wealden were exactly the same sort of indurated substances that they are now, and were yielding to the operations of some denuding agent. The alternating clays and limestones of this outlier, each of which must have been in turn an upper layer at the bottom of some lake or estuary, are abundantly fossiliferous. In some the fresh-water character of the deposit is well marked: Cyprides are so exceedingly numerous in some of the bands, that they impart to the stone an Oölitic appearance; while others of a dark-colored limestone we see strewed over, like the oozy bottom of a modern lake, with specimens of what seem Paludina, Cyclas, and Planorbus. Some of the other shells are more equivocal: a Mytilus or Modiola, which abounds in some of the bands, may have been either a sea or a fresh-water shell; and a small oyster and Astarte seem decidedly marine. Remains of fish are very abundant,—scales, plates, teeth, ichthyodorulites, and in some instances entire ichthyolites. I saw, in the collection of Mr. Duff, a small but very entire specimen of *Lepidotus minor*, with the fins spread out on the limestone, as in an anatomical preparation, and almost every plate and scale in its place. Some of his specimens of ichthyodorulites, too, are exceedingly beautiful, and of great size, resembling jaws thickly set with teeth, the apparent teeth being mere knobs ranged along the concave edge of the bone, the surface of which we see gracefully fluted and enamelled. What most struck me, however, in glancing over the drawers of Mr. Duff, was the character of the Ganoid scales of this deposit. The [Pg 225] Ganoid order in the days of the Weald was growing old; and two new orders,—the Ctenoid and Cycloid,—were on the eve of taking its place in creation. Hitherto it had comprised at least two-thirds of all the fish that had existed ever since the period in which fish first began; and almost every Ganoid fish had its own peculiar pattern of scale. But it would now seem as if well nigh all the simpler patterns were exhausted, and as if, in order to give the variety which nature loves, forms of the most eccentric types had to be resorted to. With scarce any exception save that furnished by the scales of the *Lepidotus minor,* which are plain lozenge-shaped plates, thickly japanned, the forms are strangely complex and irregular, easily expressible by the pencil, but beyond the reach of the pen. The remains of reptiles have been found occasionally, though rarely, in this outlier of the Weald,—the vertebra of a Plesiosaurus, the femur of some Chelonian reptile, and a large fluted tooth, supposed Saurian.

I would fain have visited some of the neighboring outliers of the Oölite, but time did not permit. Mr. Duff's collection, however, enabled me to form a tolerably adequate estimate of their organic contents. Viewed in the group, these present nearly the same aspect as the organisms of the Upper Lias of Pabba. There is in the same abundance large Pinnæ, and well-relieved Pectens, both ribbed and smooth; the same abundance, too, of Belemnites and Ammonites of resembling type. Both the Moray outliers and the Pabba deposit have their Terebratulæ, Gervilliæ, Plagiostoma, Cardiadæ, their bright Ganoid scales, and their imperfectly-preserved lignites. They belong apparently to nearly the same period, and must have been formed in nearly similar circumstances,—the one on the western, the other on the eastern coast of a country then covered by the vegetation of the Oölite, and now known, [Pg 226] with reference to an antiquity of but yesterday, as the ancient kingdom of Scotland. I saw among the Ammonites of these outliers at least one species, which, I believe, has not yet been found elsewhere, and which has been named, after Mr. Robertson of Inverugie, the gentleman who first discovered it, *Ammonites Robertsoni.* Like most of the genus to which it belongs, it is an exceedingly beautiful shell, with all its whorls free and gracefully ribbed, and bearing on its back, as its distinguishing specific peculiarity, a triple keel. I spent the evening of this day in visiting, with Mr. Duff, the Upper Old Red Sandstones of Scat-Craig. In Elginshire, as in Fife and elsewhere, the Upper Old Red consists of three grand divisions,—a superior bed of pale yellow sandstone, which furnishes the finest building-stone anywhere found in the north of Scotland,—an intermediate calcareous bed, known technically as the Cornstone,—and an inferior bed of sandstone, chiefly, in this locality, of a grayish-red color, and generally very incoherent in its structure. The three beds, as shown by the fossil contents of the yellow sandstones above, and of the grayish-red sandstones below, are members of the same formation,—a formation which, in Scotland at least, does not possess an organism in common with the Middle Old Red formation; that of the Cephalaspis, as developed in Forfarshire, Stirling, and Ayr, or the Lower Old Red formation; that of the Coccosteus, as developed in Caithness, Cromarty, Inverness, and Banff shires, and in so many different localities in Moray. The Sandstones at Scat-Craig belong to the grayish-red base of the Upper Old Red formation. They lie about five miles south of Elgin, not far distant from where the palæozoic deposits of the coast-side lean against the great primary nucleus of the interior. We pass from the town, through deep rich fields, carefully cultivated and well inclosed: the country, as we advance [Pg 227] on the moorlands, becomes more open; the homely cottage takes the place of the neat villa; the brown heath, of the grassy lea; and unfenced patches of corn here and there alternate with plantings of dark sombre firs, in their mediocre youth. At length we near the southern boundary of the landscape,—an undulating moory ridge, partially planted; and see where a deep gap in the outline opens a way to the upland districts of the province, a lively hill-stream descending towards the east through the bed which it has scooped out for itself in a soft red conglomerate. The section we have come to explore lies along its course: it has been the grand excavator in the densely occupied burial-ground over which it flows; but its labors have produced but a shallow scratch after all,—a mere ditch, some ten or twelve feet deep, in a deposit the entire depth of which is supposed greatly to exceed a hundred fathoms. The shallow section, however, has been well wrought; and its suit of fossils is one of the finest, both from the great specific variety which they exhibit, and their excellent state of keeping, that the Upper Old Red Sandstone has anywhere furnished.

So great is the incoherency of the matrix, that we can dig into it with our chisels, unassisted by the hammer. It reminds us of the loose gravelly soil of an ancient graveyard, partially consolidated by a night's frost,—a resemblance still further borne out by the condition and appearance of its organic contents. The numerous bones disseminated throughout the mass do not exist, as in so many of the Upper Old Red Sandstone rocks, as mere films or impressions, but in their original forms, retaining bulk as well as surface: they are true graveyard bones, which may be detached entire from the inclosing mass, and of which, were we sufficiently well acquainted with the anatomy of the long-perished races to which they [Pg 228] belonged, entire skeletons might be reconstructed. I succeeded in disinterring, during my short stay, an occipital plate of great beauty, fretted on its outer surface by numerous tubercles, confluent on its anterior part, and surrounded on its posterior portion, where they stand detached, by punctulated markings. I found also a fine scale of *Holoptychius Nobilissimus*, and a small tooth, bent somewhat like a nail that had been drawn out of its place by two opposite wrenches, and from the internal structure of which Professor Owen has bestowed on the animal to which it belonged the generic name Dendrodus. I have ascertained, however, through the indispensable assistance of Mr. George Sanderson, that the genus Holoptychius of Agassiz, named from a peculiarity in the sculpture of the scale, is the identical genus Dendrodus of Professor Owen, named from a peculiarity in the structure of the teeth. Those teeth of the genus Holoptychius, whether of the Lower or Upper Old Red, that belong to the second or *reptile* row with which the creature's jaws were furnished, present in the cross section the appearance of numerous branches, like those of trees, radiating from a centre like spokes from the nave of a wheel; and their arborescent aspect suggested to the Professor the name Dendrodus. It seems truly wonderful, when one but considers it, to what minute and obscure ramifications the variety of pattern, specific and generic, which nature so loves to preserve, is found to descend. We see great diversity of mode and style in the architecture of a city built of brick; but while the houses are different, the bricks are always the same. It is not so in nature. The bricks are as dissimilar as the houses. We find, for instance, those differences, specific and generic, that obtain among fishes, both recent and extinct, descending to even the microscopic structure of their teeth. There is more variety of pattern,—in most [Pg 229] cases of very elegant pattern,—in the sliced fragments of the teeth of the ichthyolites of a single formation, than in the carved blocks of an extensive calico-print yard. Each species has its own distinct pattern, as if in all the individuals of which it consisted the same block had been employed to stamp it; each genus has its own general *type* of pattern, as if the same inventive idea, variously altered and modified, had been wrought upon in all. In the genus Dendrodus, for instance, it is the generic type, that from a central nave there should radiate, spoke-like, a number of leafy branches; but in the several species, the branches, if I may so express myself, belong to different shrubs, and present dissimilar outlines. There are no repetitions of earlier patterns to be found among the generically different ichthyolites of other formations. We see in the world of fashion old modes of ornament continually reviving: the range of invention seems limited; and we find it revolving, in consequence, in an irregular, ever-returning cycle. But Infinite resource did not need to travel in a circle, and so we find no return or doublings in its course. It has appeared to me, that an argument against the transmutation of species, were any such needed, might be founded on those inherent peculiarities of structure that are ascertained thus to pervade the entire texture of the framework of animals. If we find one building differing from another merely in external form, we have no difficulty in conceiving how, by additions and alterations, they might be made to present a uniform appearance; transmutation, development, progression,—if one may use such terms,—seem possible in such circumstances. But if the buildings differ from each other, not only in external form, but also in every brick and beam, bolt and nail, no mere scheme of external alteration can induce a real resemblance. Every brick must be taken down, and every beam [Pg 230] and belt removed. The problem cannot be wrought by the remodelling of an old house: there is no other mode of solving it save by the erection of a new one.

Among the singularly interesting Old Red fossils of Mr. Duff's collection I saw the impression of a large ichthyolite from the superior yellow sandstone of the Upper Old Red, which had been brought him by a country diker only a few days before. In breaking open a building stone, the diker had found the inside of it, he said, covered over with curiously carved flowers; and, knowing that Mr. Duff had a turn for curiosities, he had brought the flowers to him. The supposed flowers are the sculpturings on the scales of the ichthyolite; and, true to the analogy of the diker, on at least a first glance, they may be held to resemble the rather equivocal florets of a cheap wall-paper, or of an ornamental tile. The specimen exhibits the impressions of four rows of oblong rectangular scales. One row contains seven of these, and another eight. Each scale averages about an inch and a quarter in length, by about three quarters of an inch in breadth; and the parallelogramical field which it presents is occupied by a curious piece of carving. By a sort of pictorial illusion, the device appears as if in motion: it would seem as if a sudden explosion had taken place in the middle of the field, and as if the numerous dislodged fragments, propelled all around by the central force, were hurrying to the sides. But these seeming fragments were not elevations in the original scale, but depressions. They almost seem as if they had been indented into it, in the way one sees the first heavy drops of a thunder shower indented into a platform of damp sea sand; and this last peculiarity of appearance seems to have suggested the name which this sole representative of an extinct genus has received during the course of the last few weeks from Agassiz. An Elgin gentleman forwarded [Pg 231] to Neufchatel a singularly fine calotype of the fossil, taken by Mr. Adamson of Edinburgh, with a full-size drawing of a few of the scales; and from the calotype and the drawing the naturalist has decided that the genus is entirely new, and that henceforth it shall bear the descriptive name of Stagonolepis, or drop-scale. As I looked for the first time on this broken fragment of an ichthyolite,—the sole representative and record of an entire genus of creatures that had been once called into existence to fulfil some wise purpose of the Creator long since accomplished,—I bethought me of Rogers's noble lines on the Torso,—

"And dost thou still, thou mass of breathing stone,
(Thy giant limbs to night and chaos hurled)
Still sit as on the fragment of a world,
Surviving all?"

Here, however, was a still more wonderful Torso than that of the dismembered Hercules, which so awakened the enthusiasm of the poet. Strange peculiarities of being,—singular habits, curious instincts, the history of a race from the period when the all-producing Word had spoken the first individuals into being, until, in circumstances unfitted for their longer existence, or in some great annihilating catastrophe, the last individuals perished,—were all associated with this piece of sculptured stone; but, like some ancient inscription of the desert, written in an unknown character and dead tongue, its dark meanings were fast locked up, and no inhabitant of earth possessed the key. Does that key anywhere exist, save in the keeping of Him who knows all and produced all, and to whom there is neither past nor future? Or is there a record of creation kept by those higher intelligences,—the first-born of spiritual natures,—whose existence stretches far into the eternity that has gone by, and who possess, as their [Pg 232] inheritance, the whole of the eternity to come? We may be at least assured, that nothing can be too low for angels to remember, that was not too low for God to create.

I took coach for Edinburgh on the following morning; for with my visit to Scat-Craig terminated the explorations of my Summer Ramble. During the summer of the present year I have found time to follow up some of the discoveries of the last. In the course of a hasty visit to the island of Eigg, I succeeded in finding *in situ* reptile remains of the kind which I had found along the shores in the previous season, in detached water-rolled masses. The deposit in which they occur lies deep in the Oölite. In some parts of the island there rest over it alternations of beds of trap and sedimentary strata, to the height of more than a thousand feet; but in the line of coast which intervenes between the farm-house of Keill and the picturesque shieling described in my fifth chapter, it has been laid bare by the sea immediately under the cliffs, and we may see it jutting out at a low angle from among the shingle and rolled stones of the beach for several hundred feet together, charged everywhere with the teeth, plates, and scales of Ganoid fishes, and somewhat more sparingly, with the ribs, vertebræ, and digital bones of saurians. But a full description of this interesting deposit, as its discovery belongs to the Summer Ramble of a year, the ramblings of which are not yet completed, must await some future time.

[Pg 233]

CHAPTER XIII.

SUPPLEMENTARY.

Supplementary—Isolated reptile Remains in Eigg—Small Isles revisited—The Betsey again—Storm bound—Tacking—Becalmed—Medusæ caught and described—Rain—A Shoal of Porpoises—Change of Weather—The bed-ridden Woman—The Poor Law Act for Scotland—Geological Excursion—Basaltic Columns—Oölitic Beds—Abundance of Organic Remains—Hybodus Teeth—Discovery of reptile Remains *in situ*—Musical Sand of Laig re-examined—Explanation suggested—Sail for Isle Ornsay—Anchored Clouds—A Leak sprung—Peril of the Betsey—At work with Pump and Pails—Safe in Harbor—Return to Edinburgh.

It is told of the "Spectator," on his own high authority, that having "read the controversies of some great men concerning the antiquities of Egypt, he made a voyage to Grand Cairo, on purpose to take the measure of a pyramid, and that, so soon as he had set himself right in that particular, he returned to his native country with great satisfaction." My love of knowledge has not carried me altogether so far, chiefly, I dare say, because my voyaging opportunities have not been quite so great. Ever since my ramble of last year, however, I have felt, I am afraid, a not less interest in the geologic antiquities of Small Isles than that cherished by "Spectator" with respect to the comparatively modern antiquities of Egypt; and as, in a late journey to these islands the object of my visit involved but a single point, nearly as insulated as the dimensions of a pyramid, I think I cannot do better than shelter myself under the authority of the short-faced gentleman who wrote articles in the reign of Queen Anne. I had found in Eigg, [Pg 234] in considerable abundance and fine keeping, reptile remains of the Oölite; but they had occurred in merely rolled masses, scattered along the beach. I had not discovered the bed in which they had been originally deposited, and could neither tell its place in the system, nor its relation to the other rocks of the island. The discovery was but a half-discovery,—the half of a broken medal, with the date on the missing portion. And so, immediately after the rising of the General Assembly in June last [1845], I set out to revisit Small Isles, accompanied by my friend Mr. Swanson, with the determination of acquainting myself with the burial-place of the old Oölitic reptiles, if it lay anywhere open to the light.

We found the Betsey riding in the anchoring ground at Isle Ornsay, in her foul-weather dishabille, with her topmast struck and in the yard, and her cordage and sides exhibiting in their weathered aspect the influence of the bleaching rains and winds of the previous winter. She was at once in an undress and getting old, and, as seen from the shore through rain and spray,—for the weather was coarse and boisterous,—she had apparently gained as little in her good looks from either circumstance as most other ladies do. We lay storm-bound for three days at Isle Ornsay, watching from the window of Mr. Swanson's dwelling the incessant showers sweeping down the loch. On the morning of Saturday, the gale, though still blowing right ahead, had moderated; the minister was anxious to visit this island charge, after his absence of several weeks from them at the Assembly; and I, more than half afraid that my term of furlough might expire ere I had reached my proposed scene of exploration, was as anxious as he; and so we both resolved, come what might, on doggedly beating our way adown the Sound of Sleat to Small Isles. If the wind does not fail us, said my friend, we [Pg 235] have little more than a day's work before us, and shall get into Eigg about midnight. We had but one of our seamen aboard, for John Stewart was engaged with his potato crop at home; but the minister was content, in the emergency, to rank his passenger as an able-bodied seaman; and so, hoisting sail and anchor, we got under way, and, clearing the loch, struck out into the Sound.

We tacked in long reaches for several hours, now opening up in succession the deep withdrawing lochs of the mainland, now clearing promontory after promontory in the island district of Sleat. In a few hours we had left a bulky schooner, that had quitted Isle

Ornsay at the same time, full five miles behind us; but as the sun began to decline, the wind began to sink; and about seven o'clock, when we were nearly abreast of the rocky point of Sleat, and about half-way advanced in our voyage, it had died into a calm; and for full twenty hours thereafter there was no more sailing for the Betsey. We saw the sun set, and the clouds gather, and the pelting rain come down, and nightfall, and morning break, and the noon-tide hour pass by, and still were we floating idly in the calm. I employed the few hours of the Saturday evening that intervened between the time of our arrest and nightfall, in fishing from our little boat for medusæ with a bucket. They had risen by myriads from the bottom as the wind fell, and were mottling the green depths of the water below and around far as the eye could reach. Among the commoner kinds,—the kind with the four purple rings on the area of its flat bell, which ever vibrates without sound, and the kind with the fringe of dingy brown, and the long stinging tails, of which I have sometimes borne from my swimming excursions the nettle-like smart for hours,—there were at least two species of more unusual occurrence, both of them very minute. The one, scarcely [Pg 236] larger than a shilling, bore the common umbiliferous form, but had its area inscribed by a pretty orange-colored wheel; the other, still more minute, and which presented in the water the appearance of a small hazel-nut of a brownish-yellow hue, I was disposed to set down as a species of beroe. On getting one caught, however, and transferred to a bowl, I found that the brownish-colored, melon-shaped mass, though ribbed like the beroe, did not represent the true outline of the animal; it formed merely the centre of a transparent gelatinous bell, which, though scarce visible in even the bowl, proved a most efficient instrument of motion. Such were its contractile powers, that its sides nearly closed at every stroke, behind the opaque orbicular centre, like the legs of a vigorous swimmer; and the animal, unlike its more bulky congeners,—that, despite their slow but persevering flappings, seemed greatly at the mercy of the tide, and progressed all one way,—shot, as it willed, backwards, forwards, or athwart. As the evening closed, and the depths beneath presented a dingier and yet dingier green, until at length all had become black, the distinctive colors of the acelpha,—the purple, the orange, and the brown,—faded and disappeared, and the creatures hung out, instead, their pale phosphoric lights, like the lanterns of a fleet hoisted high to prevent collision in the darkness. Now they gleamed dim and indistinct as they drifted undisturbed through the upper depths, and now they flamed out bright and green, like beaten torches, as the tide dashed them against the vessel's sides. I bethought me of the gorgeous description of Coleridge, and felt all its beauty:—

> "They moved in tracks of shining white,
> And when they reared, the elfish light
> Fell off in hoary flakes.
> [Pg 237] Within the shadow of the ship
> I watched their rich attire,—
> Blue, glassy green, and velvet black:
> They curled, and swam, and every track
> Was a flash of golden fire."

A crew of three, when there are watches to set, divides wofully ill. As there was, however, nothing to do in the calm, we decided that our first watch should consist of our single seaman, and the second of the minister and his friend. The clouds, which had been thickening for hours, now broke in torrents of rain, and old Alister got into his water-proof oil-skin and souwester, and we into our beds. The seams of the Betsey's deck had opened so sadly during the past winter, as to be no longer water-tight, and the little cabin resounded drearily in the darkness, like some dropping cave, to the ceaseless patter of the leakage. We continued to sleep, however, somewhat longer than we ought,—for Alister had been unwilling to waken the minister; but we at length got up, and, relieving watch the first from the tedium of being rained upon and doing nothing, watch the second was set to do nothing and be rained upon in turn. We had drifted during the night-time on a kindly tide, considerably nearer our island, which we could now see looming blue and indistinct through the haze some seven or eight miles away. The rain ceased a little before nine, and the clouds rose, revealing the surrounding lands, island and main,—Rum, with its abrupt mountain-peaks,—the dark Cuchullins of Skye,—and, far to the south-east, where Inverness bounds on Argyllshire, some of the tallest hills in Scotland,—among the rest, the dimly-seen Ben-Wevis. But long wreaths of pale gray cloud lay lazily under their summits, like shrouds half drawn from off the features of the dead, to be again spread over them, and we concluded that the dry weather had not yet come. A little before noon we were surrounded [Pg 238] for miles by an immense but thinly-spread shoal of porpoises, passing in pairs to the south, to prosecute, on their own behalf, the herring fishing in Lochfine or Gareloch; and for a full hour the whole sea, otherwise so silent, became vocal with long-breathed blowings, as if all the steam-tenders of all the railways in Britain were careering around us; and we could see slender jets of spray rising in the air on every side, and glossy black backs and pointed fins, that looked as if they had been fashioned out of Kilkenny marble, wheeling heavily along the surface. The clouds again began to close as the shoal passed, but we could now hear in the stillness the measured sound of oars, drawn vigorously against the gunwale in the direction of the island of Eigg, still about five miles distant, though the boat from which they rose had not yet come in sight. "Some of my poor people," said the minister, "coming to tug us ashore!" We were boarded in rather more than half an hour after,—for the sounds in the dead calm had preceded the boat by miles,—by four active young men, who seemed wonderfully glad to see their pastor; and then, amid the thickening showers, which had recommenced heavy as during the night, they set themselves to tow us into the harbor. The poor fellows had a long and fatiguing pull, and were thoroughly drenched ere, about six o'clock in the evening, we had got up to our anchoring ground, and moored, as usual, in the open tideway between *Eilan Chasteil* and the main island. There was still time enough for an evening discourse, and the minister, getting out of his damp clothes, went ashore and preached.

The evening of Sunday closed in fog and rain, and in fog and rain the morning of Monday arose. The ceaseless patter made dull music on deck and skylight above, and the slower drip, drip, through the leaky beams, drearily [Pg 239] beat time within. The roof of my bed was luckily water-tight; and I could look out from my snuggery of blankets on the desolations of the leakage, like Bacon's philosopher surveying a tempest from the shore. But the minister was somewhat less fortunate, and had no little trouble in diverting an ill-conditioned drop that had made a dead set at his pillow. I was now a full week from Edinburgh, and had seen and done nothing; and, were another week to pass after the same manner,—as, for aught that appeared, might well happen,—I might just go home again, as I had come, with my labor for my pains. In the course of the afternoon, however, the weather unexpectedly cleared up, and we set out somewhat impatiently through the wet grass, to visit a cave a few hundred yards to the west of *Naomh Fraingh*, in which it had been said the Protestants of the island might meet for the purposes of religious worship, were they to be ejected from the cottage erected by Mr. Swanson, in which they had worshipped hitherto. We reëxamined, in the passing, the pitch stone dike mentioned in a former chapter, and the charnel cave of Frances; but I found nothing to add to my former descriptions, and little to modify, save that perhaps the cave appeared less dark, in at least the outer half of its area, than it had seemed to me in the former year, when examined by torch-light, and that the straggling twilight, as it fell on the ropy sides, green with moss and mould, and on the damp bone-strewn floor, overmantled with a still darker crust, like that of a stagnant pool, seemed also to wear its tint of melancholy greenness, as if transmitted through a depth of sea-water. The cavern we had come to examine we found to be a noble arched opening in a dingy-colored precipice of augitic trap,—a cave roomy and lofty as the nave of a cathedral, and ever resounding to the dash of the sea; but though it could have amply accommodated a [Pg 240] congregation of at least five hundred, we found the way far too long and difficult for at least the weak and the elderly, and in some places inaccessible at full flood; and so we at once decided against the accommodation which it offered. But its shelter will, I trust, scarce be needed.

On our return to the Betsey, we passed through a straggling group of cottages on the hill-side, one of which, the most dilapidated and smallest of the number, the minister entered, to visit a poor old woman, who had been bed-ridden for ten years. Scarce ever before had I seen so miserable a hovel. It was hardly larger than the cabin of the Betsey, and a thousand times less comfortable. The walls and roof, formed of damp grass-grown turf, with a few layers of unconnected stone in the basement tiers, seemed to constitute one continuous hillock, sloping upwards from foundation to ridge, like one of the lesser moraines of Agassiz, save where the fabric here and there bellied outwards or inwards, in perilous dilapidation, that seemed but awaiting the first breeze. The low chinky door opened direct into the one wretched apartment of the hovel, which we found lighted chiefly by holes in the roof. The back of the sick woman's bed was so placed at the edge of the opening, that it had formed at one time a sort of partition to the portion of the apartment, some five or six feet square, which contained the fireplace; but the boarding that had rendered it such had long since fallen away, and it now presented merely a naked rickety frame to the current of cold air from without. Within a foot of the bed-ridden woman's head there was a hole in the turf-wall, which was, we saw, usually stuffed with a bundle of rags, but which lay open as we entered, and which furnished a downward peep of sea and shore, and the rocky *Eilan Chasteil*, with the minister's yacht riding in the channel hard by. The little hole in the wall had formed the poor [Pg 241] creature's only communication with the face of the external world for ten weary years. She lay under a dingy coverlet, which, whatever its original hue, had come to differ nothing in color from the graveyard earth, which must so soon better supply its place. What perhaps first struck the eye was the strange flatness of the bed-clothes, considering that a human body lay below: there seemed scarce bulk enough under them for a human skeleton. The light of the opening fell on the corpse-like features of the woman,—sallow, sharp, bearing at once the stamp of disease and of famine; and yet it was evident, notwithstanding, that they had once been agreeable,—not unlike those of her daughter, a good-looking girl of eighteen, who, when we entered, was sitting beside the fire. Neither mother nor daughter had any English; but it was not difficult to determine, from the welcome with which the minister was greeted from the sick-bed, feeble as the tones were, that he was no unfrequent visitor. He prayed beside the poor creature, and, on coming away, slipped something into her hand. I learned that not during the ten years in which she had been bed-ridden had she received a single farthing from the proprietor, nor, indeed, had any of the poor of the island, and that the parish had no session-funds. I saw her husband a few days after,—an old worn-out man, with famine written legibly in his hollow cheek and eye, and on the shrivelled frame, that seemed lost in his tattered dress; and he reiterated the same sad story. They had no means of living, he said, save through the charity of their poor neighbors, who had so little to spare; for the parish or the proprietor had never given them anything. He had once, he added, two fine boys, both sailors, who had helped them; but the one had perished in a storm off the Mull of Cantyre, and the other had died of fever when on a West India voyage; [Pg 242] and though their poor girl was very dutiful, and staid in their crazy hut to take care of them in their helpless old age, what other could she do in a place like Eigg than just share with them their sufferings? It has been recently decided by the British Parliament, that in cases of this kind the starving poor shall not be permitted to enter the law courts of the country, there to sue for a pittance to support life, until an intermediate newly-erected court, alien to the Constitution, before which they must plead at their own expense, shall have first given them permission to prosecute their claims. And I doubt not that many of the English gentlemen whose votes swelled the majority, and made it such, are really humane men, friendly to an equal-handed justice, and who hold it to be the peculiar glory of the Constitution, as well shown by De Lolme, that it has not one statute-book for the poor, and another for the rich, but the same law and the same administration of law for all. They surely could not have seen that the principle of their Poor Law Act for Scotland sets the pauper beyond the pale of the Constitution in the first instance, that he may be starved in the second. The suffering paupers of this miserable island cottage would have all their wants fully satisfied in the grave, long ere they could establish at their own expense, at Edinburgh, their claim to enter a court of law. I know not a fitter case for the interposition of our lately formed "Scottish Association for the Protection of the Poor" than that of this miserable family; and it is but one of many which the island of Eigg will be found to furnish.

After a week's weary waiting, settled weather came at last; and the morning of Tuesday rose bright and fair. My friend, whose absence at the General Assembly had accumulated a considerable amount of ministerial labor on his hands, had to employ the day professionally; and as [Pg 243] John Stewart was still engaged with his potato crop, I was necessitated to sally out on my first geological excursion alone. In passing vessel-wards, on the previous year, from the *Ru Stoir* to the farm-house of Keill, along the escarpment under the cliffs, I had examined the shores somewhat too cursorily during the one-half of my journey, and the closing evening had prevented me from exploring them during the other half at all; and I now set myself leisurely to retrace the way backwards from the farm-house to the *Stoir*. I descended to the bottom of the cliffs, along the pathway which runs between Keill and the solitary midway shieling formerly described, and found that the basaltic columns over head, which had seemed so picturesque in the twilight, lost none of their beauty when viewed by day. They occur in forms the most beautiful and fantastic; here grouped beside some blind opening in the precipice, like pillars cut round the opening of a tomb, on some rock-front in Petræa; there running in long colonnades, or rising into tall porticoes; yonder radiating in straight lines from some common centre, resembling huge pieces of fan-work, or bending out in bold curves over some shaded chasm, like rows of crooked oaks projecting from the steep sides of some dark ravine. The various beds of which the cliffs are composed, as courses of ashlar compose a wall, are of very different degrees of solidity: some are of hard porphyritic or basaltic trap; some of soft Oölitic sandstone or shale. Where the columns rest on a soft stratum, their foundations have in many places given way, and whole porticoes and colonnades hang perilously forward in tottering ruin, separated from the living rock behind by deep chasms. I saw one of these chasms, some five or six feet in width, and many yards in length, that descended to a depth which the eye could not penetrate; and another partially filled up with earth and stones, through which, [Pg 244] along a dark opening not much larger than a chimney-vent, the boys of the island find a long descending passage to the foot of the precipice, and emerge into light on the edge of the grassy talus half-way down the hill. It reminded me of the tunnel in the rock through which Imlac opened up a way of escape to Rasselas from the happy valley,—the "subterranean passage," begun "where the summit hung over the middle part," and that "issued out behind the prominence."

From the commencement of the range of cliffs, on half-way to the shieling, I found the shore so thickly covered up by masses of trap, the debris of the precipices above, that I could scarce determine the nature of the bottom on which they rested. I now, however, reached a part of the beach where the Oölitic beds are laid bare in thin party-colored strata, and at once found something to engage me. Organisms in vast abundance, chiefly shells and fragmentary portions of fishes, lie closely packed in their folds. One limestone bed, occurring in a dark shale, seems almost entirely composed of a species of small oyster; and some two or three other thin beds, of what appears to be either a species of small Mytilus or Avicula, mixed up with a few shells resembling large Paludina, and a few more of the gaper family, so closely resembling existing species, that John Stewart and Alister at once challenged them as *smurslin*, the Hebridean name for a well-known shell in these parts,—the *Mya truncata*. The remains of fishes,—chiefly Ganoid scales and the teeth of Placoids,—lie scattered among the shells in amazing abundance. On the surface of a single fragment, about nine inches by five, which I detached from one of the beds, and which now lies before me, I reckon no fewer than twenty-five teeth, and twenty-two on the area of another. They are of very various forms,—some of them squat and round, like ill-formed [Pg 245] small shot,—others spiky and sharp, not unlike flooring nails,—some straight as needles, some bent like the beak of a hawk,—some, like the palatal teeth of the Acrodus of the Lias, resemble small leeches; some, bearing a series of points ranged on a common base, like masts on the hull of a vessel, the tallest in the centre, belong to the genus Hybodus. There is a palpable approximation in the teeth of the leech-like form to the teeth with the numerous points. Some of the specimens show the same plicated structure common to both; and on some of the leech backs, if I may so speak, there are protuberant knobs, that indicate the places of the spiky points on the hybodent teeth. I have got three of each kind slit up by Mr. George Sanderson, and the internal structure appears to be the same. A dense body of bone is traversed by what seem innumerable roots, resembling those of woody shrubs laid bare along the sides of some forest stream. Each internal opening sends off on every side its myriads of close-laid filaments; and nowhere do they lie so thickly as in the line of the enamel, forming, from the regularity with which they are arranged, a sort of framing to the whole section. It is probable that the Hybodus,—a genus of shark which became extinct some time about the beginning of the chalk,—united, like the shark of Port Jackson, a crushing apparatus of palatal teeth to its lines of cutting ones. Among the other remains of these beds I found a dense fragment of bone, apparently reptilian, and a curious dermal plate punctulated with thick-set depressions, bounded on one side by a smooth band, and altogether closely resembling some saddler's thimble that had been cut open and straightened.

Following the beds downwards along the beach, I found that one of the lowest which the tide permitted me to examine,—a bed colored with a tinge of red,—was formed [Pg 246] of a denser limestone than any of the others, and composed chiefly of vast numbers of small univalves resembling Neritæ. It was in exactly such a rock I had found, in the previous year, the reptile remains; and I now set myself, with no little eagerness, to examine it. One of the first pieces I tore up contained a well-preserved Plesiosaurian vertebra; a second contained a vertebra and a rib; and, shortly after, I disinterred a large portion of a pelvis. I had at length found, beyond doubt, the reptile remains *in situ*. The bed in which they occur is laid bare here for several hundred feet along the beach, jutting out at a low angle among boulders and gravel, and the reptile remains we find embedded chiefly in its under side. It lies low in the Oölite. All the stratified rocks of the island, with the exception of a small Liasic patch, belong to the Lower Oölite, and the reptile-bed occurs deep in the base of the system,—low in its relation to the nether division, in which it is included. I found it nowhere rising to the level of high-water mark. It forms one of the foundation tiers of the island, which, as the latter rises over the sea in some places to the height of about fourteen hundred feet, its upper peaks and ridges must overlie the bones, making allowance for the dip, to the depth of at least sixteen hundred. Even at the close of the Oölitic period this sepulchral stratum must have been a profoundly ancient one. In working it out, I found two fine specimens of fish jaws, still retaining their ranges of teeth;—ichthyodorulites,—occipital plates of various forms, either reptile or ichthyic,—Ganoid scales, of nearly the same varieties of pattern as those in the Weald of Morayshire,—and the vertebræ and ribs, with the digital, pelvic, and limb-bones, of saurians. It is not unworthy of remark, that in none of the beds of this deposit did I find any of the more characteristic shells of the system,—Ammonites, Belemnites, Gryphites, or Nautili.

[Pg 247]

I explored the shores of the island on to the *Ru Stoir*, and thence to the Bay of Laig; but though I found detached masses of the reptile bed occurring in abundance, indicating that its place lay not far beyond the fall of ebb, in no other locality save the one described did I find it laid bare. I spent some time beside the Bay of Laig in reëxamining the musical sand, in the hope of determining the peculiarities on which its sonorous qualities depended. But I examined, and cross-examined it in vain. I merely succeeded in ascertaining, in addition to my previous observations, that the loudest sounds are elicited by drawing the hand slowly through the incoherent mass, in a segment of a circle, at the full stretch of the arm, and that the vibrations which produce them communicate a peculiar titillating sensation to the hand or foot by which they are elicited, extending in the foot to the knee, and in the hand to the elbow. When we pass the wet finger along the edge of an ale-glass partially filled with water, we see the vibrations thickly wrinkling the surface: the undulations which, communicated to the air, produce sound, render themselves, when communicated to the water, visible to the eye; and the titillating feeling seems but a modification of the same phenomenon acting on the nerves and fluids of the leg or arm. It appears to be produced by the wrinklings of the vibrations, if I may so speak, passing along sentient channels. The sounds will ultimately be found dependent, I am of opinion, though I cannot yet explain the principle, on the purely quartzose character of the sand, and the friction of the incoherent upper strata against under strata coherent and damp. I remained ten days in the island, and went over all my former ground, but succeeded in making no further discoveries.

On the morning of Wednesday, June 25th, we set sail for Isle Ornsay, with a smart breeze from the north-west. The [Pg 248] lower and upper sky was tolerably clear, and the sun looked cheerily down on the deep blue of the sea; but along the higher ridges of the land there lay long level strata of what the meteorologists distinguish as parasitic clouds. When every other patch of vapor in the landscape was in motion, scudding shorewards from the Atlantic before the still-increasing gale, there rested along both the Scuir of Eigg and the tall opposite ridge of the island, and along the steep peaks of Rum, clouds that seemed as if anchored, each on its own mountain-summit, and over which the gale failed to exert any propelling power. They were stationary in the middle of the rushing current, when all else was speeding before it. It has been shown that these parasitic clouds are mere local condensations of strata of damp air passing along the mountain-summits, and rendered visible but to the extent in which the summits affect the temperature. Instead of being stationary, they are ever-forming and ever-dissipating clouds,—clouds that form a few yards in advance of the condensing hill, and that dissipate a few yards after they have quitted it. I had nothing to do on deck, for we had been joined at Eigg by John Stewart; and so, after watching the appearance of the stationary clouds for some little time, I went below, and, throwing myself into the minister's large chair, took up a book. The gale meanwhile freshened, and freshened yet more; and the Betsey leaned over till her lee chain-plate lay along in the water. There was the usual combination of sounds beneath and around me,—the mixture of guggle, clunk, and splash,—of low, continuous rush, and bluff, loud blow, which forms in such circumstances the voyager's concert. I soon became aware, however, of yet another species of sound, which I did not like half so well,—a sound as of the washing of a shallow current over a rough surface; and, on the minister coming below, I asked him, tolerably well prepared for his answer, what it might [Pg 249] mean. "It means," he said, "that we have sprung a leak, and a rather bad one; but we are only some six or eight miles from the Point of Sleat, and must soon catch the land." He returned on deck, and I resumed my book. Presently, however, the rush became greatly louder; some other weak patch in the Betsey's upper works had given way, and anon the water came washing up from the lee side along the edge of the cabin floor. I got upon deck to see how matters stood with us; and the minister, easing off the vessel for a few points, gave instant orders to shorten sail, in the hope of getting her upper works out of the water, and then to unship the companion ladder, beneath which a hatch communicated with the low strip of hold under the cabin, and to bring aft the pails. We lowered our foresail; furled up the mainsail half-mast high; John Stewart took his station at the pump; old Alister and I, furnished with pails, took ours, the one at the foot, the other at the head, of the companion, to hand up and throw over; a young girl, a passenger from Eigg to the mainland, lent her assistance, and got wofully drenched in the work; while the minister, retaining his station at the helm, steered right on. But the gale had so increased, that, notwithstanding our diminished breadth of sail, the Betsey, straining hard in the rough sea, still lay in to the gunwale; and the water, pouring in through a hundred opening chinks in her upper works, rose, despite of our exertions, high over plank, and beam, and cabin-floor, and went dashing against beds and lockers. She was evidently filling, and bade fair to terminate all her voyagings by a short trip to the bottom. Old Alister, a seaman of thirty years' standing, whose station at the bottom of the cabin stairs enabled him to see how fast the water was gaining on the Betsey, but not how the Betsey was gaining on the land, was by no means the least anxious among us. Twenty years previous he had seen a vessel go down in [Pg 250] exactly similar circumstances, and in nearly the same place, and the reminiscence, in the circumstances, seemed rather an uncomfortable one. It had been a bad evening, he said, and the vessel he sailed in, and a sloop, her companion, were pressing hard to gain the land. The sloop had sprung a leak, and was straining, as if for life and death, under a press of canvas. He saw her outsail the vessel to which he belonged, but, when a few shots a-head she gave a sudden lurch, and disappeared from the surface instantaneously as a vanishing spectre, and neither sloop nor crew were ever more heard of.

There are, I am convinced, few deaths less painful than some of those untimely and violent ones at which we are most disposed to shudder. We wrought so hard at pail and pump,—the occasion, too, was one of so much excitement, and tended so thoroughly to awaken our energies,—that I was conscious, during the whole time, of an exhilaration of spirits rather pleasurable than otherwise. My fancy was active, and active, strange as the fact may seem, chiefly with ludicrous objects. Sailors tell regarding the flying Dutchman, that he was a hard-headed captain of Amsterdam, who, in a bad night and head wind, when all the other vessels of his fleet were falling back on the port they had recently quitted, obstinately swore that, rather than follow their example, he would keep beating about till the day of judgment. And the Dutch captain, says the story, was just taken at his word, and is beating about still. When matters were at the worst with us, we got under the lea of the point of Sleat. The promontory interposed between us and the roll of the sea; the wind gradually took off; and, after having seen the water gaining fast and steadily on us for considerably more than an hour, we, in turn, began to gain on the water. It came ebbing out of drawers and beds, and sunk downwards along pannels and table-legs,—a second [Pg 251] retiring deluge; and we entered Isle Ornsay with the cabin-floor all visible, and less than two feet water in the hold. On the following morning, taking leave of my friend the minister, I set off, on my return homewards, by the Skye steamer, and reached Edinburgh on the evening of Saturday.

[Pg 252]

RAMBLES OF A GEOLOGIST;

OR,

TEN THOUSAND MILES OVER THE FOSSILIFEROUS DEPOSITS OF SCOTLAND.

[Pg 255]

RAMBLES OF A GEOLOGIST;

OR,

TEN THOUSAND MILES OVER THE FOSSILIFEROUS DEPOSITS OF SCOTLAND. [10]

CHAPTER I.

Embarkation—A foundered Vessel—Lateness of the Harvest dependent on the Geological character of the Soil—A Granite Harvest and an Old Red Harvest—Cottages of Redstone and of Granite—Arable Soil of Scotland the result of a Geological Grinding Agency—Locality of the Famine of 1846—Mr. Longmuir's Fossils—Geology necessary to a Theologian—Popularizers of Science when dangerous—"Constitution of Man," and "Vestiges of Creation"—Atop of the Banff Coach—A Geologist's Field Equipment—The trespassing "Stirk"—Silurian Schists inlaid with Old Red—Bay of Gamrie how formed—Gardenstone—Geological Free-masonry illustrated—How to break an Ichthyolite Nodule—An old Rhyme mended—A raised Beach—Fossil Shells—Scotland under water at the time of the Boulder Clays.

From circumstances that in no way call for explanation, my usual exploratory ramble was thrown this year (1847) from the middle of July into the middle of September; and I embarked at Granton for the north just as the night began to count hour against hour with the day. The weather was fine, and the voyage pleasant. I saw by the way, however, at least one melancholy memorial of a hurricane which had swept the eastern coasts of the island about a fortnight before, and filled the provincial newspapers with [Pg 256] paragraphs of disaster. Nearly opposite where the Red Head lifts its mural front of Old Red Sandstone a hundred yards over the beach, the steamer passed a foundered vessel, lying about a mile and a half off the land, with but her topmast and the point of her peak over the surface. Her vane, still at the mast-head, was drooping in the calm; and its shadow, with that of the fresh-colored *spar* to which it was attached, white atop and yellow beneath, formed a well-defined undulatory strip on the water, that seemed as if ever in the process of being rolled up, and yet still retained its length unshortened. Every recession of the swell showed a patch of mainsail attached to the peak: the sail had been hoisted to its full stretch when the vessel went down. And thus, though no one survived to tell the story of her disaster, enough remained to show that she had sprung a leak when straining in the gale, and that, when staggering under a press of canvas towards the still distant shore, where, by stranding her, the crew had hoped to save at least their lives, she had disappeared with a sudden lurch, and all aboard had perished. I remembered having read, among other memorabilia of the hurricane, without greatly thinking of the matter, that "a large sloop had foundered off the Red Head,—name unknown." But the minute portion of the wreck which I saw rising over the surface, to certify, like some frail memorial in a churchyard, that the dead lay beneath, had an eloquence in it which the words wanted, and at once sent the imagination back to deal with the stern realities of the disaster, and the feelings abroad to expatiate over saddened hearths and melancholy homesteads, where for many a long day the hapless perished would be missed and mourned, but where the true story of their fate, though too surely guessed at, would never be known.

The harvest had been early; and on to the village of [Pg 257] Stonehaven, and a mile or two beyond, where the fossiliferous deposits end and the primary begin, the country presented from the deck only a wide expanse of stubble. Every farm-steading we passed had its piled stack-yard; and the fields were bare. But the line of demarcation between the Old Red Sandstone and the granitic districts formed also a separating line between an earlier and later harvest; the fields of the less kindly subsoil derived from the primary rocks were, I could see, still speckled with sheaves; and, where the land lay high, or the exposure was unfavorable, there were reapers at work. All along in the course of my journey northward from Aberdeen I continued to find the country covered with shocks, and laborers employed among them; until, crossing the Spey, I entered on the fossiliferous districts of Moray; and then, as in the south, the champaign again showed a bare breadth of stubble, with here and there a ploughman engaged in turning it down. The traveller bids farewell at Stonehaven to not only the Old Red Sandstone and the early-harvest districts, but also to the rich wheat-lands of the country, and does not again fairly enter upon them until, after travelling nearly a hundred miles, he passes from Banffshire into the province of Moray. He leaves behind him at the same line the wheat-fields and the cottages built of red stone, to find only barley and oats, and here and there a plot of rye, associated with cottages of granite and gneiss, hyperstene and mica schist; but on crossing the Spey, the red cottages reäppear, and fields of rich wheat-land spread out around them, as in the south. The circumstance is not unworthy the notice of the geologist. It is but a tedious process through which the minute lichen, settling on a surface of naked stone, forms in the course of ages a soil for plants of greater bulk and a higher order; and had Scotland been left to the exclusive operation of this slow agent, it would [Pg 258] be still a rocky desert, with perhaps here and there a strip of alluvial meadow by the side of a stream, and here and there an insulated patch of rich soil among the hollows of the crags. It might possess a few gardens for the spade, but no fields for the plough. We owe our arable land to that comparatively modern geologic agent, whatever its character, that crushed, as in a mill, the upper parts of the surface-rocks of the kingdom, and then overlaid them with their own debris and rubbish to the depth of from one to forty yards. This debris, existing in one locality as a boulder-clay more or less finely comminuted, in another as a grossly pounded gravel, forms, with few exceptions, that subsoil of the country on which the existing vegetation first found root; and, being composed mainly of the formations on which it more immediately rests, it partakes of their character,—bearing a comparatively lean and hungry aspect over the primary rocks, and a greatly more fertile one over those deposits in which the organic matters of earlier creations lie diffused. Saxon industry has done much for the primary districts of Aberdeen and Banffshires, though it has failed to neutralize altogether the effects of causes which date as early as the times of the Old Red Sandstone; but in the Highlands, which belong almost exclusively to the non-fossiliferous formations, and which were, on at least the western coasts, but imperfectly subjected to that grinding process to which we owe our subsoils, the poor Celt has permitted the consequences of the original difference to exhibit themselves in full. If we except the islands of the Inner Hebrides, the famine of 1846 was restricted in Scotland to the primary districts.

I made it my first business, on landing in Aberdeen, to wait on my friend Mr. Longmuir, that I might compare with him a few geological notes, and benefit by his knowledge of the surrounding country. I was, however, unlucky [Pg 259] enough to find that he had gone, a few days before, on a journey, from which he had not yet returned; but, through the kindness of Mrs. Longmuir, to whom I took the liberty of introducing myself, I was made free of his stone-room, and held half an hour's conversation with his Scotch fossils of the Chalk. These had been found, as the readers of the *Witness* must remember from his interesting paper on the subject, on the hill of Dudwick, in the neighborhood of Ellon, and were chiefly impressions—some of them of singular distinctness and beauty—in yellow flint. I saw among them several specimens of the Inoceramus, a thin-shelled, ponderously-hinged conchifer, characteristic of the Cretaceous group, but which has no living representative; with numerous flints, traversed by rough-edged, bifurcated hollows, in which branched sponges had once lain, a well-preserved Pecten; the impressions of spines of Echini of at least two distinct species; and the nicely-marked impression of part of a Cidaris, with the balls on which the sockets of the club-like spines had been fitted existing in the print as spherical moulds, in which shot might be cast, and with the central ligamentary depression, which in the actual fossil exists but as a minute cavity, projecting into the centre of each hollow sphere, like the wooden fusee into the centre of a bomb-shell. This latter cast, fine and sharp as that of a medal taken in sulphur, seems sufficient of itself to establish two distinct points: in the first place, that the siliceous matter of which the flint is composed, though now so hard and rigid, must, in its original condition, have been as impressible as wax softened to receive the stamp of the seal; and, in the next, that though it was thus yielding in its character, it could not have greatly shrunk in the process of hardening. I looked with no little interest on these remains of a Scotch formation now so entirely broken up, that, like those ruined cities of the [Pg 260] East which exist but as mere lines of wrought material barring the face of the desert, there has not "been left one stone of it upon another," but of which the fragments, though widely scattered, bear imprinted upon them, like the stamped bricks of Babylon, the story of its original condition, and a record of its *founders*. All Mr. Longmuir's Cretaceous fossils from the hill of Dudwick are of flint,—a substance not easily ground down by the denuding agencies.

I found several other curious fossils in Mr. Longmuir's collection. Greatly more interesting, however, than any of the specimens which it contains, is the general fact, that it should be the collection of a Free Church minister, sedulously attentive to the proper duties of his office, but who has yet found time enough to render himself an accomplished geologist; and whose week-day lectures on the science attract crowds, who receive from them, in many instances, their first knowledge of the strange revolutions of which our globe has been the subject, blent with the teachings of a wholesome theology. The present age, above all that has gone before, is peculiarly the age of physical science; and of all the physical sciences, not excepting astronomy itself, geology, though it be a fact worthy of notice, that not one of our truly accomplished geologists is an infidel, is the science of which infidelity has most largely availed itself. And as the theologian in a metaphysical age,—when skepticism, conforming to the character of the time, disseminated its doctrines in the form of nicely abstract speculations,—had, in order that the enemy might be met in his own field, to become a skilful metaphysician, he must now, in like manner, address himself to the tangibilities of natural history and geology, if he would avoid the danger and disgrace of having his flank turned by every sciolist in these walks whom he may chance to encounter. [Pg 261] It is those identical bastions and outworks that are *now* attacked, which must be *now* defended; not those which were attacked some eighty or a hundred years ago. And as he who succeeds in first mixing up fresh and curious truths, either with the objections by which religion is assailed or the arguments by which it is defended, imparts to his cause all the interest which naturally attaches to these truths, and leaves to his opponent, who passes over them after him as at second hand, a subject divested of the fire-edge of novelty, I can deem Mr. Longmuir well and not unprofessionally employed, in connecting with a sound creed the picturesque marvels of one of the most popular of the sciences, and by this means introducing them to his people, linked, from the first, with right associations. According to the old fiction, the look of the basilisk did not kill unless the creature saw before it was seen;—its mere *return* glance was harmless; and there is a class of thoroughly dangerous writers who in this respect resemble the basilisk. It is perilous to give them a first look of the public. They are formidable simply as the earliest popularizers of some interesting science, or the first promulgators of some class of curious little-known facts, with which they mix up their special contributions of error,—often the only portion of their writings that really belongs to themselves. Nor is it at all so easy to *counteract* as to *confute* them. A masterly confutation of the part of their works truly their own may, from its subject, be a very unreadable book: it can have but the insinuated poison to deal with, unmixed with the palatable pabulum in which the poison has been conveyed; and mere treatises on poisons, whether moral or medical, are rarely works of a very delectable order. It seems to be on this principle that there exists no confutation of the "Constitution of Man" in which the ordinary reader finds amusement to carry him through; whereas the work itself, full [Pg 262] of curious miscellaneous information, is eminently readable; and that the "Vestiges of Creation,"—a treatise as entertaining as the "Arabian Nights,"—bids fair, not from the amount of error which it contains, but from the amount of fresh and interestingly told truth with which the error is mingled, to live and do mischief when the various solidly-scientific replies which it has called forth are laid upon the shelf. Both the "Constitution" and the "Vestiges" had the advantage, so essential to the basilisk, of taking the first glance of the public on their respective subjects; whereas their confutators have been able to render them back but mere *return* glances. The only efficiently counteractive mode of looking down the danger, in cases of this kind, is the mode adopted by Mr. Longmuir.

There was a smart frost next morning; and, for a few hours, my seat on the top of the Banff coach, by which I travelled across the country to where the Gamrie and Banff roads part company, was considerably more cool than agreeable. But the keen morning improved into a brilliant day, with an atmosphere transparent as if there had been no atmosphere at all, through which the distant objects looked out as sharp of outline, and in as well-defined light and shadow, as if they had occupied the background, not of a Scotch, but of an Italian landscape. A few speck-like sails, far away on the intensely blue sea, which opened upon us in a stretch of many leagues, as we surmounted the moory ridge over Macduff, gleamed to the sun with a radiance bright as that of the sparks of a furnace blown to a white heat. The land, uneven of surface, and open, and abutting in bold promontories on the frith, still bore the sunny hue of harvest, and seemed as if stippled over with shocks from the ridgy hill summits, to where ranges of giddy cliffs flung their shadows across the beach. I struck off for Gamrie by a [Pg 263] path that runs eastward, nearly parallel to the shore,—which at one or two points it overlooks from dark-colored cliffs of grauwacke slate,—to the fishing village of Gardenstone. My dress was the usual fatigue suit of russet, in which I find I can work amid the soil of ravines and quarries with not only the best effect, but with even the least possible sacrifice of appearance: the shabbiest of all suits is a good suit spoiled. My hammer-shaft projected from my pocket; a knapsack, with a few changes of linen, slung suspended from my shoulders; a strong cotton umbrella occupied my better hand; and a gray maud, buckled shepherd-fashion aslant the chest, completed my equipment. There were few travellers on the road, which forked off on the hill-side a short mile away, into two branches, like a huge letter Y, leaving me uncertain which branch to choose; and I made up my mind to have the point settled by a woman of middle age, marked by a hard, *manly* countenance, who was coming up towards me, bound apparently for the Banff or Macduff market, and stooping under a load of dairy produce. She too, apparently, had her purpose to serve or point to settle; for as we met, she was the first to stand; and, sharply scanning my appearance and aspect at a glance, she abruptly addressed me. "Honest man," she said, "do you see yon house wi' the chimla?" "That house with the farm-steadings and stacks beside it?" I replied. "Yes." "Then I'd be obleeged if ye wald just stap in as ye'r gaing east the gate, and tell *our* folk that the stirk has gat fra her tether, an' 'ill brak on the wat clover. Tell them to sen' for her *that* minute." I undertook the commission, and, passing the endangered stirk, that seemed luxuriating, undisturbed by any presentiment of impending peril, amid the rich swathe of a late clover crop, still damp with the dews of the morning frost, I tapped at the door of the [Pg 264] farm-house, and delivered my message to a young good-looking girl, in nearly the words of the woman:—"The gude-wife bade me tell *them,*" I said, "to send that instant for the stirk, for she had gat fra her tether, and would brak on the wat clover." The girl blushed just a very little, and thanked me; and then, after obliging me, in turn, by laying down for me my proper route,—for I had left the question of the forked road to be determined at the farm-house,—she set off at high speed, to rescue the unconscious stirk. A walk of rather less than two hours brought me abreast of the Bay of Gamrie,—a picturesque indentation of the coast, in the formation of which the agency of the old denuding forces, operating on deposits of unequal solidity, may be distinctly traced. The surrounding country is composed chiefly of Silurian schists, in which there is deeply inlaid a detached strip of mouldering Old Red Sandstone, considerably more than twenty miles in length, and that varies from two to three miles in breadth. It seems to have been let down into the more ancient formation,—like the keystone of a bridge into the ringstones of the arch when the work is in the act of being completed,—during some of those terrible convulsions which cracked and rent the earth's crust, as if it had been an earthen pipkin brought to a red heat and then plunged into cold water. Its consequent occurrence in a lower tier of the geological edifice than that to which it originally belonged has saved it from the great denudation which has swept from the surface of the surrounding country the tier composed of its contemporary beds and strata, and laid bare the grauwacke on which this upper tier rested. But where it presents its narrow end to the sea, as the older houses in our more ancient Scottish villages present their gables to the street, the waves of the German Ocean, by incessantly charging against it, propelled [Pg 265] by the tempests of the stormy north, have hollowed it into the Bay of Gamrie, and left the more solid grauwacke standing out in bold promontories on either side, as the headlands of Gamrie and Troup.

In passing downwards on the fishing village of Gardenstone, mainly in the hope of procuring a guide to the ichthyolite beds, I saw a laborer at work with a pickaxe, in a little craggy ravine, about a hundred yards to the left of the path, and two gentlemen standing beside him. I paused for a moment, to ascertain whether the latter were not brother-workers in the geologic field. "Hilloa!—here,"—shouted out the stouter of the two gentlemen, as if, by some *clair-voyant* faculty, he had dived into my secret thought; "come here." I went down into the ravine, and found the laborer engaged in disinterring ichthyolitic nodules out of a bed of gray stratified clay, identical in its composition with that of the Cromarty fish-beds; and a heap of freshly-broken nodules, speckled with the organic remains of the Lower Old Red Sandstone,—chiefly occipital plates and scales,—lay beside him. "Know you aught of these?" said the stouter gentleman, pointing to the heap. "A little," I replied; "but your specimens are none of the finest. Here, however, is a dorsal plate of Coccosteus; and here a scattered group of scales of Osteolepis; and here the occipital plates of *Cheirolepis Cummingiæ*; and here the spine of the anterior dorsal of *Diplacanthus striatus*." My reading of the fossils was at once recognized, like the mystic sign of the freemason, as establishing for me a place among the geologic brotherhood; and the stout gentleman producing a spirit-flask and a glass, I pledged him and his companion in a bumper. "Was I not sure?" he said, addressing his friend: "I knew by the cut of his jib, notwithstanding his shepherd's plaid, that he was a wanderer of the scientific [Pg 266] cast." We discussed the peculiarities of the deposit, which, in its mineralogical character, and generically in that of its organic contents, resembles, I found, the fish-beds of Cromarty (though, curiously enough, the intervening contemporary deposits of Moray and the western parts of Banffshire differ widely, in at least their chemistry, from both); and we were right good friends ere we parted. To men who travel for amusement, incident is incident, however trivial in itself, and always worth something. I showed the younger of the two geologists my mode of breaking open an ichthyolitic nodule, so as to secure the best possible section of the fish. "Ah," he said, as he marked a style of handling the hammer which, save for the fifteen years' previous practice of the operative mason, would be perhaps less complete, — "Ah, you must have broken open a great many.' His own knowledge of the formation and its ichthyolites had been chiefly derived, he added, from a certain little treatise on the "Old Red Sandstone," rather popular than scientific, which he named. I of course claimed no acquaintance with the work; and the conversation went on.

The ill luck of my new friends, who had been toiling among the nodules for hours without finding an ichthyolite worth transferring to their bag, showed me that, without excavating more deeply than my time allowed, I had no chance of finding good specimens. But, well content to have ascertained that the ichthyolite bed of Gamrie is identical in its composition, and, generically at least, in its organisms, with the beds with which I was best acquainted, I rose to come away. The object which I next proposed to myself was, to determine whether, as at Eathie and Cromarty, the fossils here appear not only on the hill-side, but also crop out along the shore On taking leave, however, of the geologists, I was reminded by the younger of [Pg 267] what I might have otherwise forgotten, — a raised beach in the immediate neighborhood (first described by Mr. Prestwich, in his paper on the Gamrie ichthyolites), which contains shells of the existing species at a higher level than elsewhere, — so far as is yet known, — on the east coast of Scotland. And, kindly conducting me till he had brought me full within view of it, we parted. The ichthyolites which I had just been laying open occur on the verge of that Strathbogie district in which the Church controversy raged so hot and high; and by a common enough trick of the associative faculty, they now recalled to my mind a stanza which memory had somehow caught when the battle was at the fiercest. It formed part of a satiric address, published in an Aberdeen newspaper, to the not very respectable non-intrusionists who had smoked tobacco and drank whisky in the parish church at Culsalmond, on the day of a certain forced settlement there, specially recorded by the clerks of the Justiciary Court.

> "Tobacco and whisky cost siller,
> And meal is but scanty at hame;
> But gang to the stane-mason M— —r,
> Wi' Old Red Sandstone fish he'll fill your wame."

Rather a dislocated line that last, I thought, and too much in the style in which Zachary Boyd sings "Pharaoh and the Pascal." And as it is wrong to leave the beast of even an enemy in the ditch, however long its ears, I must just try and set it on its legs. Would it not run better thus?

> "Tobacco and whisky cost siller,
> An' meal is but scanty at hame;
> But gang to the stane-mason M— —r,"
> He'll pang wi' ichth'ólites your wame,—
> Wi' *fish*!! as Agassiz has ca'ed 'em,
> In Greek, like themsel's, *hard* an' *odd*,
> That were baked in stane pies afore Adam
> Gaed names to the haddocks and cod.

[Pg 268]

Bad enough as rhyme, I suspect; but conclusive as evidence to prove that the animal spirits, under the influence of the bracing walk, the fine day, and the agreeable recounter at the fish-beds,—not forgetting the half-gill bumper,—had mounted very considerably above their ordinary level at the editorial desk.

The raised beach may be found on the slopes of a grass-covered eminence, once the site of an ancient hill-fort, and which still exhibits, along the rim-like edge of the flat area atop, scattered fragments of the vitrified walls. A general covering of turf restricted my examination of the shells to one point, where a land-slip on a small scale had laid the deposit bare; but I at least saw enough to convince me that the debris of the shell-fish used of old as food by the garrison had not been mistaken for the remains of a raised beach,—a mistake which in other localities has occurred, I have reason to believe, oftener than once. The shells, some of them exceedingly minute, and not of edible species, occur in layers in a siliceous stratified sand, overlaid by a bed of bluish-colored silt. I picked out of the sand two entire specimens of a full-grown Fusus, little more than half an inch in length,—the *Fusus turricola*; and the greater number of the fragments that lay bleaching at the foot of the broken slope, in a state of chalky friability, seemed to be fragments of those smaller bivalves, belonging to the genera *Donax*, *Venus*, and *Mactra*, that are so common on flat sandy shores. But when the sea washed over these shells, they could have been the denizens of at least no *flat* shore. The descent on which they occur sinks downwards to the existing beach, over which it is elevated at this point two hundred and thirty feet, at an angle with the horizon of from thirty-five to forty degrees. Were the land to be now submerged to where they appear on the hill-side, the bay of Gamrie, as abrupt in its slopes [Pg 269] as the upper part of Loch Lomond or the sides of Loch Ness, would possess a depth of forty fathoms water at little more than a hundred yards from the shore. I may add, that I could trace at this height no marks of such a continuous terrace around the sides of the bay as the waves would have infallibly excavated in the diluvium, had the sea stood at a level so high, or, according to the more prevalent view, had the land stood at a level so low, for any considerable time; though the green banks which sweep around the upper part of the inflection, unscarred by the defacing plough, would scarce have failed to retain some mark of where the surges had broken, had the surges been long there. Whatever may in this special case be the fact, however, I cannot doubt that in the comparatively modern period of the boulder clays, Scotland lay buried under water to a depth at least five times as great as the space between this ancient sea-beach and the existing tide-line.

[Pg 270]

CHAPTER II.

Character of the Rocks near Gardenstone—A Defunct Father-lasher—A Geological Inference—Village of Gardenstone—The drunken Scot—Gardenstone Inn—Lord Gardenstone—A Tempest threatened—The Author's Ghost Story—The Lady in Green—Her Appearance and Tricks—The Rescued Children—The murdered Peddler and his Pack—Where the Green Dress came from—Village of Macduff—Peculiar Appearance of the Beach at the Mouth of the Deveron—Dr. Emslie's Fossils—*Pterichthys quadratus*—Argillaceous Deposit of Blackpots—Pipe-laying in Scotland—Fossils of Blackpots Clay—Mr. Longmuir's Description of them—Blackpots Deposit a Re-formation of a Liasic Patch—Period of its Formation.

I lingered on the hill-side considerably longer than I ought; and then, hurrying downwards to the beach, passed eastwards under a range of abrupt, mouldering precipices of red sandstone, to the village. From the lie of the strata, which, instead of inclining coast-wise, dip towards the interior of the country, and present in the descent seawards the outcrop of lower and yet lower deposits of the formation, I found it would be in vain to look for the ichthyolite beds along the shore. They may possibly be found, however, though I lacked time to ascertain the fact, along the sides of a deep ravine, which occurs near an old ecclesiastical edifice of gray stone, perched, nest-like, half-way up the bank, on a green hummock that overlooks the sea. The rocks, laid bare by the tide, belong to the bed of coarse-grained red sandstone, varying from eighty to a hundred and fifty feet in thickness, which lies between the lower fish-bed and the great conglomerate, and which, in not a few of its strata, passes itself into a species of conglomerate, different only from that which it overlies, in [Pg 271] being more finely comminuted. The continuity of this bed, like that of the deposit on which it rests, is very remarkable. I have found it occurring at many various points, over an area at least ten thousand square miles in extent, and bearing always the same well-marked character of a more thoroughly ground-down conglomerate than the great conglomerate on which it reposes. The underlying bed is composed of broken fragments of the rocks below, crushed, as if by some imperfect rudimentary process, like that which in a mill merely breaks the grain; whereas, in the bed above, a portion of the previously-crushed materials seems to have been subjected to some further attritive process, like that through which, in the mill, the broken grain is ground down into meal or flour.

As I passed onwards, I saw, amid a heap of drift-weed stranded high on the beach by the previous tide, a defunct father-lasher, with the two defensive spines which project from its opercles stuck fast into little cubes of cork, that had floated its head above water, as the tyro-swimmer floats himself upon bladders; and my previous acquaintance with the habits of a fishing village enabled me at once to determine why and how it had perished. Though almost never used as food on the eastern coast of Scotland, it had been inconsiderate enough to take the fisherman's bait, as if it had been worthy of being eaten; and he had avenged himself for the trouble it had cost him, by mounting it on cork, and sending it off, to wander between wind and water, like the Flying Dutchman, until it died. Was there ever on earth a creature save man that could have played a fellow-mortal a trick at once so ingeniously and gratuitously cruel? Or what would be the proper inference, were I to find one of the many-thorned ichthyolites of the Lower Old Red Sandstone with the spines of its pectorals similarly fixed on cubes of lignite?—that there had existed in these [Pg 272] early ages not merely *physical death*, but also *moral evil*; and that the being who perpetrated the evil could not only inflict it simply for the sake of the pleasure he found in it, and without prospect of advantage to himself, but also by so adroitly reversing, fiend-like, the purposes of the benevolent Designer, that the weapons given for the defence of a poor harmless creature should be converted into the instruments of its destruction. It was not without meaning that it was forbidden by the law of Moses to seethe a kid in its mother's milk.

A steep bulwark in front, against which the tide lashes twice every twenty-four hours,—an abrupt hill behind,—a few rows of squalid cottages built of red sandstone, much wasted by the keen sea-winds,—a wilderness of dunghills and ruinous pig-styes,—women seated at the doors, employed in baiting lines or mending nets,—groups of men lounging lazily at some gable-end fronting the sea,—herds of ragged children playing in the lanes,—such are the components of the fishing village of Gardenstone. From the identity of name, I had associated the place with that Lord Gardenstone of the

Court of Sessions who published, late in the last century, a volume of "Miscellanies in Prose and Verse," containing, among other clever things, a series of tart criticisms on English plays, transcribed, it was stated in the preface, from the margins and fly-leaves of the books of a "small library kept open by his Lordship" for the amusement of travellers at the inn of some village in his immediate neighborhood; and taking it for granted, somehow, that Gardenstone was the village, I was looking around me for the inn, in the hope that where his Lordship had opened a library I might find a dinner. But failing to discern it, I addressed myself on the subject to an elderly man in a pack-sheet apron, who stood all alone, looking out upon the sea, like Napoleon, in the print, from a projection [Pg 273] of the bulwark. He turned round, and showed, by an unmistakable expression of eye and feature, that he was what the servant girl in "Guy Mannering" characterizes as "very particularly drunk,"—not stupidly, but happily, funnily, conceitedly drunk, and full of all manner of high thoughts of himself. "It'll be an awfu' coorse nicht," he said, "fra the sea." "Very likely," I replied, reiterating my query in a form that indicated some little confidence of receiving the needed information; "I daresay you could point me out the public-house here?" "Aweel, I wat, that I can; but what's that?" pointing to the straps of my knapsack;—"are ye a sodger on the Queen's account, or ye'r ain?" "On my own, to be sure; but have ye a public-house here?" "Ay, twa; ye'll be a traveller?" "O yes, great traveller, and very hungry: have I passed the best public-house?" "Ay; and ye'll hae come a gude stap the day?" A woman came up, with spectacles on nose, and a piece of white seam-work in her hand; and, cutting short the dialogue by addressing myself to her, she at once directed me to the public-house. "Hoot, gude-wife," I heard the man say, as I turned down the street, "we suld ha'e gotten mair oot o' him. He's a great traveller yon, an' has a gude Scots tongue in his head."

Travellers, save when, during the herring season, an occasional fish-curer comes the way, rarely bait at the Gardenstone inn; and in the little low-browed room, with its windows in the thatch, into which, as her best, the land-lady ushered me, I certainly found nothing to identify the *locale* with that chosen by the literary lawyer for his open library. But, according to Ferguson, though "learning was scant, provision was good;" and I dined sumptuously on an immense platter of fried flounders. There was a little bit of cold pork added to the fare; but, aware from previous experience of the pisciverous habits of the swine [Pg 274] of a fishing village, I did what I knew the defunct pig must have very frequently done before me,—satisfied a keenly-whetted appetite on fish exclusively. I need hardly remind the reader that Lord Gardenstone's inn was not that of Gardenstone, but that of Laurence-kirk,—the thriving village which it was the special ambition of this law-lord of the last century to create; and which, did it produce only its famed snuff-boxes, with the invisible hinges, would be rather a more valuable boon to the country than that secured to it by those law-lords of our own days, who at one fell blow disestablished the national religion of Scotland, and broke off the only handle by which their friends the politicians could hope to *manage* the country's old vigorous Presbyterianism. Meanwhile it was becoming apparent that the man with the apron had as shrewdly anticipated the character of the coming night as if he had been soberer. The sun, ere its setting, disappeared in a thick leaden haze, which enveloped the whole heavens; and twilight seemed posting on to night a full hour before its time. I settled a very moderate bill, and set off under the cliffs at a round pace, in the hope of scaling the hill, and gaining the high road atop which leads to Macduff, ere the darkness closed. I had, however, miscalculated my distance; I, besides, lost some little time in the opening of the deep ravine to which I have already referred as that in which possibly the fish-beds may be found cropping out; and I had got but a little beyond the gray ecclesiastical ruin, with its lonely burying-ground, when the tempest broke and the night fell.

One of the last objects which I saw, as I turned to take a farewell look of the bay of Gamrie, was the magnificent promontory of Troup Head, outlined in black on a ground of deep gray, with its two terminal stacks standing apart in the sea. And straightway, through one of those tricks of [Pg 275] association so powerful in raising, as if from the dead, buried memories of things of which the mind has been oblivious for years, there started up in recollection the details of an ancient ghost-story, of which I had not thought before for perhaps a quarter of a century. It had been touched, I suppose, in its obscure, unnoted corner, as Ithuriel touched the toad, by the apparition of the insulated stacks of Troup, seen dimly in the thickening twilight over the solitary burying-ground. For it so chances that one of the main incidents of the story bears reference to an insulated sea-stack; and it is connected altogether, though I cannot fix its special locality, with this part of the coast. The story had been long in my mother's family, into which it had been originally brought by a great-grandfather of the writer, who quitted some of the seaport villages of Banffshire for the northern side of the Moray Frith, about the year 1718; and, when pushing on in the darkness, straining as I best could, to maintain a sorely-tried umbrella against the capricious struggles of the tempest, that now tatooed furiously upon its back as if it were a kettle-drum, and now got underneath its stout ribs, and threatened to send it up aloft like a balloon, and anon twisted it from side to side, and strove to turn it inside out, like a Kilmarnock night-cap,—I employed myself in arranging in my mind the details of the narrative, as they had been communicated to me half an age before by a female relative.

The opening of the story, though it existed long ere the times of Sir Walter Scott or the Waverly novels, bears some resemblance to the opening in the "Monastery," of the story of the White Lady of Avenel. The wife of a Banffshire proprietor of the minor class had been about six months dead, when one of her husband's ploughmen, returning on horseback from the smithy, in the twilight of an autumn evening, was accosted, on the banks of a small [Pg 276] stream, by a stranger lady, tall and slim, and wholly attired in green, with her face wrapped up in the hood of her mantle, who requested to be taken up behind him on the horse, and carried across. There was something in the tones of her voice that seemed to thrill through his very bones, and to insinuate itself, in the form of a chill fluid, between his skull and the scalp. The request, too, appeared a strange one; for the rivulet was small and low, and could present no serious bar to the progress of the most timid traveller. But the man, unwilling ungallantly to offend a lady, turned his horse to the bank, and she sprang up lightly behind him. She was, however, a personage that could be better seen than felt; she came in contact with the ploughman's back, he said, as if she had been an ill-filled sack of wool; and when, on reaching the opposite side of the streamlet, she leaped down as slightly as she had mounted, and he turned fearfully round to catch a second glimpse of her, it was in the conviction that she was a creature considerably less earthly in her texture than himself. She had opened, with two pale, thin arms, the enveloping hood, exhibiting a face equally pale and thin, which seemed marked, however, by the roguish, half-humorous expression of one who had just succeeded in playing off a good joke. "My dead mistress!!" exclaimed the ploughman. "Yes, John, *your mistress*," replied the ghost. "But ride home, my bonny man, for it's growing late: you and I will be better acquainted ere long." John accordingly rode home and told his story.

Next evening, about the same hour, as two of the laird's servant-maids were engaged in washing in an out-house, there came a slight tap to the door. "Come in," said one of the maids, and the lady entered, dressed, as on the previous night, in green. She swept past them to the inner part of the washing-room; and, seating herself on a low [Pg 277] bench, from which, ere her death, she used occasionally to superintend their employment, she began to question them, as if still in the body, about the progress of their work. The girls, however, were greatly too frightened to make any reply. She then visited an old woman who had nursed the laird, and to whom she used to show, ere her departure, greatly more kindness than her husband. And she now seemed as much interested in her welfare as ever. She inquired whether the laird was kind to her, and looking round her little smoky cottage, regretted she should be so indifferently lodged, and that her cupboard, which was rather of the emptiest at the time, should not be more amply furnished. For nearly a twelvemonth after, scarce a day passed in which she was not seen by some of the domestics; never, however, except on one occasion, after the sun had risen, or before it had set. The maids could see her, in the gray of the morning flitting like a shadow round their beds, or peering in upon them at night through the dark window-panes, or at half-open doors. In the evening she would glide into the kitchen or some of the out-houses,—one of the most familiar and least dignified of her class that ever held intercourse with mankind,—and inquire of the girls how they had been employed during the day; often, however, without obtaining an answer, though from a cause different from that which had at first tied their tongues. For they had become so regardless of her presence, viewing her simply as a troublesome mistress, who had no longer any claim to be heeded, that when she entered, and they had dropped their conversation, under the impression that their visitor was a creature of flesh and blood like themselves, they would again resume it, remarking that the entrant was "only the green lady." Though always cadaverously pale, and miserable looking, she affected a joyous disposition, and was frequently heard to laugh, [Pg 278] even when invisible. At one time, when provoked by the studied silence of a servant girl, she flung a pillow at her head, which the girl caught up and returned; at another, she presented her first acquaintance, the ploughman, with what seemed to be a handful of silver coin, which he transferred to his pocket, but which, on hearing her laugh, he drew out, and found to be merely a handful of slate shivers. On yet another occasion, the man, when passing on horseback through a clump of wood, was repeatedly struck from behind the trees by little pellets of turf; and, on riding into the thicket, he found that his assailant was the green lady. To her husband she never appeared; but he frequently heard the tones of her voice echoing from the lower apartments, and the faint peal of her cold, unnatural laugh.

One day at noon, a year after her first appearance, the old nurse was surprised to see her enter the cottage; as all her previous visits had been made early in the morning or late in the evening; whereas now,—though the day was dark and lowering, and a storm of wind and rain had just broken out,—still it *was* day. "Mammie," she said, "I cannot open the heart of the laird, and I have nothing of my own to give you; but I think I can do something for you now. Go straight to the White House [that of a neighboring proprietor], and tell the folk there to set out with all the speed of man and horse for the black rock in the sea, at the foot of the crags, or they'll rue it dearly to their dying day. Their bairns, foolish things, have gone out to the rock, and the tide has flowed around them; and, if no help reach them soon, they'll be all scattered like sea-ware on the shore ere the fall of the sea. But if you go and tell your story at the White House, mammie, the bairns will be safe for an hour to come, and there will be something done by their mother to better you, for the news." The woman went, as directed, and told her story; [Pg 279] and the father of the children set out on horseback in hot haste for the rock,—a low, insulated skerry, which, lying on a solitary part of the beach, far below the line of flood, was shut out from the view of the inhabited country by a wall of precipices, and covered every tide by several feet of water. On reaching the edge of the cliffs, he saw the black rock, as the woman had described, surrounded by the sea, and the children clinging to its higher crags. But, though the waves were fast rising, his attempts to ride out through the surf to the poor little things were frustrated by their cries, which so frightened his horse as to render it unmanageable; and so he had to gallop on to the nearest fishing village for a boat. So much time was unavoidably lost in consequence, that nearly the whole beach was covered by the sea, and the surf had begun to lash the feet of the precipices behind; but until the boat arrived, not a single wave dashed over the black rock; though immediately after the last of the children had been rescued, an immense wreath of foam rose twice a man's height over its topmost pinnacle.

The old nurse, on her return to the cottage, found the green lady sitting beside the fire. "Mammie," she said, "you have made friends to yourself to-day, who will be kinder to you than your foster-son. I must now leave you. My time is out, and you'll be all left to yourselves; but I'll have no rest, mammie, for many a twelvemonth to come. Ten years ago, a travelling peddler broke into our garden in the fruit season, and I sent out our old ploughman, who is now in Ireland, to drive him away. It was on a Sunday, and everybody else was in church. The men struggled and fought, and the peddler was killed. But though I at first thought of bringing the case before the laird, when I saw the dead man's pack, with its silks and its velvets, and this unhappy piece of green satin (shaking [Pg 280] her dress), my foolish heart beguiled me, and I made the ploughman bury the peddler's body under our ash tree, in the corner of our garden, and we divided his goods and money between us. You must bid the laird raise his bones, and carry them to the churchyard; and the gold, which you will find in the little bowl under the tapestry in my room, must be sent to a poor old widow, the peddler's mother, who lives on the shore of Leith. I must now away to Ireland to the ploughman; and I'll be e'en less welcome to him, mammie, than at the laird's; but the hungry blood cries loud against us both,—him and me,—and we must suffer together. Take care you look not after me till I have passed the knowe." She glided away, as she spoke, in a gleam of light; and when the old woman had withdrawn her hand from her eyes, dazzled by the sudden brightness, she saw only a large black gray-hound crossing the moor. And the green lady was never afterwards seen in Scotland. The little hoard of gold pieces, however, stored in a concealed recess of her former apartment, and the mouldering ruins of the peddler under the ash tree, gave evidence to the truth of her narrative. The story was hardly wild enough for a night so drear and a road so lonely; its ghost-heroine was but a homely ghost-heroine, too little aware that the same familiarity which, according to the proverb, breeds contempt when exercised by the denizens of this world, produces similar effects when too much indulged in by the inhabitants of another. But the arrangement and restoration of the details of the tradition,—for they had been scattered in my mind like the fragments of a broken fossil,—furnished me with so much amusement, when struggling with the storm, as to shorten by at least one-half the seven miles which intervene between Gamrie and Macduff. Instead, however, of pressing on to Banff, as I had at first intended, [Pg 281] I baited for the night at a snug little inn in the latter village, which I reached just wet enough to enjoy the luxury of a strong clear fire of Newcastle coal.

Mrs. Longmuir had furnished me with a note of introduction to Dr. Emslie of Banff, an intelligent geologist, familiar with the deposits of the district; and, walking on to his place of residence next morning, in a rain as heavy as that of the previous night, I made it my first business to wait on him, and deliver the note. Ere, however, crossing the Deveron, which flows between Banff and Macduff, I paused for a few minutes in the rain, to mark the peculiar appearance presented by the beach where the river disembogues into the frith. Occurring as a rectangular spit in the line of the shore, with the expanded stream widening into an estuary on its upper side, and the open sea on the lower, it marks the scene of an obstinate contest between antagonist forces,—the powerful sweep of the torrent, and the not less powerful waves of the stormy north-east; and exists, in consequence, as a long gravelly prism, which presents as steep an angle of descent to the waves on the one side as to the current on the other. It is a true river bar, beaten in from its proper place in the sea by the violence of the surf, and fairly stranded. Dr. Emslie obligingly submitted to my inspection his set of Gamrie fossils, containing several good specimens of Pterichthys and Coccosteus, undistinguishable, like those I had seen on the previous day, in their state of keeping, and the character of the nodular matrices in which they lie, from my old acquaintance the Cephalaspians of Cromarty. The animal matter which the bony plates and scales originally contained has been converted, in both the Gamrie and Cromarty ichthyolites, into a jet-black bitumen; and in both, the inclosing nodules consist of a smoke-colored argillaceous limestone, which formed around the [Pg 282] organisms in a bed of stratified clay, and at once exhibits, in consequence, the rectilinear lines of the stratification, mechanical in their origin, and the radiating ones of the sub-crystalline concretion, purely a trick of the chemistry of the deposit. A Pterichthys in Dr. Emslie's collection struck me as different in its proportions from any I had previously seen, though, from its state of rather imperfect preservation, I hesitated to pronounce absolutely upon the fact. I cannot now doubt, however, that it belonged to a species not figured nor described at the time; but which, under the name of *Pterichthys quadratus*, forms in part the subject of a still unpublished memoir, in which Sir Philip Egerton, our first British authority on fossil fish, has done me the honor to associate my humble name with his own; and which will have the effect of reducing to the ranks of the Pterichthyan genus the supposed genera *Pamphractus* and *Homothorax*. A second set of fossils, which Dr. Emslie had derived from his tile-works at Blackpots, proved, I found, identical with those of the Eathie Lias. As this Banffshire deposit had formed a subject of considerable discussion and difference among geologists, I was curious to examine it; and the Doctor, though the day was still none of the best, kindly walked out with me, to bring under my notice appearances which, in the haste of a first examination, I might possibly overlook, and to show me yet another set of fossils which he kept at the works. He informed me, as we went, that the Grauwacke (Lower Silurian) deposits of the district, hitherto deemed so barren, had recently yielded their organisms in a slate quarry at Gamrie-head; and that they belong to that ancient family of the Pennatularia which, in this northern kingdom, seems to have taken precedence of all the others. Judging from what now appears, the Graptolite must be regarded as the first settler who [Pg 283] squatted for a living in that deep-sea area of undefined boundary occupied at the present time by the bold wave-worn headlands and blue hills of Scotland; and this new Banffshire locality not only greatly extends the range of the fossil in reference to the kingdom, but also establishes, in a general way, the fossiliferous identity of the Lower Silurian deposits to the north of the Grampians with that of Peebles-shire and Galloway in the south,—so far as I know, the only other two Scottish districts in which this organism has been found.

The argillaceous deposit of Blackpots occupies, in the form of a green swelling bank, a promontory rather soft than bold in its contour, that projects far into the sea, and forms, when tipped with its slim column of smoke from the tile-kiln, a pleasing feature in the landscape. I had set it down on the previous day, when it first caught my eye from the lofty cliffs of Gamrie-head, at the distance of some ten or twelve miles, as different in character from all the other features of the prospect. The country generally is moulded on a framework of primary rock, and presents headlands of hard, sharp outline, to the attrition of the waves; whereas this single headland in the midst,—soft-lined, undulatory, and plump,—seems suited to remind one of Burns's young Kirk Alloway beauty disporting amid the thin old ladies that joined with her in the dance. And it *is* a greatly younger beauty than the Cambrian and mica-schist protuberances that encroach on the sea on either side of it. The sheds and kilns of a tile-work occupy the flat terminal point of the promontory; and as the clay is valuable, in this tile-draining age, for the facility with which it can be moulded into pipe-tiles (a purpose which the ordinary clays of the north of Scotland, composed chiefly of re-formations of the Old Red Sandstone, are what is technically termed too *short* to serve), it is gradually [Pg 284] retreating inland before the persevering spade and mattock of the laborer. The deposit has already been drawn out into many hundred miles of cylindrical pipes, and is destined to be drawn out into many thousands more,—such being one of the strange metamorphoses effected in the geologic formations, now that that curious animal the Bimana has come upon the stage; and at length it will have no existence in the country, save as an immense system of veins and arteries underlying the vegetable mould. Will these veins and arteries, I marvel, form, in their turn, the *fossils* of another period, when a higher platform than that into which they have been laid will be occupied to the full by plants and animals specifically different from those of the present scene of things,—the existences of a happier and more finished creation? My business to-day, however, was with the fossils which the deposit now contains,—not with those which it may ultimately form.

The Blackpots clay is of a dark-bluish or greenish-gray color, and so adhesive, that I now felt, when walking among it, after the softening rains of the previous night and morning, as if I had got into a bed of bird-lime. It is thinly charged with rolled pebbles, septaria, and pieces of a bituminous shale, containing broken Belemnites, and sorely-flattened Ammonites, that exist as thin films of a white chalky lime. The pebbles, like those of the boulder-clay of the northern side of the Moray Frith, are chiefly of the primary rocks and older sandstones, and were probably in the neighborhood, in their present rolled form, long ere the re-formation of the inclosing mass; while the shale and the septaria are, as shown by their fossils, decidedly Liasic. I detected among the conchifers a well-marked species of our northern Lias, figured by Sowerby from Eathie specimens,—the *Plagiostoma concentrica*; and among the [Pg 285] Cephalopoda, though considerably broken, the *Belemnite elongatus* and *Belemnite lanceolata*, with the *Ammonite Kœnigi* (*mutabilis*),—all Eathie shells. I, besides, found in the bank a piece of a peculiar-looking quartzose sandstone, traversed by hard jaspedeous veins of a brownish-gray color, which I have never found, in Scotland at least, save associated with the Lias of our north-eastern coasts. Further, my attention was directed by Dr. Emslie to a fine Lignite in his collection, which had once formed some eighteen inches or two feet of the trunk of a straight slender pine,—probably the *Pinites Eiggensis*,—in which, as in most woods of the Lias and Oölite, the annual rings are as strongly marked as in the existing firs or larches of our hill-sides. [11] The Blackpots deposit is evidently a re-formation [Pg 286] of a Liasic patch, identical, both in mineralogical character and in its organic remains, with the lower beds of the Eathie Lias; while the fragments of shale which it contains belong chiefly to an upper Liasic bed. So rich is [Pg 287] the dark-colored tenacious argil of the Inferior Lias of Eathie, that the geologist who walks over it when it is still moist with the receding tide would do well to look to his footing;—the mixture of soap and grease spread by the [Pg 288] ship-carpenter on his launch-slips, to facilitate the progress of his vessel seawards, is not more treacherous to the tread: while the Upper Liasic deposit which rests over it is composed of a dark slaty shale, largely charged with bitumen. And of a Liasic deposit of this compound character, consisting in larger part of an inferior argillaceous bed, and in lesser part of a superior one of dark shale, the tile-clay of Blackpots has been formed.

I had next to determine whether aught remained to indicate the period of its re-formation. The tile-works at the point of the promontory rest on a bed of shell-sand, composed exclusively, like the sand so abundant on the western coast of Scotland, of fragments of existing shells. These, however, are so fresh and firm, that, though the stratum which they form seems to underlie the clay at its edges, I cannot regard them as older than the most modern of our ancient sea-margins. They formed, in all probability, in the days of the old coast line, a white shelly beach, under such a precipitous front of the dark clay as argillaceous deposits almost always present to the undermining wear of the waves. On the recession of the sea, however, to its present line, the abrupt, steep front, loosened by the frosts and washed by the rains, would of course gradually moulder down over them into a slope; and there would thus be communicated to the shelly stratum, at least at its edges, an underlying character. The true period of the re-formation of the deposit was, I can have no doubt, that of the boulder-clay. I observed that the septaria and larger masses of shale which the bed contains, bear, on roughly-polished surfaces, in the line of their larger axes, the mysterious groovings and scratchings of this period,—marks which I have never yet known to fail in their chronological evidence. It may be mentioned, too, simply as a fact, though one of less value than [Pg 289] the other, that the deposit occurs in its larger development exactly where, in the average, the boulder-clays also are most largely developed,—a little over that line where the waves for so many ages charged against the coast, ere the last upheaval of the land or the recession of the sea sent them back to their present margin. There had probably existed to the west or north-west of the deposit, perhaps in the middle of the open bay formed by the promontory on which it rests,—for the small proportion of other than Liasic materials which it contains serves to show that it could be derived from no great distance,—an outlier of the Lower Lias. The icebergs of the cold glacial period, propelled along the submerged land by some arctic current, or caught up by the gulf-stream, gradually grated it down, as a mason's laborer grates down the surface of the sandstone slab which he is engaged in polishing; and the comminuted debris, borne eastwards by the current, was cast down here. It has been stated that no Liasic remains have been found in the boulder-clays of Scotland. They are certainly rare in the boulder-clays of the northern shores of the Moray Frith; for there the nearest Lias, bearing in a western direction from the clay, is that of Applecross, on the other side of the island; and the materials of the boulder-deposits of the north have invariably been derived in the line, westerly in its general bearing, of the grooves and scratches of the iceberg era. But on the southern shore of the frith, where that westerly line passed athwart the Liasic beds of our eastern coast, organisms of the Lias are comparatively common in the boulder-clays; and here, at Blackpots, we find an extensive deposit of the same period formed of Liasic materials almost exclusively. Fragments of still more modern rocks occur in the boulder-clays of Caithness. My friend Mr. Robert Dick, of Thurso, to whose persevering [Pg 290] labors and interesting discoveries in the Old Red Sandstone of his locality I have had frequent occasion to refer, has detected in a blue boulder-clay, scooped into precipitous banks by the river Thorsa, fragments both of chalk-flints and a characteristic conglomerate of the Oölite. He has, besides, found it mottled from top to bottom, a full hundred feet over the sea-level, and about two miles inland, with comminuted fragments of existing shells. But of this more anon.

[Pg 291]

CHAPTER III.

From Blackpots to Portsoy—Character of the Coast—Burn of Boyne—Fever Phantoms—Graphic Granite—Maupertuis and the Runic Inscription—Explanation of the *quo modo* of Graphic Granite—Portsoy Inn—Serpentine Beds—Portsoy Serpentine unrivalled for small ornaments—Description of it—Significance of the term *serpentine*—Elizabeth Bond and her "Letters"—From Portsoy to Cullen—Attritive Power of the Ocean illustrated—The Equinoctial—From Cullen to Fochabers—The Old Red again—The old Pensioner—Fochabers—Mr. Joss, the learned Mail-guard—The Editor a sort of Coach-guard—On the Coach to Elgin—Geology of Banffshire—Irregular paging of the Geologic Leaves—Geologic Map of the County like Joseph's Coat—Striking Illustration.

I parted from Dr. Emslie, and walked on along the shore to Portsoy,—for three-fourths of the way over the prevailing grauwacke of the county, and for the remaining fourth over mica schist, primary limestone, hornblende slate, granitic and quartz veins, and the various other kindred rocks of a primary district. The day was still gloomy and gray, and ill suited to improve homely scenery; nor is this portion of the Banff coast nearly so striking as that which I had travelled over the day before. It has, however, its spots of a redeeming character,—rocky recesses on the shore, half-beach, half-sward, rich in wild-flowers and shells,—where one could saunter in a calm sunny morning, with one's *bairns* about one, very delightfully; and the interior is here and there agreeably undulated by diluvial hillocks, that, when the sun falls low in the evening, must chequer the landscape with many a pleasing alternation of light and shadow. The Burn of Boyne,—which separates, about two miles from Portsoy, a grauwacke from a mica-schist district,—with its bare, [Pg 292] open valley, its steep limestone banks, and its gray, melancholy castle, long since roofless and windowless, and surrounded by a few stunted trees, bears a deserted and solitary shagginess about it, that struck me as wildly agreeable. It is such a valley as one might expect to meet a ghost in, in some still, dewy evening, as gloamin was darkening into uncertainty the outlines of the ancient ruin, and the newly-kindled stars looked down upon the stream.

It so happened, however, that my only story connected with either ruin or valley was as little a ghost story as might be. I remember that, when lying ill of fever on one occasion,—indisposed enough to see apparition after apparition flitting across the bed-curtains, like the figures of a magic lantern posting along the darkened wall, and yet self-possessed enough to know that they were but mere pictures in the eye, and to watch them as they rose,—I set myself to determine whether they were in any degree amenable to the will, or connected by the ordinary associative links of the metaphysician. Fixing my mind on a certain object, I strove to call it up in the character, not of an image of the conceptive faculty, but of a fever-vision on the retina. The image which I pictured to myself was that of a death's head, yellow and grim, and lighted up, as if from within, amid the darkness of a burial vault. But the death's head obstinately refused to rise. I had no control, I found, over the fever imagery. And the picture that rose instead, uncalled and unexpected, was that of a coal-fire burning brightly in a grate, with a huge tea-kettle steaming cheerily over it.

In traversing the bare height which, rising on the western side of the valley of the Boyne, owes its comparatively bold relief in the landscape to the firmness of the primary rock which composes it, I picked up a piece of graphic [Pg 293] granite, bearing its inlaid characters of dark quartz on a ground of cream-colored feldspar. This variety, however, though occasionally found in rolled boulders in the neighborhood of Portsoy, is not the graphic granite for which the locality is famous, and which occurs in a vein in the mica schist of the eminence I was now traversing, about a mile to the east of the town. The prevailing ground of the granite of the vein is a flesh-colored feldspar; and the thickly-marked quartzose characters with which it is set, greatly smaller and paler than in the cream-colored stone, bear less the antique Hebraic look, and would scarce deceive even the most credulous antiquary. Antiquarians, however, *have* been sometimes deceived by weathered specimens of this graphic rock, in which the characters were of considerable size, and restricted to thin veins, covering the surface of a schistose groundwork. Maupertuis, during his famous journey to Lapland, undertaken in 1737, to establish, from actual measurement, that the degrees of latitude are longer towards the pole than at the equator, and which demonstrated, of consequence, the true figure of the earth, travelled thirty leagues out of his way, through a wild country covered with snow, to examine an ancient monument, of which, he says, "the Fins and Laplanders frequently spoke, as containing in its inscription the knowledge of everything of which they were ignorant." He found it on the side of a mountain, buried in snow; and ascertained, after kindling a great fire around it, in order to lay it bare, that it was a stone of irregular form, composed of various layers of unequal hardness, and that the characters, which were rather more than an inch in length, were written on "a layer of a species of flint," chiefly in two lines, with a few scattered signs beneath, while the rest of the mass was composed of a rock more soft and foliated. Graphic granite, it may be mentioned, [Pg 294] generally occurs, not in masses, but in veins and layers. The inscription had been described in a previously published dissertation of immense erudition, as Runic; but a Runic scholar of the party found he could make nothing of it. The philosopher himself was struck by the frequent repetition of characters of nearly the same form on the stone; but he was ingenious enough to get over the difficulty, by remembering that in our notation, after the Arabic manner, characters shaped exactly alike may be very frequently repeated,—nay, as in some of the lines of the Lapland inscription, may succeed each other, as in the sums I. II. III. IIII. or X. XX. XXX.,—and yet very distinct and definite ideas attach to them all. Still, however, he could not, he says, venture on authoritatively deciding whether the inscription was a work of man or a sport of nature. He stood between his two conclusions, like our Edinburgh antiquarians between the two fossil Maries of Gueldres; and, richer in eloquence than most of the philosophers his contemporaries, was quite prepared, in his uncertainty, to give gilded mounting and a purple pall to both.

"Should it be no other than a sport of nature," he concludes, "the reputation which the stone bears in this country deserves that we should have given a description of it. If, on the other hand, what is on it be an inscription, though it certainly does not possess the beauty of the sculpture of Greece or Rome, it very possibly has the advantage of being the oldest in the universe. The country in which it is found is inhabited only by a race of men who live like beasts in the forests. We cannot imagine that they can have ever had any memorable event to transmit to posterity, nor, if ever they had had, that they could have invented the means. Nor can it be conceived that this country, with its present aspect, ever possessed more civilized inhabitants. [Pg 295] The rigor of the climate and the barrenness of the land have destined it for the retreat of a few miserable wretches, who know no other. It seems, therefore, that the inscription must have been cut at a period when the country was situated in a different climate, and before some one of those great revolutions which, we cannot doubt, have taken place on our globe. The position that the earth's axis holds at present with respect to the ecliptic, occasions Lapland to receive the sun's rays very obliquely: it is therefore condemned to a long winter, adverse to man, as well as to all the productions of nature. No great movement, possibly, in the heavens was necessary, however, to cause all its misfortunes. These regions may formerly have been those on which the sun shone most favorably; the polar circles may have been what now the tropics are, and the torrid zone have filled the place occupied by the temperate." Pretty well, Monsieur, for a philosopher! The various attempts made to unriddle the real history of graphic granite are, however, scarce less curious than the speculations connected with what may be termed its romance. It seems to be generally held, since the days of old Hutton, who, in his "Theory of the Earth," discussed the subject with his usual ingenuity, that the feldspathic basis of the stone first crystallized, leaving interstices between the crystals, partaking of a certain regularity of form,—a consequence of the regularity of the crystals themselves,—and of a certain irregularity from the eccentric dispositions which these manifest in their position and relations to each other; and that these interstices, being afterwards filled up with quartz, form the characters of the rock,—characters partaking enough of the first element of *regularity* to present their peculiar graphic appearance, and enough of the second element of *irregularity* to exhibit forms of an alphabet-like variety of outline. The chemist, however, in cross-questioning the [Pg 296] explanation, has his puzzle to propound regarding it. Quartz, he says, being considerably less fusible than feldspar, would naturally consolidate first, and so would give form to the more fusible substance, instead of deriving form from it. On what principle, then, is it that, reversing its ordinary character, it should have been the last of the two substances to consolidate in the graphic granite?—a query to which there seems to be no direct reply, but which as little affects the fact that it *was* the substance which last consolidated, and which took form from the other, as the decision of the learned Strasburgers, which determined the impossibility of the long nose in Slawkenbergius's Tale, affected the actual existence of that remarkable feature. "It happens *to be,* notwithstanding your objection," said the controversialists on the pro-nose side of the question. "But it *ought not,*" replied their opponents.

The rain again returned as I was engaged in examining the graphic granite of the Portsoy vein; the breeze from the sea heightened into a gale, that soon fringed the coast with a broad border of foam; and I entered the town, which looked but indifferently well in its gray dishabille of haze and spray, tolerably wet and worn, with but the prospect before me of being weather-bound for the rest of the day. I found an old-fashioned inn, kept by somewhat old-fashioned people, who had lately come from the country to "open a public;" and ensconced myself by the fireside, in a huge many-windowed room, that must have witnessed the county dinners of at least a century ago. Soon wearying, however, of hearing the rain beating mad-like ratans upon the panes, and availing myself of a comparatively "lucid interval," I sallied out, wrapped up in my plaid, to examine the serpentine beds in the neighborhood, which produce what is so extensively known as the Portsoy marble. The *beds* or *veins* of this substance,—for it is still a moot point whether [Pg 297] they occur here as mere insulated masses of contemporary origin with the primary formations which surround them, or as Plutonic dykes injected into fissures at a later period,—are of very considerable extent, one of them measuring about twenty-five yards across, and another considerably more than a quarter of a mile; and, had they but the solidity of the true marbles, they would scarce fail to be regarded as valuable quarries of a highly ornamental stone, admirably suited for the interior decorations of the architect. But they are unluckily what the quarrier would term rubbly,—traversed by an infinity of cracks and fissures; and it is rare indeed to find a continuous mass out of which a chimney-jamb or lintel could be fashioned. The serpentine was wrought here considerably more than a century and a half ago, and exported to France for the magnificent Palace of Versailles; which, though regarded by the French nation, says Voltaire, as "a favorite without merit," Louis the Fourteenth persisted at the time in lavishly beautifying, and looked as for abroad as Portsoy for materials with which to adorn it. I have, however, seen it stated that the greater part of a ship's cargo, brought afterwards to Paris on speculation, was suffered to lie unwrought for years in the stone-dealer's yard, and was ultimately disposed of as rubbish,—a consequence, probably, of its unfitness, from its shaky texture, for ornamental purposes on a large scale, though for ornaments of the smaller kind, such as boxes, vases, and plates, it has been pronounced unrivalled. "At Zöblitz, in Upper Saxony," says Professor Jamieson, "several hundred people are employed in quarrying, cutting, turning, and polishing the serpentine which occurs in that neighborhood; and the various articles into which it is manufactured are carried all over Germany. The serpentine of Portsoy," he adds, "is, however, far superior to that of Zöblitz, in color, hardness, and transparency, and, when cut, is very beautiful."

[Pg 298]

It is really a pretty stone; and, bad as the evening was, it was by no means one of the worst of evenings for seeing it to advantage *in situ*, or among the rolled pebbles on the shore. The varnish-like gloss of the wet imparted to the undressed masses all the effect of polish, and brought out in their proper variegations of color, every cloud, streak, and vein. Viewed in the mass, the general hue is green; so much so, that an insulated stack, which stands abreast of one of the beds, a stone-cast in the sea, has greatly the appearance, at a little distance, of an immense mass of verdigris. But red, gray, and brown are also prevailing colors in the rock; occasional veins and blotches of white give lightness to the darker portions; and veins of hematitic and deep umbry tints, variety to the portions that are lighter. The greens vary from the palest olive to the deepest black-green of the mineralogist; the reds and browns, from blood-red to dark chocolate, and from wood-brown to brownish-black; and, thus various in shade, they occur in almost every possible variety of combination and form,—dotted, spotted, clouded, veined,—so that each separate pebble on the shore seems the representative of a rock different from the rocks represented by almost all the others. Though not much of a mineralogist, I could have spent considerably more time than the weather permitted me to employ this evening, in admiring the beauties of this beach of *marbles*, or rather,—as the real name, derived from those gorgeous, many-colored cloudings, that impart a terrible splendor to the skins of the snake and viper family, is not only the more correct, but also the more poetical of the two,—this beach of *serpentines*. I had, however, to compromise matters between the fierce wind and rain and the pretty rocks and pebbles, by adjourning to the workshop of the Portsoy lapidary, Mr. Clark, and examining under cover his polished specimens, of which I purchased for a few shillings a characteristic [Pg

299] and elegant little set. Portsoy is peculiarly rich in minerals; and hence it reckons among its mechanics of the ordinary class, what perhaps no other village in Scotland of the same size and population possesses,—a skilful lapidary. Mr. Clark's collection of the graphic granites, serpentines, and talcose and mica schists, of the district, with their associated minerals, such as schorl, talc, asbestos, amianthus, mountain cork, steatite, and schiller spar, will be found eminently worthy a visit by the passing traveller.

I made several inquiries in the village, though not, as it proved, in the right direction, regarding a poor old lady, several years dead, of whom I had known a very little considerably more than a quarter of a century before, and whose grave I would have visited, bad as the night was, had I met any one who could have pointed it out to me. But ungrateful Portsoy seemed to have forgotten poor Miss Bond, who, in all her printed letters and little stories, so rarely forgot *it*. Have any of my readers ever seen the work (in two slim volumes), "Letters of a Village Governess," published in 1814 by Elizabeth Bond, and dedicated to Sir Walter Scott? If not, and should they chance to see, as I lately did, a copy on a stall (with uncut leaves, alas! and selling dog cheap), they might possibly do worse things than buy it. [12]

With better weather I could have spent a day or two very agreeably in Portsoy and its neighborhood; but the rain dashed unceasingly, and made exploration under the cover of the umbrella somewhat resemble that of a sea-bottom under cover of the diving-bell. I could see but little at a time, and the little imperfectly. Miss Bond, in her "Letters," refers, in her light, pleasing style, to what [Pg 300] in more favorable circumstances *might* be seen. "My troop of *light infantry*," she says, "keeps me so well employed here during the day, that the silence and repose of the evening is very delightful. In fine weather I walk by the sea-side, and scramble among the rugged rocks, many of which are inaccessible to human feet, forming a fine retreat for foxes. These animals often may be seen from the heights, sporting with their cubs in perfect safety. This day I went to see the works of an old *virtuoso*, who turns in marble, or rather granite [serpentine] all kinds of chimney-piece ornaments, rings, ear-rings, etc. Several specimens of his work, which must have cost him a vast deal of trouble, I thought very beautiful. It was in this neighborhood that the celebrated Ferguson spent so much of his time. The globular stones on the gate of Durn are still to be seen, on which he mapped out the figuring of the terrestrial and celestial globes. I was told it was forbidden ground to approach the premises of Durn; but I could not resist the temptation of visiting the spot where the young philosopher had shown such early proofs of his genius; and I accordingly paid the forfeit of an *impertinent*, for the gentleman who resides there caught the prowler, and in genteel terms bade her go about her business, and never return. How ungracious! She was doing no harm."

The morning arose as gloomily as the evening had fallen; and I walked on in the rain to Cullen, fully disposed to sympathize by the way with the "hardy Byron,"—he of the "Narrative,"—who, from his ill-luck in weather, went among his sailors by the name of "Foul-weather Jack." In the sandy bay of Cullen, where the road, after inflecting inland for some five or six miles, comes again upon the sea, I found the surf charging home in long white lines six waves deep,—

> [Pg 301] "Each stepping where his comrade stood,
> The instant that he fell."

The appearance was such as to impart no inadequate idea of the vast attritive power of ocean in wearing down the land. When pausing for a little abreast of the fishing village, partially sheltered by an old boat, to mark the fierce turmoil, it suddenly occurred to me,—as the tempest weltered around reef and skerry, and roared wildly, mile after mile, along the beach,—that the day and night were now just equal, and that it was the customary equinoctial storm that had broken out to accompany me on my journey. And so, calculating on a few days more of it, instead of waiting on in the hope of a fair afternoon to examine the outlier of Old Red which occurs in the neighborhood of Cullen, I was content to see at a distance its mural-sided cliffs rising like broken walls through the flat sand; and, taking the road for Fochabers, with the intention of leaving exploration till fairer weather set in, I resolved on posting straight on, to join my relatives on the opposite side of the Frith. The deep-red color of the boulder-clay, as exhibited by the way-side, in the water-courses and the water,—for every runnel was tumbling down big and turbid with the rains,—intimated, when, after leaving Cullen some six or seven miles behind me, I passed from a bare moory region of quartz rock into a region of woods and fields, that I was again upon my ancient acquaintance, the Old Red Sandstone. And the section furnished by the Burn of Tynet showed me shortly after that the intimation was a correct one, and how generally it may be laid down as a rule, that at least the more impalpable portions of the boulder-clay are derived from the rocks on which it rests. The ichthyolite beds appear in the course of the burn. They have furnished several good specimens,—among the others, the specimen of [Pg 302] Coccosteus figured by Mr. Patrick Duff in his "Sketches of the Geology of Moray;" and they are, besides, curious, as being the first to exhibit to the traveller who explores from Gamrie westwards, that peculiar style of coloring which characterizes the Old Red ichthyolites of the shires of Moray and Nairn, and which differs so strikingly from the more sombre style exhibited by the other ichthyolites of Banffshire, with those of Cromarty, Ross, Caithness, and Orkney. Instead of bearing, like these, one uniform hue, as if deeply shaded with Indian ink, they are gorgeously attired, especially when newly laid open, in white, red, purple, and blue. The day, however, was ill-suited for fishing Pterichthyes and Osteolepi out of the Tynet: the red water was roaring from bank to brae; here eddying along the half-submerged furze,—there tearing down the boulder-days in raw, red land-slips; and so, casting but one eager glance at the bed where the fish lay, I travelled on, and entered the tall woods to the east of Fochabers. The rain ceased for a time; and I met in the woods an old pensioner, who had been evidently weather-bound in some public-house, and had now taken the opportunity of the fair interval to stagger to his dwelling. He was eminently, exuberantly happy,—there could not be two opinions on that head,—full of all manner of bright sunshiny thoughts and imaginations, rendered just a little tremulous and uncertain by the *summer-heat* exhalations of the imbibed moisture, like distant objects in a hot noonday landscape in July seen through volumes of rising vapor; and a sheep's head and trotters, which he carried under his arm, was, I saw, to serve as a peace-offering to his wife at home. True, he had been taking a dram, but he was mindful of the family for all that. He confronted me with the air of an old acquaintance; gave the military salute; and then, laying hold of a corner of my plaid [Pg 303] with his thumb and forefinger,—"I know you," he said, "I know *your kind* well; ye're a Highland-Donald. Od, I've seen ye in the *thick o't*. Ye're *reugh* fellows when ye're bluid's up!" He had taken me for a grenadier of the 42d; and I lacked the moral courage to undeceive him. I met nothing further on my way worthy of record, save and except a sheep's trotter, dropped by the old pensioner in one of his zigzaggings to the extreme left; but having no particular use for the trotter at the time and in the circumstances, I left it to benefit the next passer-by. I finished my journey of eighteen miles in capital style, and was within five minutes' walk of Fochabers when the horn of the mail-guard was sounding up the street. And, entering the village, I found the vehicle standing opposite the inn door, minus the horses.

The *insides* and *outsides* were sitting down to dinner together as I entered the inn; and I felt, after my long walk, that it would be rather an agreeable matter to join with them. But in the hope of meeting my old friend Mr. Joss, I requested to be shown, not into the passengers' room, but into that of the coachman and guard; and with them I dined. It so chanced, however, that Mr. Joss was not *out* that day; and the man in the red long coat was a stranger whom I had never seen before. I inquired of him regarding Mr. Joss,—one of perhaps the most remarkable mail-guards in Europe. I have at least never heard of another who, like him, amuses his leisure on the coach-top with the "Principia" of Newton, and understands it. And the man, drawing his inference from the interest in Mr. Joss which my queries evinced, asked me whether I myself was not a coach-guard. "No," I rather thoughtlessly replied, "I am not a coach-guard." Half a minute's consideration, however, led me to doubt whether I had given the right answer. "I am not sure," I said to myself, [Pg 304] on second thoughts, "but the man has cut pretty fairly on the point;—I daresay *I am* a sort of coach-guard. I have to mount my twice-a-week coach in all weathers, like any mail-guard among them all; I have to start at the appointed hour, whether the vehicle be empty or full; I have to keep a sharp eye on the opposition coaches; I am responsible, like any other mail-guard, for all the parcels carried, however little I may have had to do with the making of them up; I have always to keep my blunderbuss full charged to the muzzle,—not wishing harm to any one, but bound in duty to let drive at all and sundry who would make war upon the passengers, or attempt running the conveyance off the road; and, finally, as my friend Mr. Joss takes the "Principia" to *his* coach-top, I take pockets full of fossils to the top of mine, and amuse myself in fine days by working out, as I best can, the problems which they furnish. Yes, I rather think *I am* a coach-guard." And so, taking my seat beside my red-coated brother, who had guessed the true nature of my occupation so much more shrewdly than myself, I rode on to Elgin, where I passed the night.

It is difficult to arrange in the mind the geologic formations of Banffshire in their character as a series of deposits. The pages of the stony record which the county composes, like those of an unskilfully-folded pamphlet, have been strangely mixed together, so that page last succeeds in some places to page first, and, of the intermediate pages, some appear at the beginning of the work, and some at the end. It is not until we reach the western confines of the county, some two or three miles short of the river Spey, its terminal boundary in this direction, that we find the beds comparatively little disturbed, and arranged chronologically in their original places. In the eastern and southern parts of the shire, rocks widely separated by the date of their formation have been set down side by side in [Pg 305] patches, occasionally of but inconsiderable extent. Now the traveller passes over a district of grauwacke, now over a reformation of the Lias; anon he finds himself on a primary limestone,—gneiss, syenite, clay-slate, or quartz-rock; and yet anon amid the fossils of some outlier of the Old Red. The geological map of the county is, like Joseph's coat, of many colors. I remember seeing, when a boy, more years ago than I am inclined to specify, some workmen engaged in pulling down what had been a house-painter's shop, a full century before. The painter had been in the somewhat slovenly habit of cleaning his brushes by rubbing them against a hard-cast wall, which was covered, in consequence, by a many-colored layer of paint, a full half-inch in thickness, and as hard as a stone. Taking a little bit home with me, I polished it by rubbing the upper surface smooth; and, lo! a geological map. The *strata* of variously hued pigment, spread originally over the surface of the hard-cast wall, were cut open, by the *denudation* of the grindstone, into all manner of fantastic forms, and seemed thrown into all sorts of strange neighborhoods. The *map* lacked merely the additional perplexity of a few bold *faults*, with here and there a decided *dike*, in order to render it on a small scale a sort of miniature transcript of the geology of Banff; and I have very frequently found my thoughts reverting to it, in connection with deposits of this broken character. On a rough *hard-cast* basis of granite I have laid down in imagination, as if by way of priming, coat after coat of the primary rocks,—gneiss, and stratified hornblend, and mica-schist, and quartz-rock, and day-slate; and then, after breaking the coatings well up, and rubbing them well down, and so spoiling and crumpling up the work as to make their original order considerably a puzzle, I have begun anew to paint over the rough surface with thick coatings of grauwacke and grauwacke-slate. [Pg 306] When this part of the operation was completed, I have again begun to break up and grind down,—here letting a tract of grauwacke sink into the broken primary,—there wearing it off the surface altogether,—yonder elevating the original granitic *hard-cast* till it rose over all the coatings, Primary and Palæozoic. And then I have begun to paint yet a third time with thick Old Red Sandstone pigment; and yet again to break up and wear down,—here to insert a tenon of the Old Red deep into a mortise of the grauwacke, as at Gamrie,—there to dovetail it into the clay-slate, as at Tomantoul,—yonder, after laying it across the upturned quartz-rock, as at Cullen, to rub by much the greater part of it away again, leaving but mere remainder-patches and fragments, to mark where it had been. Lastly, if I had none of the superior Palæozoic or Secondary formations to deal with, I have brushed over the whole, by way of finish, with the variously-derived coatings of the superficial deposits; and thus, as I have said, I have often completed, in idea, after the chance suggestion of the old painter's shop, my portable models of the geology of disturbed districts like the Banffshire one. The deposits of Moray are greatly less broken. Denudation has partially worn them down; but they seem to have almost wholly escaped the previous crumpling process.

[Pg 307]

CHAPTER IV.

Yellow-hued Houses Of Elgin—Geology of the Country indicated by the coloring of the Stone Houses—Fossils of Old Red north of the Grampians different from those of Old Red south—Geologic Formations at Linksfield difficult to be understood—Ganoid Scales of the Wealden—Sudden Reaction, from complex to simple, in the Scales of Fishes—Pore-covered Scales—Extraordinary amount of Design exhibited in Ancient Ganoid Scales—Holoptychius Scale illustrated by Cromwell's "fluted pot"—Patrick Duff's Geological Collection—Elgin Museum—Fishes of the Ganges—Armature of Ancient Fishes—Compensatory Defences—The Hermit-crab—Spines of the Pimelodi—Ride to Campbelton—Theories of the formation of Ardersier and Fortrose Promontories—Tradition of their construction by the Wizard, Michael Scott—A Region of Legendary Lore.

The prevailing yellow hue of the Elgin houses strikes the eye of the geologist who has travelled northwards from the Frith of Forth. He takes leave of a similar stone at Cupar-Fife,—a warmly-tinted yellow sandstone, peculiarly well-suited for giving effect to architectural ornament; and after passing along the deep-red sandstone houses of the shires of Angus and Kincardine, and the gneiss, granite, hyperstene, and mica-schist houses of Aberdeen and Banff shires, he again finds houses of a deep red on crossing the Spey, and houses of a warm yellow tint on reaching Elgin,—geologically the Cupar-Fife of the north. And the story that the colored buildings tell him is, that he has been passing, though by a somewhat circuitous route of a hundred and fifty miles, over an anticlinal geological section,—*down* in the scale till he reached Aberdeen and had gone a little beyond it, and then *up* again, until at Elgin he arrives at the same superior yellow bed of Old Red Sandstone which he had quitted at Cupar-Fife. Both beds contain the same [Pg 308] organisms. The Holoptychius of Dura Den, near Cupar, must have sprung from the same original as the Holoptychius of the Hospital and Bishop-Mill quarries near Elgin; and it seems not improbable that the two beds, thus identical in their character and contents, may have existed, ere the upheaval of the Grampians broke their continuity, as an extended deposit, at the bottom of the same sea. But with this last and newest of the formations of the Old Red Sandstone the identity of the deposits to the south and north ceases. The strata which in the south overlie the yellow bed of the Holoptychius represent the Carboniferous period, the overlying strata in the north represent the Oölitic one. On the one side the miner sinks his shaft, and finds a true coal, composed of the Stigmaria, Calamites, Club-mosses, Ferns, and Araucarians of the Palæozoic era; he sinks his shaft on the other side, and finds but thin seams of an imperfect lignite, composed of the Cycadeæ, Pines, Sphenopteri, and Clathraria of the Secondary period. The flora which found its subsoil in the Old Red Sandstone north of the Grampians, belonged to a scene of things so much more modern than the flora which found its subsoil in the Old Red Sandstone of the south, that all its productions were green and flourishing, waving beside lake, river, and sea, at a time when the productions of the other were locked up, as now, in sand and shale, lime and clay,—the dead mummies of ages long departed.

Another thoroughly wet morning! varied only from the morning of the preceding day by the absence of wind, and the greater weight of the persevering vertical rain, that leaped upwards in myriads of little dancing pyramids from the surface of every pool. I walked out under cover of my umbrella, to renew my acquaintance with the outlier of the Weald at Linksfield, and ascertain what sort of section it now presented under the quarrying operations of the limeburners. [Pg 309] There was, however, little to be seen; the bands of green and blue clays, alternating with strata of fossiliferous limestone, and layers of a gray shade, thickly charged with minute shells of Cypris, were sadly blurred this morning by the trail of numerous slips from above, which had fallen during the rains, and softened into mud as they rushed downwards athwart the face of the quarry: and the arched band of boulder-clay which so mysteriously underlies the deposit was, save in a few parts, wholly covered up by the debris. The occurrence of the clay here as an inferior bed, with but the cornstone of the Old Red beneath, and all the beds of the Weald resting over it, forms a riddle somewhat difficult of solution; but it is palpably not reading it aright to regard the deposit, with at least one geologist who has written on the subject, as older than the rocks above. It is, on the contrary, as a vast amount of various and unequivocal evidence demonstrates, incalculably more modern; nay, we find proof of the fact here in that very bed which has been instanced as rendering it doubtful; the clay of which the interpolation is composed is found to contain fragments, not only of the cornstone on which it rests, but also of the Wealden limestone and shales which it underlies. It forms the mere filling up of a flat-roofed cavern, or rather of two flat-roofed caverns,—for the limestone roof dipped in the centre to the cornstone floor,—which, previous to the times of the boulder-clay, had lain open in what was then, as now, an old-world deposit, charged with long extinct organisms, but which, during the iceberg period, was penetrated and occupied by the clay, as run lime penetrates and occupies the interstices of a dry-stone wall. It was no day for gathering fossils. I saw a few ganoid scales, washed by the rain from the investing rubbish, glittering on fragments of the limestone, with a few of the characteristic shells of the deposit, chiefly Unionidæ; but nothing worth bringing [Pg 310] away. The adhesive clay of the Weald, widely scattered by the workmen, and wrought into mortar by the beating rains, made it a matter of some difficulty for the struggling foot to retain the shoe, and, sticking to my soles by pounds at a time, rendered me obnoxious to the old English nickname of "rough-footed Scot." And so, after traversing the heaps, somewhat like a fly in treacle, I had to yield to the rain above and the mud beneath, and to return to do in Elgin what cannot be done equally well in almost any other town of its size in Scotland,—pursue my geological inquiries under cover.

On this, as on other occasions, I was struck by the complex and very various forms assumed by the ganoid scales of the Wealden. Throughout the Oölitic system generally, including the Lias, there obtains a singular complexity of type in these little glittering tiles of enamelled bone, which contrasts strongly with the greatly more simple style which obtained among the ganoids of the Palæozoic period. In many of these last, as in the Cœlacanth family, including the genera Holoptychius, Asterolepis, and Glyptolepis, in all their many species, with at least one genus of Dipterians, the genus Dipterus, the external outline and arrangement of scale was as simple as in any of the Cycloid family of the present time. Like slates on a roof, each single scale covered two, and was covered by two in turn; and the only point of difference which existed in relation to the *laying down* of these massy *slates* of *bone*, and the laying down of the very thin ones of *horn* which cover fish such as the carp or salmon, was, that in the massier *slates*, the sides, or *cover*,—nicely bevelled, in order to preserve an equability of thickness throughout,—were so adjusted, that two scales at their edges, where they lay the one over the other, were not thicker than one scale at its centre. Even in the other ganoids, their contemporaries, such as the Osteolepis and [Pg 311] Diplopterus, where the scales were ranged more in the tile fashion, side by side, there was, with much ingenious carpentry in the fitting, a general simplicity of form. It would almost appear, however, that ere the ganoid order reached the times of the Weald, the simple forms had been exhausted, and that nature, abhorring repetition, and ever stamping upon the scales some specific characteristic of the creature that bore them, was obliged to have recourse to forms of a more complex and involved outline. These latter-day scales send out nail-like spikes laterally and atop, to lay hold upon their neighbors, and exhibit in their undersides grooves that accommodated the nails sent out, in turn, by their neighbors, to lay hold upon *them*. Their forms, too, are indescribably various and fantastic. It seems curious enough, that immediately after this extremely *artificial* state of things, if I may so speak, the two prevailing orders of the fish of the present day, the Cycloids and Ctenoids, should have been ushered upon the scene, and more than the original simplicity of scale restored. There took place a sudden reäction, from the fantastic and the complex to the simple and the plain.

It is further worthy of notice, that though many of the ganoid scales of the Secondary systems, including those of the Wealden, glitter as brightly in burnished enamel as the more splendent scales of the Old Red Sandstone and Coal Measures, there is a curious peculiarity exhibited in the structure of many of the older scales of the highly enamelled class, which, so far as I have yet seen, does not extend beyond the Palæozoic period. The outer layer of the scale, which lies over a middle layer of a cellular cancellated structure, and corresponds, apparently, with that scarf-skin which in the human subject overlies the *rete mucosum*, is thickly set over with microscopic pores, funnel-shaped in the transverse section, and which, examined [Pg 312] by a good glass, in the horizontal one resemble the puncturings of a sieve. The Megalichthys of the Coal Measures, with its various carboniferous congeners, with the genera Diplopterus, Dipterus, and Osteolepis of the Old Red Sandstone,—

all brilliantly enamelled fish,—are thickly pore-covered. But whatever purpose these pores may have served, it seems in the Secondary period to have been otherwise accomplished, if, indeed, it continued to exist. It is a curious circumstance, that in no case do the pores seem to pass *through* the scale. Whatever their use, they existed merely as communications between the cells of the middle cancellated layer and the surface. In a fish of the Chalk,—*Macropoma Mantelli*,—the exposed fields of the scales are covered over with apparently hollow, elongated cylinders, as the little tubes in a shower-bath cover their round field of tin, save that they lie in a greatly flatter angle than the tubes; but I know not that, like the pores of the Dipterians and the Megalichthys, they communicated between the interior of the scale and its external surface. Their structure is at any rate palpably different, and they bear no such resemblance to the pores of the human skin as that which the Palæozoic pores present.

The amount of design exhibited in the scales of some of the more ancient ganoids,—design obvious enough to be clearly read,—is very extraordinary. A single scale of *Holoptychius Nobilissimus*,—fast locked up in its red sandstone rock,—laid by, as it were, for ever,—will be seen, if we but set ourselves to unravel its texture, to form such an instance of nice adaptation of means to an end as might of itself be sufficient to confound the atheist. Let me attempt placing one of these scales before the reader, in its character as a flat counter of bone, of a nearly circular form, an inch and a half in diameter, and an eighth-part [Pg 313] of an inch in thickness; and then ask him to bethink himself of the various means by which he would impart to it the greatest possible degree of strength. The human skull consists of two tables of solid bone, an inner and an outer, with a spongy cellular substance interposed between them, termed the *diploe*; and such is the effect of this arrangement, that the blow which would fracture a continuous wall of bone has its force broken by the spongy intermediate layer, and merely injures the outer table, leaving not unfrequently the inner one, which more especially protects the brain, wholly unharmed. Now, such also was the arrangement in the scale of the *Holoptychius Nobilissimus*. It consisted of its two well-marked tables of solid bone, corresponding in their dermal character, the outer to the cuticle, the inner to the true skin, and the intermediate cellular layer to the *rete mucosum*; but bearing an unmistakable analogy also, as a mechanical contrivance, to the two plates and the *diploe* of the human skull. To the strengthening principle of the two tables, however, there were two other principles added. Cromwell, when commissioning for a new helmet, his old one being, as he expresses it, "ill set," ordered his friend to send him a "*fluted pot*," *i.e.*, a helmet ridged and furrowed on the surface, and suited to break, by its protuberant lines, the force of a blow, so that the vibrations of the stroke would reach the body of the metal deadened and flat. Now, the outer table of the scale of the Holoptychius was a "fluted pot." The alternate ridges and furrows which ornamented its surface served a purpose exactly similar with that of the flutes and fillets of Cromwell's helmet. The inner table was strengthened on a different but not less effective principle. The human stomach consists of three coats; and two of these, the outermost or peritoneal coat, and the middle or muscular coat, are so [Pg 314] arranged, that the fibres of the one cross at nearly right angles those of the other. The violence which would tear the compact sides of this important organ along the fibres of the outer coat, would be checked by the transverse arrangement of the fibres of the middle coat, and *vice versa*. We find the cotton manufacturer weaving some of his stronger fabrics on a similar plan;—they also are made to consist of two *coats*; and what is technically termed the *tear* of the upper is so disposed that it lies at an angle of forty-five degrees with the *tear* of the coat which lies underneath. Now, the inner table of the scale of the Holoptychius was composed, on this principle, of various layers or coats, arranged the one over the other, so that the fibres of each lay at right angles with the fibres of the others in immediate contact with it. In the inner table of one scale I reckon nine of these alternating, variously-disposed layers; so that any application of violence, which, in the language of the lath-splitter, would *run lengthwise along the grain* of four of them, would be checked by the *cross grain* in five. In other words, the line of the *tear* in five of the layers was ranged at right angles with the line of the *tear* in four. There were thus in a single scale, in order to secure the greatest possible amount of strength,—and who can say what other purposes may have been secured besides?—three distinct principles embodied,—the principle of the two tables and *diploe* of the human skull,—the principle of the variously arranged coats of the human stomach,—and the principle of Oliver Cromwell's "fluted pot." There have been elaborate treatises written on those ornate flooring-tiles of the classical and middle ages, that are occasionally dug up by the antiquary amid monastic ruins, or on the sites of old Roman stations. But did any of them ever tell a story half so instructive or so strange as that told by the incalculably [Pg 315] more ancient ganoid *tiles* of the Palæozoic and Secondary periods?

I called, on my way back from Linksfield, upon my old friend Mr. Patrick Duff, and was introduced once more to his exquisite collection, with its unique ichthyolites of at least two genera of fishes of the Old Red,—the *Stagonolepis* and *Placothorax* of Agassiz,—which up to the present time are to be seen nowhere else; and various other fine specimens of rare species, which, having sat for their portraits, have their forms preserved in the great work of the naturalist of Neufchatel. He showed me, with some triumph, one of his later acquisitions,—a fine specimen of Holoptychius from the upper yellow sandstone of Bishop-Mill, which exhibits the dorsal ridge covered with a line of large overlapping scales, not at all unlike those overlapping plates which cover the tail of the lobster; for which, by the way, they were mistaken by the workman who first laid the fossil open. I examined, too, with some interest, fragments of a gigantic species of Pterichthys, belonging to an inferior division of the same Upper Old Red formation as the yellow stone, designated by Agassiz *Pterichthys major*, which must have attained to at least thrice the size, linearly, of even its bulkier congeners of the Lower formation of the Coccosteus. After examining many a drawer, stored, from the deposits of the neighborhood, with characteristic fossils of the Lias, the Weald, and the Oölite, and of the Upper and Lower Old Red, we set out together to expatiate amid the treasures of the Town Museum.

Among other recent additions to the Museum, there is an interesting set of the fishes of the Ganges, the donation of a gentleman long resident in India, to which Mr. Duff called my attention, as illustrative, in some of the specimens, of the more characteristic ichthyolites of the Old [Pg 316] Red Sandstone. One numerous family, the Pimelodi, abundantly represented in the Gangetic region, in not only the rivers, but also the ponds, tanks, and estuaries of the district, is certainly worthy the careful study of the geologist. It approaches nearer, in some of its more strongly-marked genera, to the Coccosteus of the Lower Old Red, than any other tribe of existing fishes which I have yet seen. The body of the Pimelodus, from the anterior dorsal downwards, is as naked as that of the eel; whereas the head, and in several of the species the back, is armed with strong plates of naked bone, curiously fretted, as in many of the ichthyolites of the Lower, and more especially of the Upper Old Red Sandstone, into ridges of confluent tubercles, that radiate from the centre to the edges of the plates. The dorsal plate, too, when detached, as in many of the species, from the plates of the head, bears upon its inner side a strong central ridge, that deepens as it descends, till it abruptly terminates a little short of the termination of the plate, exactly as in the dorsal plate of Coccosteus, which sunk its central ridge deep into the back of the animal. The point of resemblance to be mainly noticed, however, is the contrast furnished by the powerful armature of the head and back, with the unprotected nakedness of the posterior portions of the creature;—a point specially noticeable in the Coccosteus, and apparent also, though in a lesser degree, in some of the other genera of the Old Red, such as the Pterichthyes and Asterolepides. From the snout of the Coccosteus down to the posterior termination of the dorsal plate, the creature was cased in strong armor, the plates of which remain as freshly preserved in the ancient rocks of the country as those of the Pimelodi of the Ganges on the shelves of the Elgin Museum; but from the pointed termination of the plate immediately over the dorsal fin, to the tail, comprising [Pg 317] more than one half the entire length of the animal, all seems to have been exposed, without the protection of even a scale, and there survives in the better specimens only the internal skeleton of the fish and the ray-bones of the fins. It was armed, like a French dragoon, with a strong helmet and a short cuirass; and so we find its remains in the state in which those of some of the soldiers of Napoleon's old guard, that had been committed unstripped to the earth, may be dug up in the future on the fatal field of Borodino, or along the banks of the Dwina or the Wap. The cuirass lies still attached to the helmet, but we find only the naked skeleton attached to the cuirass. The Pterichthys to its strong helmet and cuirass added a posterior armature of comparatively feeble scales, as if, while its upper parts were shielded with plate armor, a lighter covering of ring or scale armor sufficed for the less vital parts beneath. In the Asterolepis the arrangement was somewhat similar, save that the plated cuirass was wanting: it was a strongly helmed warrior in slight scale armor; for the disproportion between the strength of the plated head-piece and that of the scaly coat was still greater than in the Pterichthys. The occipital star-covered plates are, in some of the larger specimens, fully three-quarters of an inch in thickness, whereas the thickness of the delicately-fretted scales rarely exceeds a line.

Why this disproportion between the strength of the armature in different parts of the same fish should have obtained, as in Pterichthys and Asterolepis, or why, while one portion of the animal was strongly armed, another portion should have been left, as in Coccosteus, wholly exposed, cannot of course be determined by the mere geologist. His rocks present him with but the fact of the disproportion, without accounting for it. But the natural [Pg 318] history of existing fish, in which, as in the Pimelodi, there may be detected a similar peculiarity of armature, may perhaps throw some light on the mystery. In Hamilton's "Fishes of the Ganges" I find but little reference made to the instincts and habits of the animals described: their deep-river haunts lie, in many cases, beyond the reach of observation; and of the observations actually made, the descriptive naturalist, intent often on mere peculiarities of structure, is not unfrequently too careless. Hamilton describes the habitats of the various Indian species of Pimelodi, whether brackish estuaries, ponds, or rivers, but not their characteristic instincts. Of the Silurus, however, a genus of the same great family, I read elsewhere that some of the species, such as the *Silurus glanis*, being unwieldy in their motions, do not pursue their prey, which consists of small fishes, but lie concealed among the mud, and seize on the chance stragglers that come their way. And of the *Pimelodus gulio*, a little, strongly-helmed fish, with a naked body, I was informed by Mr. Duff, on the authority of the gentleman who had presented the specimens to the Museum, that it burrowed in the holes of muddy banks, from which it shot out its armed head, and arrested, as they passed, the minute animals on which it preyed. The animal world is full of such compensatory defences: there is a half-suit of armor given to shield half the body, and a wise instinct to protect the rest. The *Pholas crispata* cannot shut its valves so as to protect its anterior parts, without raising them from off those parts which lie behind: like the Irishman in the haunted house, who attempted lengthening his blanket by cutting strips from the top and sewing them on to the bottom, it loses at the one end what it gains at the other; but, hemmed round by the solid walls of the recess which it is its nature to hollow out for itself in shale or stone, the anterior parts, [Pg 319] though uncovered by the shell, are not exposed. By closing its valves anteriorly, it shuts the door of its little house, made like that of the coney-folk of Scripture, in the rock; and then, of the entire cell in which it dwells so secure, what is not shut door is impregnable wall. The remark of Paley, that the "human animal is the only one which is naked, and the only one which can clothe itself," is by no means quite correct. One half the hermit crab is as naked as the "human animal," and even less fitted for exposure; for it consists of a thin-skinned, soft, unmuscular bag, filled with delicate viscera; but not even the human animal is more skilful in clothing himself in the spoils of other animals than the hermit crab in wrapping up its naked bag in the strong shell of some dead fusus or buccinum, which it carries about with it in all its peregrinations, as at once clothes, armor, and house. Nature arms its front, and it is itself wise enough to arm its rear. Now, it seems not improbable that the half-armed Coccosteus, a heavy fish, indifferently furnished with fins, may have burrowed, like the recent *Silurus glanis* or *Pimelodus gulio*, in a thick mud,—of the existence of which in vast quantity, during the times of the Old Red Sandstone, the dark Caithness flagstones, the fetid breccia of Strathpeffer, and the gray stratified clays of Cromarty, Moray, and Banff, unequivocally testify: and that it may have thus not only succeeded in capturing many of its light-winged contemporaries, which it would have vainly pursued in open sea, but may have been enabled also to present to its enemies, when assailed in turn, only its armed portions, and to protect its unarmed parts in its burrow. It is further worthy of notice, that many of the Pimelodi are furnished with spines, not, like those ichthyodorulites which occur so frequently in the older Secondary and Palæozoic divisions, unfinished in appearance at their lower extremity, as if, [Pg 320] like the spines of the ancient Acanthodi, or those of the recent dog-fish (*Spinax acanthias*), they had been simply embedded in the flesh, but bearing, like the wings of the Pterichthys, an articulated aspect. Those of the *Pimelodus rita* and *Pimelodus gagata* are of singular beauty; and when the creatures have no further use for them, and the mud of the Ganges has been consolidated into shale or baked into flagstone around them, they will make very exquisite fossils. A correct drawing of the plates and spines of some of the members of the Pimelodi family, with a portion of the internal skeletons, arranged in their proper places, but divested of those more destructible parts to which they are attached, would serve admirably to show what strange forms fish not greatly removed from the ordinary type may assume in the fossil state, and might throw some light on the extraordinary appearance assumed, as ichthyolites, by the old family of the Cephalaspians.

The geological department of the Elgin Museum is not yet very complete. The private collections of the locality, by forestalling, greatly restrict the supply from the rich deposits in the neighborhood, and have an unquestioned right to do so. The Museum contains, however, several interesting organisms. I saw, among the others, a specimen of Diplopterus, that showed the form and position of the fins of this rather rare ichthyolite much better than any of the Morayshire specimens portrayed by Agassiz in his great work; and beside it, one of the two specimens of *Pterichthys oblongus* which he figures, and on which he establishes the species. The other individual,—a Cromarty specimen,—graces my little collection. The gloomy day passed pleasantly in deciphering, with so accomplished a geologist as Mr. Duff, these curious hieroglyphics of the old world, that tell such wonderful stories, and in comparing [Pg 321] *viva voce*, as we were wont to do long years before in lengthy epistles, our respective notions regarding the true key for laying open their more occult meanings. And, after sharing with him in his family dinner, I again took my seat on the mail, as a chill, raw evening was falling, and rode on, some six or eight and twenty miles, to Campbelton. The rain pattered drearily through the night on my bed-room window; and as frequent exposure to the wet had begun to tell on a constitution not altogether so strong as it had once been, I awakened oftener than was quite comfortable, to hear it. The morning, however, was dry, though gray and sunless; and, taking an early breakfast at the inn, I traversed the flat gravelly points of Ardersier and Fortrose, that, projecting like moles far into the Frith, narrow the intervening ferry to considerably less than one-third the width which it would present were they away. The origin of these long detrital promontories, which form, when viewed from the heights on either side, so peculiar a feature in the landscape, and which, were they directly opposite, instead of being set down a mile awry, would shut up the opening altogether, has not yet been satisfactorily accounted for. One special theory assigns their formation to the agency of the descending tide, striking in zig-gig style, in consequence of some peculiarity of the coast-line or of the bottom, from side to side of the Frith, and depositing a long trail of sand and gravel, at nearly right angles with the beach, first on the one shore and then on the other. But why the tide, which runs in various zig-zag crossings in the course of the Frith, should have the effect here, and nowhere else, of raising two vast mounds, each a full mile and a quarter in length, with an average breadth of from two to five furlongs, is by no means very apparent. Certainly the present tides of the Frith could not have formed them, nor could they have [Pg 322] been elevated to their present average height of ten or twelve feet over the flood-line in a sea standing at the existing level. If they in reality originated in this cause, it must have been ere the latter upheavals of the land or recessions of the sea, when the great Caledonian Valley existed as a narrow ocean sound, swept by powerful currents. Upon another and entirely different hypothesis, these flat promontories have been regarded as the remains, levelled by the waves, and gapped direct in the middle by the tide, of a vast transverse morain of the great valley, belonging to the same glacial age as the lateral morains some ten or fifteen miles higher up, that extend from the immediate neighborhood of Inverness to the mansion-house of Dochfour. But this hypothesis, like the other, is not without its difficulties. Why, for instance, should the promontories be a mile awry? There is, however, yet another mode of accounting for their formation, which I am not in the least disposed to criticise.

They were constructed, says tradition, through the agency of the arch-wizard Michael Scott. Michael had called up the hosts of Faery to erect the cathedral of Elgin and the chanonry kirk of Fortrose, which they completed from foundation to ridge, each in a single night,—committing, in their hurry, merely the slight mistake of locating the building intended for Elgin in Fortrose, and that intended for Fortrose in Elgin; but, their work over and done, and when the magician had no further use for them, they absolutely refused to be *laid*; and, like a *posse* of Irish laborers thrown out of a job, came thronging round him, clamoring for more employment. Fearing lest he should be torn in pieces,—a catastrophe which has not unfrequently happened in such circumstances in the olden time, and of which those recent philanthropists who engage themselves in finding work for the unemployed may have perhaps [Pg 323] entertained some little dread in our own days,—he got rid of them for the time by setting them off in a body to run a mound across the Moray Frith from Fortrose to Ardersier. Toiling hard in the evening of a moonlight night, they had proceeded greatly more than two-thirds towards the completion of the undertaking, when a luckless Highlander passing by bade God-speed the work, and, by thus breaking the charm, arrested at once and forever the construction of the mound, and saved the navigation of Inverness.

I stood for a few seconds at the Burn of Rosemarkie undecided whether I should take the Scarfs-Craig road,—a break-neck path which runs eastwards along the cliffs, and which, though the rougher, is the more direct Cromarty line of the two,—or the considerably better though longer line of the White Bog, which strikes upwards along the burn in a westerly direction, and joins the Cromarty and Inverness highway on the moor of the Maolbuie. I had got into a part of the country where every little locality, and every more striking feature in the landscape, has its associated tradition; and the pause of a few moments at the two roads recalled to my memory the details of a ghost-story, long regarded in the district in which it was best known as one of the most authentic of its class, but which seems by no means inexplicable on natural principles. [13]

[Pg 324]

CHAPTER V.

Rosemarkie and its Scaurs—Kaes' Craig—A Jackdaw Settlement—"Rosemarkie Kaes" and "Cromarty Cooties"—"The Danes," a Group of Excavations—At Home in Cromarty—The Boulder-clay of Cromarty "begins to tell its story"—One of its marked Scenic Peculiarities—Hints to Landscape Painters—"Samuel's Well"—A Chain of

Bogs geologically accounted for—Another Scenic Peculiarity—"*Ha-has* of Nature's digging"—The Author's earliest Field of Hard Labor—Picturesque Cliff of Boulder-clay—Scratchings on the Sandstone—Invariable Characteristic of true Boulder-clay—Scratchings on Pebbles in the line of the longer axis—Illustration from the Boulder-clay of Banff.

Rosemarkie, with its long narrow valley and its red abrupt *scaurs*, [14] is chiefly interesting to the geologist for its vast beds of the boulder-clay. I am acquainted with no other locality in the kingdom where this deposit is hollowed into ravines so profound, or presents precipices so imposing and lofty. The clay lies thickly over most part of the Black Isle and the peninsula of Easter Ross,—both soft sandstone districts,—bearing everywhere an obvious relation, as a deposit, to both the form and the conditions of exposure of the existing land,—just as the accumulated snow of a long-lying snow-storm, exposed to the drifting wind, bears relation to the heights and hollows of the tracts which it covers. On the higher eminences the clay forms a comparatively thin stratum, and in not a few instances it has been wholly worn away; while on the lower grounds, immediately over the old coast line, and in the sides of hollow valleys,—exactly such places as we might expect to see the [Pg 325] snow occupying most deeply after a night of drift,—we find it accumulated in vast beds of from eighty to an hundred feet in thickness. One of these occurs in the opening of the narrow valley along which my course this morning lay, and is known far and wide,—for it forms a marked feature in the landscape, and harbors in its recesses a countless multitude of jackdaws,—as the "Kaes' Craig of Rosemarkie." It presents the appearance of a hill that had been cut sheer through the middle from top to base, and exhibits in its abrupt front a broad red perpendicular section of at least a hundred feet in height, barred transversely by thin layers of sand, and scored vertically by the slow action of the rains. Originally it must have stretched its vanished limb across the opening like some huge snow-wreath accumulated athwart a frozen rivulet; but the incessant sweep of the stream that runs through the valley has long since amputated and carried it away; and so only half the hill now remains. The Kaes' Craig resembles in form a lofty chalk cliff, square, massy, abrupt, with no sloping fillet of vegetation bound across its brow, but precipitous direct from the hill-top. The little ancient village of Rosemarkie stretches away from its base on the opposite side of the stream; and on its summit and along its sides, groups of chattering jackdaws, each one of them as reflective and philosophic as the individual immortalized by Cowper, look down high over the chimneys into the streets. The clay presents here, more than in almost any other locality with which I am acquainted, the character of a stratified deposit; and the numerous bands of sand by which the cliff is horizontally streaked from top to bottom we find hollowed, as we approach, into a multitude of circular openings, like shot-holes in an old tower, which form breeding-places for the daw and the sand-martin. The biped inhabitants of the cliff are greatly more numerous than the biped inhabitants [Pg 326] of the quiet little hamlet below; and on Fortrose fair-days, when, in virtue of an old feud, the Rosemarkie boys were wont to engage in formidable bickers with the boys of Cromarty, I remember, as one of the invading belligerents, that, in bandying names with them in the fray, we delighted to bestow upon them, as their hereditary sobriquet, given, of course, in allusion to their feathered neighbors, the designation of the "*Rosemarkie kaes*." Cromarty, however, is two-thirds surrounded by the waters of a frith abounding in sea-fowl; and the little fellows of Rosemarkie, indignant at being classed with their *kaes*, used to designate us with hearty emphasis, in turn, as the "*Cromarty cooties*," *i.e.*, coots.

A little higher up the valley, on the western side, there occurs in the clay what may be termed a *group* of excavations, composing a piece of scenery ruinously broken and dreary, and that bears a specific character of its own which scarce any other deposit could have exhibited. The excavations are of considerable depth and extent,—hollows out of which the materials of pyramids might have been taken. The precipitous sides are fretted by jutting ridges and receding inflections, that present in abundance their diversified alternations of light and shadow. The steep descents form cycloid curves, that flatten at their bases, and over which the ferruginous stratum of mould atop projects like a cornice. Between neighboring excavations there stand up dividing walls, tall and thin as those of our city buildings, and in some cases broken at their upper edges into rows of sharp pinnacles or inaccessible turf-coped turrets; while at the bottom of the hollows, washed by the runnels which, in the slow lapse of years, have been the architects of the whole, we find cairn-like accumulations of water-rolled stones,—the disengaged pebbles and boulders of the deposit. The boulders and pebbles project also from the steep sides, at all [Pg 327] heights and of all sizes, like the primary masses inclosed in our ancient conglomerates, when exhibited in wave-worn precipices,—forcing upon the mind the conclusion that the boulder-clay is itself but an unconsolidated conglomerate of the later periods, which occupies nearly the same relative position to the existing vegetable mould, with all its recent productions, that the great conglomerate of the Old Red Sandstone occupies in relation to the lower ichthyolite beds of that system, with their numerous extinct organisms. But its buried stones are fretted with hieroglyphic inscriptions, in the form of strange scratchings and polishings, grooves, ridges, and furrows,—always associated with the boulder-clays,—which those of the more ancient conglomerates want, and which, though difficult to read, seem at length to be yielding up the story which they record. Of this, however, more anon. Viewed by moonlight, when the pale red of the clay where the beam falls direct is relieved by the intense shadows, these excavations of the valley of Rosemarkie form scenes of strange and ghostly wildness: the projecting, buttress-like angles,—the broken walls,—the curved inflections,—the pointed pinnacles,—the turrets, with their masses of projecting coping,—the utter lack of vegetation, save where the heath and the furze rustle far above,—all combine to form assemblages of dreary ruins, amid which, in the solitude of night, one almost expects to see spirits walk. These excavations have been designated, from time immemorial, by the neighboring town's-people, as "the Danes;" but whether the name be, as is most probable, merely a corruption of an appropriate enough Saxon word, "the dens," or derived, as a vague tradition is said to testify, from the ages of Danish invasion, it is not quite the part of the geologist to determine. It may be worth mentioning, however, from its bearing on the point, that there are two excavations in the boulder-clay [Pg 328] near Cromarty, one of which has been long known by the name of "the Morial's Den," while the other, greatly smaller in size, rejoices in the double diminutive of "the Little Dennie." For an hour or so the Danes proved agreeable though somewhat silent companions; and then, climbing the opposite side of the valley, I gained the high road, and, walking on to Cromarty, found myself once more among "the old familiar faces."

In a few days the storm blew by; and as the prolonged rains had cleared out the deep ravines of the district, and given to the boulder-clay in which they are scooped a freshness in its section analogous to fresh fracture in rocks of harder consistency, I availed myself of the facilities afforded me in consequence, for exploring it once more. It has long constituted one of the hardest of the many riddles with which our Scottish deposits exercise the patience and ingenuity of the geologist. I remember a time when, after passing a day under its barren *scaurs*, or hid in its precipitous ravines, I used to feel in the evening as if I had been travelling under the cloud of night, and had seen nothing. It was a morose and taciturn companion, and had no speculation in it. I might stand in front of its curved precipices, red, yellow or gray, according to the prevailing average color of the rocks on which it rests, and mark their water-rolled boulders, of all qualities and sizes, sticking out in bold relief from the surface, like the rock-like protuberances that roughen the rustic basements of the architect, from the line of the wall; but I had no *open sesame* to form vistas through them into the recesses of the past. I saw merely the stiff pastry matrix of which they are composed, and the inclosed pebbles. But the boulder-clay has of late become more sociable; and, though with much hesitancy and irresolution, like old Mr. Spectator on the first formal opening of his mouth,—a consequence, doubtless, in both cases of [Pg 329] previous habits of silence long indulged,—it begins to tell its story. And a most curious story it is.

The morning was clear, but just a little chill; and a soft covering of snow, that had fallen during the storm on the flat summit of Ben-Wevis, and showed its extreme tenuity by the paleness of its tint of watery blue, was still distinctly visible at the distance of full twenty miles. The sun, low in the sky,—for the hour was early,—cast its slant rays athwart the prospect, giving to each nearer bank and hillock, and to the more distant protuberances on the mountain-sides, those well-defined accompaniments of shadow that serve by throwing the minor features of a landscape upon the eye in bold relief, to impart to it an air of higher finish and more careful filling up than it ever bears under a more vertical light. I took the road which, leading westward from the town towards Invergordon Ferry, skirts the Frith on the one hand, and runs immediately under the noble escarpment of green bank formed by the old coast line on the other. Fully two-thirds of the entire height of the rampart here, which rises in all about a hundred feet over the sea-level, is formed of the boulder-clay; and I am acquainted with no locality in which the deposit presents more strongly, for at least the first half mile, one of its marked scenic peculiarities. It is furrowed vertically on the slope, as if by enormous flutings in the more antique Doric style; and the ridges by which these are separated,—each from a hundred to a hundred and fifty feet in length, and from five-and-twenty to thirty feet in average height,—resemble those burial mounds with which the sexton frets the churchyard turf; with this difference, however, that they seem the burial mounds of giants, tall and bulky as those that of old warred against the gods. They are striking enough to have caught the eye of the children of the place, and are known among them as the Giants' [Pg 330] Graves. I could fain have taken their portrait in a calotype this morning, as they lay against the green bank,—their feet to the shore, and their heads on the top of the escarpment,—like patients on a reclining bed, and strongly marked, each by its broad bar of yellow light and of dark shadow, like the ebon and ivory buttresses of the poet. This little vignette, I would have said to the landscape painter, represents the boulder-clay, after its precipitous banks—worn down, by the frosts and rains of centuries, into parallel runnels, that gradually widened into these hollow grooves—had sunk into the angle of inclination at which the disintegrating agents ceased to operate, and the green sward covered all up. You must be studying these peculiarities of aspect more than ever you studied them before. There is a time coming when the connoisseur will as rigidly demand the specific character of the various geologic rocks and deposits in your hills, *scaurs*, and precipices, as he now demands specific character in your shrubs and trees.

It is worthy the notice of the young geologist, who has just set himself to study the various effects produced on the surface of a country by the deposits which lie under it, that for about a quarter of a mile or so, the base of the escarpment here is bordered by a line of bogs, that bear in the driest weather their mantling of green. They are fed with a perennial supply of water, by a range of deep-seated springs, that come bursting out from under the boulder-clay; and one of their number, which bears I know not why, the name of Samuel's Well, and yields its equable flow at an equable temperature, summer and winter, into a stone trough by the way-side, is not a little prized by the town's-people, and the seamen that cast anchor in the opposite roadstead, for the lightness and purity of its water. What is specially worthy of notice in the case is, the very definite [Pg 331] beginning and ending of the chain of bogs. All is dry at the base of the escarpment, up to the point at which they commence; and then all is equally dry at the point at which they terminate. And of exactly the same extent,—beginning where the bogs begin, and ending where they end,—we may trace an ancient stratum of pure sand,—of considerable thickness, intercalated between the base of the clay and the superior surface of the Old Red Sandstone. It is through this permeable sand that the profoundly seated springs find their way to the surface,—for the clay is impermeable; and where it comes in contact with the rock on either side of the arenaceous stratum, the bogs cease. The chain of green bogs is a consequence of the stratum of permeable sand. I have in vain sought this ancient layer of sand,—decidedly of the same era with the argillaceous bed which overlies it,—for aught organic. A single shell, so unequivocally of the period of the boulder-clay as to occur at the base of the deposit, would be worth, I have said, whole drawerfuls of fossils furnished by the better-known deposits. But I have since seen in abundance shells of the boulder-clay.

There is another scenic peculiarity of the clay, which the neighborhood of Cromarty finely illustrates, and of which my walk this morning furnished numerous striking instances. The Giants' Graves—to borrow from the children of the place—occur on the steep slopes of the old coast line, or in the sides of ravines, where the clay, as I have said, had once presented a precipitous front, but had been gradually moulded, under the attritive influences of the elements, into series of alternating ridges and furrows, which, when they had flattened into the proper angle, the green sward covered up from further waste. But the deep dells and narrow ravines in which many ranges of these graves occur are themselves peculiarities of the deposit. [Pg 332] Wherever the boulder-clay lies thick and continuous, as in the parish of Cromarty, on a sloping table-land, every minute streamlet cuts its way to the solid rock at the bottom, and runs through a deep dell, either softened into beauty by the disintegrating process, or with all its precipices standing up raw and abrupt over the stream. Four of these ravines, known as the "Old Chapel Burn," the "Ladies' Walk," the "Morial's Den," and the "Red Burn," each of them cutting the escarpment of the ancient coast line from top to base, and winding far into the interior, occur in little more than a mile's space; and they lie still more thickly farther to the west. These dells of the boulder clay, in their lower windings,—for they become shallower and tamer as they ascend, till they terminate in the uplands in mere *drains*, such as a ditcher might excavate at the rate of a shilling or two per yard,—are eminently picturesque. On those gentler slopes where the vegetable mould has had time and space to accumulate, we find not a few of the finest and tallest trees of the district. There is a bosky luxuriance in their more sheltered hollows, well known to the schoolboy what time the fern begins to pale its fronds, for their store of hips, sloes, and brambles; and red over the foliage we may see, ever and anon as we wend upwards, the abrupt frontage of some precipitous *scaur*, suited to remind the geologist, from its square form and flat breadth of surface, of the cliffs of the chalk. When viewed from the sea, at the distance of a few miles, these ravines seem to divide the sloping tracts in which they occur into large irregular fields, laid out considerably more in accordance with the principles of the landscape gardener than the stiffly squared rectilinear fields of the agriculturist. They are *ha-has* of Nature's digging; and their bottom and sides in this part of the country we still find occupied in a few cases—though in many more they have been ravaged [Pg 333] by the wasteful axe—by noble forest-*hedges*, tall enough to overtop, in at least their middle reaches, the tracts of table-land which they divide.

I passed, a little farther on, the quarry of Old Red Sandstone, with a huge bank of boulder-clay resting over it, in which I first experienced the evils of hard labor, and first set myself to lessen their weight by becoming an observer of geological phenomena. It had been deserted apparently for many years; and the debris of the clay partially covered up, in a sloping talus, the frontage of rock beneath. Old Red Sandstone and boulder-clay, a broad bar of each!—such was the compound problem which the excavation propounded to me when I first plied the tool in it,—a problem equally dark at the time in both its parts. I have since got on a very little way with the Old Red portion of the task; but alas for the boulder-clay portion of it! A bar of impenetrable shadow has rested long and obstinately over the newer deposit; and I scarce know whether the light which is at length beginning to play on its pebbly front be that of the sun or of a delusive meteor. But courage, patient hearts! the boulder-clay will one day yield up *its* secret too. Still further on by a few hundred yards, I could have again found use for the calotype, in transferring to paper the likeness of a protuberant picturesque cliff, which, like the Giants' Graves, could have belonged, of all our Scotch deposits, to only the boulder-clay. It stands out, on the steep acclivity of a furze-covered bank, abrupt as a precipice of solid rock, and yet seamed by the rain into numerous divergent channels, with pyramidal peaks between; and, combining the perpendicularity of a true cliff with the water-scooped furrows of a yielding clay, it presents a peculiarity of aspect which strikes, by its grotesqueness, eyes little accustomed to detect the picturesque in landscape. I remember [Pg 334] standing to gaze upon it when a mere child; and the fisher children of the neighboring town still tell that "*it has been prophesied*" it will one day fall, "and kill a man and a horse on the road below,"—a legend which shows it must have attracted *their* notice too.

I selected as the special scene of exploration this morning, a deep ravine of the boulder-clay, which had been recently deepened still more by the waters of a mill-pond, that had burst during a thunder-shower, and, after scooping out for themselves a bed in the clay some twelve or fifteen feet deep, where there had been formerly merely a shallow drain, had then tumbled into the ravine, and bared it to the rock. The sandstones of the district, soft and not very durable, show the scratched and polished surfaces but indifferently well, and, when exposed to the weather, soon lose them; but in the bottom of the runnel by which the ravine is swept I found them exceedingly well marked,—the polish as decided as the soft red stone could receive, and the lines of scratching running in their general bearing due east and west, at nearly right angles with the course of the stream. Wherever the rock had been laid bare during the last few months, *there* were the markings; wherever it had been laid bare for a few twelvemonths, they were gone. I next marked a circumstance which has now for several years been attracting my attention, and which I have found an invariable characteristic of the true boulder-clay. Not only do the rocks on which the deposit rests bear the scratched and polished surfaces, but in every instance the fragments of stone which it incloses bear the scratchings also, if from their character capable of receiving and retaining such markings, and neither of too coarse a grain nor of too hard a quality. If of limestone, or of a coherent shale, or of a close, finely-grained sandstone, or of a yielding trap, they [Pg 335] are scratched and polished,—invariably on one, most commonly on both their sides; and it is a noticeable circumstance, that the lines of the scratchings occur, in at least nine cases out of every ten, in the lines of their longer axes. When decidedly oblong or spindle-shaped, the scratchings run lengthwise, preserving in most cases, on the under and upper sides, when both surfaces are scratched, a parallelism singularly exact; whereas, when of a broader form, so that the length and breadth nearly approximate,—though the lines generally find out the longer axis, and run in that direction,—they are less exact in their parallelism, and are occasionally traversed by cross furrows. Of such certain occurrence is this longitudinal lining on the softer and finer-grained pebbles of the boulder-clay, that I have come to regard it as that special characteristic of the deposit on which I can most surely rely for purposes of identification. I am never quite certain of the boulder-clay when I do not detect it, nor doubtful of the true character of the deposit when I do. When examining, for instance, the accumulation of broken Liasic materials in the neighborhood of Banff, I made it my first care to ascertain whether the bank inclosed fragments of stone or shale bearing the longitudinal markings; and felt satisfied, on finding that it did, that I had discovered the period of its re-formation.

[Pg 336]

CHAPTER VI.

Organisms of the Boulder-clay not unequivocal—First Impressions of the Boulder-clay—Difficulty of accounting for its barrenness of Remains—Sir Charles Lyell's reasoning—A Fact to the contrary—Human Skull dug from a Clay-bank—The Author's Change of Belief respecting Organic Remains of the Boulder-clay—Shells from the Clay at Wick—Questions respecting them settled—Conclusions confirmed by Mr. Dick's Discoveries at Thurso—Sir John Sinclair's Discovery of Boulder-clay Shells in 1802—Comminution of the Shells illustrated—*Cyprina islandica*—Its Preservation in larger Proportions than those of other Shells accounted for—Boulder-clays of Scotland reformed during the existing Geological Epoch—Scotland in the Period of the Boulder-clay "merely three detached groups of Islands"—Evidence of the Subsidence of the Land in Scotland—Confirmed by Rev. Mr. Cumming's conclusion—High-lying Granite Boulders—Marks of a succeeding elevatory Period—Scandinavia now rising—Autobiography of a Boulder desirable—A Story of the Supernatural.

For the greater part of a quarter of a century I had been finding organisms in abundance in the boulder-clay, but never anything organic that unequivocally belonged to its own period. I had ascertained that it contains in Ross and Cromarty nodules of the Old Red Sandstone, which bear inside, like so many stone coffins, their well laid out skeletons of the dead; but then the markings on their surface told me that when the boulder-clay was in the course of deposition, they had been exactly the same kind of nodules that they are now. In Moray, it incloses, I had found, organisms of the Lias; but *they* also testify that they present an appearance in no degree more ancient at the present time than they did when first enveloped by the clay. In East and West Lothian too, and in the neighborhood of Edinburgh, I had detected in it occasional organisms [Pg 337] of the Mountain Limestone and the Coal Measures; but these, not less surely than its Liasic fossils in Moray, and its Old Red ichthyolites in Cromarty and Ross, belonged to an incalculably more ancient state of things than itself; and—like those shrivelled manuscripts of Pompeii or Herculaneum, which, whatever else they may record, cannot be expected to tell aught of the catastrophe that buried them up—they throw no light whatever on the deposit in which they occur. I at length came to regard the boulder-clay—for it is difficult to keep the mind in a purely blank state on any subject on which one thinks a good deal—as representative of a chaotic period of death and darkness, introductory, mayhap, to the existing scene of things.

After, however, I had begun to mark the invariable connection of the clay, as a deposit, with the dressed surfaces on which it rests, and the longitudinal linings of the pebbles and boulders which it incloses, and to associate it, in consequence, with an ice-charged sea and the Great Gulf Stream, it seemed to me extremely difficult to assign a reason why it should be thus barren of remains. Sir Charles Lyell states, in his "Elements," that the "stranding of ice-islands in the bays of Iceland since 1835 has driven away the fish for several successive seasons, and thereby caused a famine among the inhabitants of the country;" and he argues from the fact, "that a sea habitually infested with melting ice, which would chill and freshen the water, might render the same uninhabitable by marine mollusca." But then, on the other hand, it is equally a fact that half a million of seals have been killed in a single season on the meadow-ice a little to the north of Newfoundland, and that many millions of cod, besides other fish, are captured yearly on the shores of that island, though grooved and furrowed by ice-floes almost every spring. Of the seal family it is specially [Pg 338] recorded by naturalists, that many of the species "are from choice inhabitants of the margins of the frozen seas towards both poles; and, of course, in localities in which many such animals live, some must occasionally die." And though the grinding process would certainly have disjointed, and might probably have worn down and partially mutilated, the bones of the amphibious carnivora of the boulder period, it seems not in the least probable, judging from the fragments of loose-grained sandstone and soft shale which it has spared, that it would have wholly destroyed them. So it happened, however, that from North Berwick to the Ord Hill of Caithness, I had never found in the boulder-clay the slightest trace of an organism that could be held to belong to itself; and as it seems natural to build on negative evidence, if very extensive, considerably more than mere negative evidence, whatever the circumstances, will carry, I became somewhat skeptical regarding the very existence of boulder-fossils,—a skepticism which the worse than doubtful character of several supposed discoveries in the deposit served considerably to strengthen. The clay forms, when cut by a water-course, or assailed on the coast by some unusually high tide, a perpendicular precipice, which in the course of years slopes into a talus; and as it exhibits in most instances no marks of stratification, the clay of the talus—a mere re-formation of fragments detached by the frosts and rains from the exposed frontage—can rarely be distinguished from that of the original deposit. Now, in these consolidated slopes it is not unusual to find remains, animal and vegetable, of no very remote antiquity. I have seen a human skull dug out of the reclining base of a clay-bank once a precipice, fully six feet from under the surface. It might have been deemed the skull of some long-lived contemporary of Enoch,—one of the accursed race, mayhap,

[Pg 339] "Who sinned and died before the avenging flood."

But, alas! the laborer dug a little further, and struck his pickaxe against an old rybat that lay deeper still. There could be no mistaking the character of the champfered edge, that still bore the marks of the tool, nor that of the square perforation for the lock-bolt; and a rising theory, that would have referred the boulder-clay to a period in which the polar ice, set loose by the waters of the Noachian deluge, came floating southwards over the foundered land, straightway stumbled against it, and fell. Both rybat and skull had come from an ancient burying-ground, that occupies a projecting angle of the table-land above. I must now state, however, that my skepticism has thoroughly given way; and that, slowly yielding to the force of positive evidence, I have become as assured a believer in the *comminuted recent shells* of the boulder-clay as in the belemnites of the Oölite and Lias, or the ganoid ichthyolites of the Old Red Sandstone.

I had marked, when at Wick, on several occasions, a thick boulder-clay deposit occupying the southern side of the harbor, and forming an elevated platform, on which the higher parts of Pulteneytown are built; but I had noted little else regarding it than that it bears the average dark-gray color of the flagstones of the district, and that some of the granitic boulders which protrude from its top and sides are of vast size. On my last visit, however, rather more than two years ago, when sauntering along its base, after a very wet morning, awaiting the Orkney steamer, I was surprised to find, where a small slip had taken place during the rain, that it was mottled over with minute fragments of shells. These I examined, and found, so far as, in their extremely broken condition, I dared determine the point, that they belonged in such large proportion to [Pg 340] one species,—the *Cyprina islandica* of Dr. Fleming,—that I could detect among them only a single fragment of any other shell,—the pillar, apparently, of a large specimen of *Purpura lapillus*. Both shells belong to that class of old existences,—long descended, without the pride of ancient descent,—which link on the extinct to the recent scenes of being. *Cyprina islandica* and *Purpura lapillus* not only exist as living molluscs in the British seas, but they occur also as crag-shells, side by side with the dead races that have no place in the present fauna. At this time, however, I could but think of them simply in their character as recent molluscs; and as it seemed quite startling enough to find them in a deposit which I had once deemed representative of a period of death, and still continued to regard as obstinately unfossiliferous, I next set myself to determine whether it really *was* the boulder-clay in which they occurred. Almost the first pebble which I disengaged from the mass, however, settled the point, by furnishing the evidence on which for several years past I have been accustomed to settle it;—it bore in the line of its longer axis, on a polished surface, the freshly-marked grooves and scratchings of the iceberg era. Still, however, I had my doubts, not regarding the deposit, but the shells. Might they not belong merely to the talus of this bank of boulder-clay?—a re-formation, in all probability, not *more* ancient than the elevation of the most recent of the old coast lines,—perhaps greatly less so. Meeting with an intelligent citizen of Wick, Mr. John Cleghorn, I requested him to keep a vigilant eye on the shells, and to ascertain for me, when opportunity offered, whether they occurred deep in the deposit, or were restricted to merely the base of its exposed front. On my return from Orkney, he kindly brought me a small collection of fragments, exclusively, so far as I could judge, of *Cyprina islandica*, picked up in [Pg 341] fresh sections of the clay; at the same time expressing his belief that they really belonged to the deposit as such, and were not accidental introductions into it from the adjacent shore. And at this point for nearly two years the matter rested, when my attention was again called to it by finding, in the publication of Mr. Keith Johnston's admirable Geological Map of the British Islands, edited by Professor Edward Forbes, that other eyes than mine had detected shells in the boulder-clay of Caithness. "Cliffs of Pleistocene," says the Professor, in one of his notes attached to the map, "occur at Wick, containing boreal shells, especially *Astarte borealis*."

I had seen the boulder-clay characteristically developed in the neighborhood of Thurso; but, during a rather hurried visit, had lacked time to examine it. The omission mattered the less, however, as my friend Mr. Robert Dick is resident in the locality; and there are few men who examine more carefully or more perseveringly than he, or who can enjoy with higher relish the sweets of scientific research. I wrote him regarding Professor Forbes's decision on the boulder-clay of Wick and its shells; urging him to ascertain whether the boulder-clay of Thurso had not its shells also. And almost by return of post I received from him, in reply, a little packet of comminuted shells, dug out of a deposit of the boulder-clay, laid open by the river Thorsa, a full mile from the sea, and from eighty to a hundred feet over its level. He had detected minute fragments of shell in the clay about a twelvemonth before; but a skepticism somewhat similar to my own, added to the dread of being deceived by mere surface shells, recently derived from the shore in the character *of* shell-sand, or of the edible species carried inland for food, and then transferred from the ash-pit to the fields, had not only prevented him from following up the discovery, [Pg 342] but even from thinking of it as such. But he eagerly followed it up now, by visiting every bank of the boulder-clay in his locality within twenty miles of Thurso, and found them all charged, from top to bottom, with comminuted shells, however great their distance from the sea, or their elevation over it. The fragments lie thick along the course of the Thorsa, where the encroaching stream is scooping out the clay for the first time since its deposition, and laying bare the scratched and furrowed pebbles. They occur, too, in the depths of solitary ravines far amid the moors, and underlie heath, and moss, and vegetable mould, on the exposed hill-sides. The farm-house of Dalemore, twelve miles from Thurso as the crow flies, and rather more than thirteen miles from Wick, occupies, as nearly as may be, the centre of the county; and yet there, as on the sea-shore, the boulder-clay is charged with its fragments of marine shells. Though so barren elsewhere on the east coast of Scotland, the clay is everywhere in Caithness a shell-bearing deposit; and no sooner had Mr. Dick determined the fact for himself, at the expense of many a fatiguing journey, and many an hour's hard digging, than he found that it had been ascertained long before, though, from the very inadequate style in which it had been recorded, science had in scarce any degree benefited by the discovery. In 1802 the late Sir John Sinclair, distinguished for his enlightened zeal in developing the agricultural resources of the country, and for originating its statistics, employed a mineralogical surveyor to explore the underground treasures of the district; and the surveyor's journal he had printed under the title of "Minutes and Observations drawn up in the course of a Mineralogical Survey of the County of Caithness, ann. 1802, by John Busby, Edinburgh." Now, in this journal there are frequent references made to the occurrence of marine [Pg 343] shells in the blue clay. Mr. Dick has copied for me the two following entries,—for the work itself I have never seen:—"1802, Sept. 7th.—Surveyed down the river [Thorsa] to Geize; found blue clay-marl, *intermixed with marine shells* in great abundance." "Sept. 12th.—Set off this morning for Dalemore. Bored for shell-marl in the 'grass-park;' found it in one of the quagmires, but to no great extent. Bored for shell-marl in the 'house-park.' Surveyed by the side of the river, and found blue clay-marl in great plenty, *intermixed with marine shells, such as those found at Geize*. This place is supposed to be about twenty miles from the sea; and is one instance, among many in Caithness, of *the ocean's covering the inland country at some former period of time*."

The state of keeping in which the boulder-shells of Caithness occur is exactly what, on the iceberg theory, might be premised. The ponderous ice-rafts that went grating over the deep-sea bottom, grinding down its rocks into clay, and deeply furrowing its pebbles, must have borne heavily on its comparatively fragile shells. If rocks and pebbles did not escape, the shells must have fared but hardly. And very hardly they have fared: the rather unpleasant casualty of being crushed to death must have been a greatly more common one in those days than in even the present age of railways and machinery. The reader, by passing half a bushel of the common shells of our shores through a barley-mill, as a preliminary operation in the process, and by next subjecting the broken fragments thus obtained to the attritive influence of the waves on some storm-beaten beach for a twelvemonth or two, as a finishing operation, may produce, when he pleases, exactly such a water-worn shelly debris as mottles the blue boulder-clays of Caithness. The proportion borne by the fragments of one species of shell to that of [Pg 344] all the others is very extraordinary. The *Cyprina islandica* is still by no means a rare mollusc on our Scottish shores, and may, on an exposed coast, after a storm, be picked up by dozens, attached to the roots of the deep-sea tangle. It is greatly less abundant, however, than such shells as *Purpura lapillus, Mytilus edule, Cardium edule, Littorina littorea,* and several others; whereas in the boulder-clay it is, in the proportion of at least ten to one, more abundant than all the others put together. The great strength of the shell, however, may have in part led to this result; as I find that its stronger and massier portions,—those of the umbo and hinge-joint,—are exceedingly numerous in proportion to its slimmer and weaker fragments. "The *Cyprina islandica,*" says Dr. Fleming, in his "British Animals," "is the largest British bivalve shell, measuring sometimes thirteen inches in circumference, and, exclusively of the animal, weighing upwards of nine ounces." Now, in a collection of fragments of Cyprina sent me by Mr. Dick, disinterred from the boulder-clay in various localities in the neighborhood of Thurso, and weighing in all about four ounces, I have detected the broken remains of no fewer than *sixteen* hinge joints. And on the same principle through which the stronger fragments of Cyprina were preserved in so much larger proportion than the weaker ones, may Cyprina itself have been preserved in much larger proportion than its more fragile neighbors. Occasionally, however,—escaped, as if by accident,—characteristic fragments are found of shells by no means very strong,—such as *Mytilus, Tellina,* and *Astarte*. Among the univalves I can distinguish *Dentalium entale, Purpura lapillus, Turritella terebra,* and *Littorina littorea,* all existing shells, but all common also to at least the later deposits of the Crag. And among the bivalves Mr. Dick enumerates,—besides [Pg 345] the prevailing *Cyprina islandica,—Venus casina, Cardium edule, Cardium echinatum, Mytilus edule, Astarte danmoniensis* (*sulcata*), and *Astarte compressa,* with a *Mactra, Artemis,* and *Tellina*. [15] All the determined species here, with the exception of *Mytilus edule,* have, with many others, been found by the Rev. Mr. Cumming in the boulder-clays of the Isle of Man; and all of them are living shells at the present day on our Scottish coasts. It seems scarce possible to fix the age of a deposit so broken in its organisms, on the principle that would first seek to determine its per centage of extinct shells as the data on which to found. One has to search sedulously and long ere a fragment turns up sufficiently entire for the purpose of specific identification, even when it belongs to a well-known living shell; and did the clay contain some six or eight per cent. of the extinct in a similarly broken condition (and there is no evidence that it contains a single per cent. of extinct shells), I know not how, in the circumstances, the fact could ever be determined. A lifetime might be devoted to the task of fixing their real proportion, and yet be devoted to it in vain. All that at present can be said is, that, judging from what appears, the boulder-clays of Caithness, and with them the boulder-clays of Scotland generally, and of the Isle of Man,—for they are all palpably connected with the same iceberg phenomena, and occur along the same zone in reference to the sea-level,—were formed during the *existing* geological epoch.

These details may appear tediously minute; but let the reader mark how very much they involve. The occurrence of recent shells largely diffused throughout the boulder-clays of Caithness, at all heights and distances from the sea [Pg 346] at which the clay itself occurs, and not only connected with the iceberg phenomena by the closest juxtaposition, but also testifying distinctly to its agency by the extremely comminuted state in which we find them, tell us, not only according to old John Busby, "that the ocean covered the inland country at some former period of time," but that it covered it to a great height at a time geologically recent, when our seas were inhabited by exactly the same mollusca as inhabit them now, and so far as yet appears, by none others. I have not yet detected the boulder-clay at more than from six to eight hundred feet over the level of the sea; but the travelled boulders I have often found at more than a thousand feet over it; and Dr. John Fleming, the correctness of whose observations few men acquainted with the character of his researches or of his mind will be disposed to challenge, has informed me that he has detected the dressed and polished surfaces at least four hundred feet higher. There occurs a greenstone boulder, of from twelve to fourteen tons weight, says Mr. M'Laren, in his "Geology of Fife and the Lothians," on the south side of Black Hill (one of the Pentland range), at about fourteen hundred feet over the sea. Now fourteen or fifteen hundred feet, taken as the extreme height of the dressings, though they are said to occur greatly higher, would serve to submerge in the iceberg ocean almost the whole agricultural region of Scotland. The common hazel (*Corylus avellana*) ceases to grow in the latitude of the Grampians, at from one thousand two hundred to one thousand five hundred feet over the sea level; the common bracken (*Pteris aquilina*) at about the same height; and corn is never successfully cultivated at a greater altitude. Where the hazel and bracken cease to grow, it is in vain to attempt growing corn. [16] In the period [Pg 347] of the boulder-clay, then, when the existing shells of our coasts lived in those inland sounds and friths of the country that now exist as broad plains or fertile valleys, the sub-aërial superficies of Scotland was restricted to what are now its barren and mossy regions, and formed, instead of one continuous land, merely three detached groups of islands,—the small Cheviot and Hartfell group,—the greatly larger Grampian and Ben Nevis group,—and a group intermediate in size, extending from Meal-fourvonny, on the northern shores of Loch Ness, to the Maiden Paps of Caithness.

The more ancient boulder-clays of Scotland seem to have been formed when the land was undergoing a slow process of subsidence, or, as I should perhaps rather say, when a very considerable area of the earth's surface, including the sea-bottom, as well as the eminences that rose over it, was the subject of a gradual depression; for little or no alteration appears to have taken place at the time in the *relative* levels of the higher and lower portions of the sinking area: the features of the land in the northern part of the kingdom, from the southern flanks of the Grampians to the Pentland Frith, seemed to have been fixed in nearly the existing forms many ages before, at the close, apparently, of the Oölitic period, and at a still earlier age in the Lammermuir district, to the south. And so the sea around our shores must have deepened in the ratio in which the hills sank. The evidence of this process of subsidence is of a character tolerably satisfactory. The dressed surfaces occur in Scotland, most certainly, as I have already stated on the authority of Dr. Fleming, at the height of fourteen hundred feet over the present sea-level; it has been even said, at [Pg 348] fully twice that height, on the lofty flanks of Schehallion,—a statement, however, which I have had hitherto no opportunity of verifying. They may be found, too, equally well marked, under the existing high-water line; and it is obviously impossible that the dressing process could have been going on at the higher and lower levels at the same time. When the icebergs were grating along the more elevated rocks, the low-lying ones must have been buried under from three to seven hundred fathoms of water,—a depth from three to seven times greater, be it remembered, than that at which the most ponderous iceberg could possibly have grounded, or have in any degree affected the bottom. The dressing process, then, must have been a bit-and-bit process, carried on during either a period of elevation, in which the rising land was subjected, zone after zone, to the sweep of the armed ice from its higher levels *downwards*, or during a period of subsidence, in which it was subjected to the ice, zone after zone, from its lower levels *upwards*. And that it was the lower, not the higher levels, that were first dressed, appears evident from the circumstance, that though on these lower levels we find the rocks covered up by continuous beds of the boulder-clay, varying generally from twenty to a hundred feet in thickness, they are, notwithstanding, as completely dressed under the clay as on the heights above. Had it been a rising land that was subjected to the attrition of the icebergs, the debris and dressings of the higher rocks would have protected the lower from the attrition; and so the thick accumulation of boulder-clay which overlies the old coast line, for instance, would have rested, not on dressed, but on undressed surfaces. The barer rocks of the lower levels might of course exhibit their scratchings and polishings, like those of the higher; but wherever these scratchings and polishings occurred in the inferior zones, no thick protecting stratum [Pg 349] of boulder-clay would be found overlying them; and, *vice versa*, wherever in these zones there occurred thick beds of boulder-clay, there would be detected on the rock beneath no scratchings and polishings. In order to *dress* the entire surface of a country from the sea-line and under it to the tops of its hills, and at the same time to cover up extensive portions of its low-lying rocks with vast deposits of clay, it seems a necessary condition of the process that it should be carried on piece-meal from the lower level upwards,—not from the higher downwards.

It interested me much to find, that while from one set of appearances I had been inferring the gradual subsidence of the land during the period of the boulder-clay, the Rev. Mr. Cumming of King William's College had arrived, from the consideration of quite a different class of phenomena, at a similar conclusion. "It appears to me highly probable," I find him remarking, in his lately published "Isle of man," "that at the commencement of the boulder period there was a gradual sinking of this area [that of the island]. Successively, therefore, the points at different degrees of elevation were brought within the influence of the sea, and exposed to the rake of the tides, charged with masses of ice which had been floated off from the surrounding shores, and bearing on their under surfaces, mud, gravel, and fragments of hard rock." Mr. Cumming goes on to describe, in his volume, some curious appearances, which seem to bear direct on this point, in connection with a boss of a peculiarly-compounded granite, which occurs in the southern part of the island, about seven hundred feet over the level of the sea. There rise on the western side of the boss two hills, one of which attains to the elevation of nearly seven hundred, and the other of nearly eight hundred feet over it; and yet both hills to their summits are mottled over with granite boulders, [Pg 350] furnished by the comparatively low-lying boss. One of these travelled masses, fully two tons in weight, lies not sixty feet from the summit of the loftier hill, at an altitude of nearly fifteen hundred feet over the sea. Now, it seems extremely difficult to conceive of any other agency than that of a rising sea or of a subsiding land, through which these masses could have been rolled up the steep slopes of the hills. Had the boulder period been a period of elevation, or merely a stationary period, during which the land neither rose nor sank, the travelled boulders would not now be found resting at higher levels than that of the parent rock whence they were derived. We occasionally meet on our shores, after violent storms from the sea, stones that have been rolled from their place at low ebb to nearly the line of flood; but we always find that it was by the waves of the rising, not of the falling tide, that their transport was effected. For whatever removals of the kind take place during an ebbing sea are invariably in an opposite direction;—they are removals, not from lower to higher levels, but from higher to lower.

The upper subsoils of Scotland bear frequent mark of the elevatory period which succeeded this period of depression. The boulder-clay has its numerous intercalated arenaceous and gravelly beds, which belong evidently to its own era; but the numerous surface-beds of stratified sand and gravel by which in so many localities it is overlaid belong evidently to a later time. When, after possibly a long protracted period, the land again began to rise, or the sea to fall, the superior portions of the boulder-clay must have been exposed to the action of the tides and waves; and the same process of separation of parts must have taken place on a large scale, which one occasionally sees taking place in the present time on a comparatively small one, in ravines of the same clay swept by a streamlet. [Pg 351] After every shower, the stream comes down red and turbid with the finer and more argillaceous portions of the deposit; minute accumulations of sand are swept to the gorge of the ravine, or cast down in ripple-marked patches in its deeper pools; beds of pebbles and gravel are heaped up in every inflection of its banks; and boulders are laid bare along its sides. Now, a separation, by a sort of washing process of an analogous character, must have taken place in the materials of the more exposed portions of the boulder-clay, during the gradual emergence of the land; and hence, apparently, those extensive beds of sand and gravel which in so many parts of the kingdom exist, in relation to the clay, as a superior or upper subsoil; hence, too, occasional beds of a purer clay than that beneath, divested of a considerable portion of its arenaceous components, and of almost all its pebbles and boulders. This *washed* clay,—a re-formation of the boulder deposit, cast down, mostly in insulated beds in quiet localities, where the absence of currents suffered the purer particles held in suspension by the water to settle,—forms, in Scotland at least, with, of course, the exception of the ancient fire-clays of the Coal Measures, the true brick and tile clays of the agriculturist and architect.

It is to these superior beds that all the recent shells yet found above the existing sea-level in Scotland, from the Dornoch Frith and beyond it, to beyond the Frith of Forth, seem to belong. Their period is much less remote than that of the shells of the boulder-clay, and they rarely occur in the same comminuted condition. They existed, it would appear, not during the chill twilight period, when the land was in a state of subsidence, but during the after period of cheerful dawn, when hill-top after hill-top was emerging from the deep, and the close of each passing century witnessed a broader area of dry land in what is [Pg 352] now Scotland, than the close of the century which had gone before. Scandinavia is similarly rising at the present day, and presents with every succeeding age a more extended breadth of surface. Many of the boulder-stones seem to have been cast down where they now lie, during this latter time. When they occur, as in many instances, high on bare hill-tops, from five to fifteen hundred feet over the sea-level, with neither gravel nor boulder-clay beside them, we of course cannot fix their period. They may have been dropped by ice-floes or shore-ice, where we now find them, at the commencement of the period of elevation, after the clay had been formed; or they may have been deposited by more ponderous icebergs during its formation, when the land was yet sinking, though during the subsequent rise the clay may have been washed from around them to lower levels. The boulders, however, which we find scattered over the plains and less elevated hill-sides, with beds of the washed gravel or sand interposed between them and the clay, must have been cast down where they lie, during the elevatory ages. For, had they been washed out of the clay, they would have lain, not *over* the greatly lighter sands and gravels, but *under* them. Would that they could write their own histories! The autobiography of a single boulder, with notes on the various floras which had sprung up around it, and the various classes of birds, beasts, and insects by which it had been visited, would be worth nine-tenths of all the autobiographies ever published, and a moiety of the remainder to boot.

A few hundred yards from the opening of this dell of the boulder-clay, in which I have so long detained the reader, there is a wooded inflection of the bank, formed by the old coast line, in which there stood, about two centuries ago, a meal-mill, with the cottage of the miller, and [Pg 353] which was once known as the scene of one of those supernaturalities that belong to the times of the witch and the fairy. The upper anchoring-place of the bay lies nearly opposite the inflection. A shipmaster, who had moored his vessel in this part of the roadstead, some time in the latter days of the first Charles, was one fine evening sitting alone on deck, awaiting the return of his seamen, who had gone ashore, and amusing himself in watching the lights that twinkled from the scattered farm-houses, and in listening, in the extreme stillness of the calm, to the distant lowing of cattle, or the abrupt bark of the herdsman's dog. As the hour wore later, the sounds ceased, and the lights disappeared,—all but one solitary taper, that twinkled from the window of the miller's cottage. At length, however, it also disappeared, and all was dark around the shores of the bay, as a belt of black velvet. Suddenly a hissing noise was heard overhead; the shipmaster looked up, and saw what seemed to be one of those meteors known as falling stars, slanting athwart the heavens in the direction of the cottage, and increasing in size and brilliancy as it neared the earth, until the wooded ridge and the shore could be seen as distinctly from the ship-deck as by day. A dog howled piteously from one of the outhouses,—an owl whooped from the wood. The meteor descended until it almost touched the roof, when a cock crew from within; its progress seemed instantly arrested; it stood still, rose about the height of a ship's mast, and then began again to descend. The cock crew a second time; it rose as before; and, after mounting considerably higher than at first, again sank in the line of the cottage, to be again arrested by the crowing of the cock. It mounted yet a third time, rising higher still; and, in its last descent, had almost touched the roof, when the faint clap of wings was heard as if whispered over the [Pg 354] water, followed by a still louder note of defiance from the cock. The meteor rose with a bound, and, continuing to ascend until it seemed lost among the stars, did not again appear. Next night, however, at the same hour, the same scene was repeated in all its circumstances: the meteor descended, the dog howled, the owl whooped, the cock crew. On the following morning the shipmaster visited the miller's, and, curious to ascertain how the cottage would fare when the cock was away, he purchased the bird; and, sailing from the bay before nightfall, did not return until about a month after.

On his voyage inwards, he had no sooner doubled an intervening headland, than he stepped forward to the bows to take a peep at the cottage: it had vanished. As he approached the anchoring ground, he could discern a heap of blackened stones occupying the place where it had stood; and he was informed on going ashore, that it had been burnt to the ground, no one knew how, on the very night he had quitted the bay. He had it re-built and furnished, says the story, deeming himself what one of the old schoolmen perhaps term the *occasional* cause of the disaster. He also returned the cock,—probably a not less important benefit,—and no after accident befel the cottage. About fifteen years ago there was a human skeleton dug up near the scene of the tradition, with the skull, and the bones of the legs and feet, lying close together, as if the body had been huddled up twofold in a hole; and this discovery led to that of the story, which, though at one time often repeated and extensively believed, had been suffered to sleep in the memories of a few elderly people for nearly sixty years.

[Pg 355]

CHAPTER VII.

Relation of the deep red stone of Cromarty to the Ichthyolite Beds of the System—Ruins of a Fossil-charged Bed—Journey to Avoch—Red Dye of the Boulder-clay distinct from the substance itself—Variation of Coloring in the Boulder-clay Red Sandstone accounted for—Hard-pan how formed—A reformed Garden—An ancient Battle-field—Antiquity of Geologic and Human History compared—Burn of Killein—Observation made in boyhood confirmed—Fossil-nodules—Fine Specimen of *Coccosteus decipiens*—Blank strata of Old Red—New View respecting the Rocks of Black Isle—A Trip up Moray and Dingwall Friths—Altered color of the Boulder-clay—Up the Auldgrande River—Scenery of the great Conglomerate—Graphic Description—Laidlaw's Boulder—*Vaccinium myrtillus*—Profusion of Travelled Boulders—The Boulder *Clach Malloch*—Its zones of Animal and Vegetable Life.

The ravine excavated by the mill-dam showed me what I had never so well seen before,—the exact relation borne by the deep red stone of the Cromarty quarries to the ichthyolite beds of the system. It occupies the same place, and belongs to the same period, as those superior beds of the Lower Old Red Sandstone which are so largely developed in the cliffs of Dunnet Head in Caithness, and of Tarbet Ness in Ross-shire, and which were at one time regarded as forming, north of the Grampians, the analogue of the New Red Sandstone. I paced it across the strata this morning, in the line of the ravine, and found its thickness over the upper fish-beds, though I was far from reaching its superior layers, which are buried here in the sea, to be rather more than five hundred feet. The fossiliferous beds occur a few hundred yards below the dwelling-house of Rose Farm. They are not quite uncovered in the ravine; but we find their places indicated by heaps of gray [Pg 356] argillaceous shale, mingled with their characteristic ichthyolitic nodules, in one of which I found a small specimen of Cheiracanthus. The projecting edge of some fossil-charged bed had been struck, mayhap, by an iceberg, and dashed into ruins, just as the subsiding land had brought the spot within reach of the attritive ice; and the broken heap thus detached had been shortly afterwards covered up, without mixture of any other deposit, by the red boulder-clay. On the previous day I had detected the fish-beds in another new locality,—one of the ravines of the lawn of Cromarty House,—where the gray shale, concealed by a covering of soil and sward for centuries, had been laid bare during the storm by a swollen runnel, and a small nodule, inclosing a characteristic plate of Pterichthys, washed out. And my next object in to-day's journey, after exploring this ravine of the boulder-clay, was to ascertain whether the beds did not also occur in a ravine of the parish of Avoch, some eight or nine miles away, which, when lying a-bed one night in Edinburgh, I remembered having crossed when a boy, at a point which lies considerably out of the ordinary route of the traveller. I had remarked on this occasion, as the resuscitated recollection intimated, that the precipices of the Avoch ravine bore, at the unfrequented point, the peculiar aspect which I learned many years after to associate with the ichthyolitic member of the system; and I was now quite as curious to test the truth of a sort of vignette landscape, transferred to the mind at an immature period of life, and preserved in it for full thirty years, as desirous to extend my knowledge of the fossiliferous beds of a system to the elucidation of which I had peculiarly devoted myself.

As the traveller reaches the flat moory uplands of the parish, where the water stagnates amid heath and moss over a thin layer of peaty soil, he finds the underlying [Pg 357] boulder-clay, as shown in the chance sections, spotted and streaked with patches of a grayish-white. There is the same mixture of arenaceous and aluminous particles in the white as in the red portions of the mass; for, as we see so frequently exemplified in the spots and streaks of the Red

Sandstone formations, whether Old or New, the coloring matter has been discharged without any accompanying change of composition in the substance which it pervaded;—evidence enough that the red dye must be something distinct from the substance itself, just as the dye of a handkerchief is a thing distinct from the silk or cotton yarn of which the handkerchief has been woven. The stagnant water above, acidulated by its various vegetable solutions, seems to have been in some way connected with these appearances. In every case in which a crack through the clay gives access to the oozing moisture, we see the sides bleached, for several feet downwards, to nearly the color of pipe-clay; we find the surface, too, when it has been divested of the vegetable soil, presenting for yards together the appearance of sheets of half-bleached linen: the red ground of the clay has been acted upon by the percolating fluid, as the red ground of a Bandanna handkerchief is acted upon through the openings in the perforated lead, by the discharging chloride of lime. The peculiar chemistry through which these changes are effected might be found, carefully studied, to throw much light on similar phenomena in the older formations. There are quarries in the New Red Sandstone in which almost every mass of stone presents a different shade of color from that of its neighboring mass, and quarries in the Old Red the strata of which we find streaked and spotted like pieces of calico. And their variegated aspect seems to have been communicated, in every instance, not during deposition, nor after they had been hardened into stone but when, like the [Pg 358] boulder-clay, they existed in an intermediate state. Be it remarked, too, that the red clay here,—evidently derived from the abrasion of the red rocks beneath,—is in dye and composition almost identical with the substance on which, as an unconsolidated sandstone, the bleaching influences, whatever their character, had operated in the Palæozoic period, so many long ages before;—it is a repetition of the ancient experiment in the Old Red, that we now see going on in the boulder-clay. It is further worthy of notice, that the bleached lines of the clay exhibit, viewed horizontally, when the overlying vegetable mould has been removed, and the whitened surface in immediate contact with it paired off, a polygonal arrangement, like that assumed by the cracks in the bottom of clayey pools dried up in summer by the heat of the sun. Can these possibly indicate the ancient rents and fissures of the boulder-clay, formed, immediately after the upheaval of the land, in the first process of drying, and remaining afterwards open enough to receive what the uncracked portions of the surface excluded,—the acidulated bleaching fluid?

The kind of ferruginous pavement of the boulder-clay known to the agriculturist as *pan*, which may be found extending in some cases its iron cover over whole districts,—sealing them down to barrenness, as the iron and brass sealed down the stump of Nebuchadnezzar's tree,—is, like the white strips and blotches of the deposit, worthy the careful notice of the geologist. It serves to throw some light on the origin of those continuous bands of clayey or arenaceous ironstone, which in the older formations in which vegetable matter abounds, whether Oölitic or Carboniferous, are of such common occurrence. The *pan* is a stony stratum, scarcely less indurated in some localities than sandstone of the average hardness, that rests like a pavement on the surface of the boulder-clay, and that generally [Pg 359] bears atop a thin layer of sterile soil, darkened by a russet covering of stunted heath. The binding cement of the *pan* is, as I have said, ferruginous, and seems to have been derived from the vegetable covering above. Of all plants, the heaths are found to contain most iron. Nor is it difficult to conceive how, in comparatively flat tracts of heathy moor, where the surface water sinks to the stiff subsoil, and on which one generation of plants after another has been growing and decaying for many centuries, the minute metallic particles, disengaged in the process of decomposition, and carried down by the rains to the impermeable clay, should, by accumulating there, bind the layer on which they rest, as is the nature of ferruginous oxide, into a continuous stony crust. Wherever this *pan* occurs, we find the superincumbent soil doomed to barrenness,—arid and sun-baked during the summer and autumn months, and, from the same cause, overcharged with moisture in winter and spring. My friend Mr. Swanson, when schoolmaster of Nigg, found a large garden attached to the school-house so inveterately sterile as to be scarce worth cultivation; a thin stratum of mould rested on a hard impermeable pavement of *pan*, through which not a single root could penetrate to the tenacious but not unkindly subsoil below. He set himself to work in his leisure hours, and bit by bit laid bare and broke up the pavement. The upper mould, long divorced from the clay on which it had once rested, was again united to it; the piece of ground began gradually to alter its character for the better; and when I last passed the way, I found it, though in a state of sad neglect, covered by a richer vegetation than it had ever borne under the more careful management of my friend. This ferruginous pavement of the boulder-clay may be deemed of interest to the geologist, as a curious instance of deposition in a dense medium, [Pg 360] and as illustrative of the changes which may be effected on previously existing strata, through the agency of an overlying vegetation.

I passed, on my way, through the ancient battle-field to which I have incidentally referred in the story of the Miller of Resolis. [17] Modern improvement has not yet marred it by the plough; and so it still bears on its brown surface many a swelling tumulus and flat oblong mound, and—where the high road of the district passes along its eastern edge—the huge gray cairn, raised, says tradition, over the body of an ancient Pictish king. But the contest of which it was the scene belongs to a profoundly dark period, ere the gray dawn of Scottish history began. As shown by the remains of ancient art occasionally dug up on the moor, it was a conflict of the times of the stone battle-axe, the flint arrow-head, and the unglazed sepulchral urn, unindebted for aught of its symmetry to the turning-lathe,—times when there were heroes in abundance, but no scribes. And the cairn, about a hundred feet in length and breadth, by about twenty in height, with its long hoary hair of overgrown lichen waving in the breeze, and the trailing club-moss shooting upwards from its base along its sides, bears in its every lineament full mark of its great age. It is a mound striding across the stream of centuries, to connect the past with the present. And yet, after all, what a mere matter of yesterday its extreme antiquity is! My explorations this morning bore reference to but the later eras of the geologist; the portion of the geologic volume which I was attempting to decipher and translate formed the few terminal paragraphs of its concluding chapter. And yet the *finis* had been [Pg 361] added to them for thousands of years ere this latter antiquity began. The boulder-clay had been formed and deposited; the land, in rising over the waves, had had many a huge pebble washed out of its last formed red stratum, or dropped upon it by ice-floes from above; and these pebbles lay mottling the surface of this barren moor for mile after mile, bleaching pale to the rains and the sun, as the meagre and mossy soil received, in the lapse of centuries, its slow accessions of organic matter, and darkened around them. And then, for a few brief hours, the heath, no longer solitary, became a wild scene of savage warfare,—of waving arms and threatening faces,—and of human lives violently spilled, gushing forth in blood; and, when all was over, the old weathered boulders were heaped up above the slain, and there began a new antiquity in relation to the pile in its gathered state, that bore reference to man's short lifetime, and to the recent introduction of the species. The child of a few summers speaks of the events of last year as long gone by; while his father advanced into middle life, regards them as still fresh and recent.

I reached the Burn of Killein,—the scene of my purposed explorations,—where it bisects the Inverness road; and struck down the rocky ravine, in the line of the descending strata and the falling streamlet, towards the point at which I had crossed it so many years before. First I passed along a thick bed of yellow stone,—next over a bed of stratified clay. "The little boy," I said, "took correct note of what he saw, though without special aim at the time, and as much under the guidance of a mere observative instinct as Dame Quickly, when she took note of the sea-coal fire, the round table, the parcel-gilt goblet, and goodwife Keech's dish of prawns dressed in vinegar, as adjuncts of her interview with old Sir John [Pg 362] when he promised to marry her. These are unequivocally the ichthyolitic beds, whether they contain ichthyolites or no." The first nodule I laid open presented inside merely a pale oblong patch in the centre, which I examined in vain with the lens, though convinced of its organic origin, for a single scale. Proceeding farther down the stream, I picked a nodule out of a second and lower bed, which contained more evidently its organism,—a finely-reticulated fragment, that at first sight reminded me of some delicate festinella of the Silurian system. It proved, however, to be part of the tail of a Cheiracanthus, exhibiting—what is rarely shown—the interior surfaces of those minute rectangular scales which in this genus lie over the caudal fin, ranged in right lines. A second nodule presented me with the spines of *Diplacanthus striatus*; and still farther down the stream,—for the beds are numerous here, and occupy in vertical extent very considerable space in the system,—I detected a stratum of bulky nodules charged with fragments of Coccosteus, belonging chiefly to two species,—*Coccosteus decipiens* and *Coccosteus cuspidatus*. All the specimens bore conclusive evidence regarding the geologic place and character of the beds in which they occur; and in one of the number, a specimen of *Coccosteus decipiens*, sufficiently fine to be transferred to my knapsack, and which now occupies its corner in my little collection, the head exhibits all its plates in their proper order, and the large dorsal plate, though dissociated from the nail-like attachment of the nape, presents its characteristic breadth entire. It was the plates of this species, first found in the flagstones of Caithness, which were taken for those of a fresh-water tortoise; and hence apparently its specific name, *decipiens*;—it is the *deceiving* Coccosteus. I disinterred, in the course of my explorations, as many nodules as lay within reach,—now and then longing [Pg 363] for a pickaxe, and a companion robust and persevering enough to employ it with effect; and after seeing all that was to be seen in the bed of the stream and the precipices, I retraced my steps up the dell to the highway. And then, striking off across the moor to the north,—ascending in the system as I climbed the eminence, which forms here the central ridge of the old Maolbuie Common,—I spent some little time in a quarry of pale red sandstone, known, from the moory height on which it has been opened, as the quarry of the Maolbuie. But here, as elsewhere, the folds of that upper division of the Lower Old Red in which it has been excavated contain nothing organic. Why this should be so universally the case,—for in Caithness, Orkney, Cromarty, and Ross, wherever, in short, this member of the system is unequivocally developed, it is invariably barren of remains,—cannot, I suspect, be very satisfactorily explained. Fossils occur both over and under it, in rocks that seem as little favorable to their preservation; but during that intervening period which its blank strata represent, at least the *species* of all the ichthyolites of the system seem to have changed, and, so far as is yet known, the *genus* Coccosteus died out entirely.

The Black Isle has been elaborately described in the last Statistical Account of the Parish of Avoch as comprising at least the analogues of three vast geologic systems. The Great Conglomerate, and the thick bed of coarse sandstone of corresponding character that lies over it, compose all which is not primary rock of that south-eastern ridge of the district which forms the shores of the Moray Frith; and *they* are represented in the Account as Old Red Sandstone proper. Then, next in order,—forming the base of a parallel ridge,—come those sandstone and argillaceous bands to which the ichthyolite beds belong; and these [Pg 364] though at the time the work appeared their existence in the locality could be but guessed at, are described as representatives of the Coal Measures. Last of all there occur those superior sandstones of the Lower Old Red formation in which the quarry of the Maolbuie has been opened, and which are largely developed in the central or *backbone* ridge of the district. "And these," says the writer, "we have little hesitation in assigning to the *New* Red, or variegated Sandstone formation." I remember that some thirteen years ago,—in part misled by authority, and in part really afraid to represent beds of such an enormous aggregate thickness as all belonging to one inconsiderable formation,—for such was the character of the Old Red Sandstone at the time,—I ventured, though hesitatingly, and with less of detail, on a somewhat similar statement regarding the sandstone deposits of the parish of Cromarty. But true it is, notwithstanding, that the stratified rocks of the Black Isle are composed generally, not of the analogues of three systems, but of merely a fractional portion of a single system,—a fact previously established in other parts of the district, and which my discovery of this day in the Burn of Killein served yet farther to confirm in relation to that middle portion of the tract in which the parish of Avoch is situated. The geologic records, unlike the Sybilline books, grow in volume and number as one pauses and hesitates over them; demanding, however, with every addition to their bulk, a larger and yet larger sum of epochs and of ages.

The sun had got low in the western sky, and I had at least some eight or nine miles of rough road still before me; but the day had been a happy and not unsuccessful one, and so its hard work had failed to fatigue. The shadows, however, were falling brown and deep on the bleak Maolbuie, as I passed, on my return, the solitary cairn; [Pg 365] and it was dark night long ere I reached Cromarty. Next morning I quitted the town for the upper reaches of the Frith, to examine yet further the superficial deposits and travelled boulders of the district.

I landed at Invergordon a little after noon, from the Leith steamer, that, on its way to the upper ports of the Moray and Dingwall Friths, stops at Cromarty for passengers every Wednesday; and then passing direct through the village, I took the western road which winds along the shore towards Strathpeffer, skirting on the right the ancient province of the Munroes. The day was clear and genial; and the wide-spreading woods of this part of the country, a little touched by their autumnal tints of brown and yellow, gave a warmth of hue to the landscape, which at an earlier season it wanted. A few slim streaks of semi-transparent mist, that barred the distant hill-peaks, and a few towering piles of intensely white cloud, that shot across the deep blue of the heavens, gave warning that the earlier part of the day was to be in all probability the better part of it, and that the harvest of observation which it was ultimately to yield might be found to depend on the prompt use made of the passing hour. What first attracts the attention of the geologist, in journeying westwards, is the altered color of the boulder-clay, as exhibited in ditches by the way-side, or along the shore. It no longer presents that characteristic red tint,—borrowed from the red sandstone beneath,—so prevalent over the Black Isle, and in Easter Ross generally; but is of a cold leaden hue, not unlike that which it wears above the Coal Measures of the south, or over the flagstones of Caithness. The altered color here is evidently a consequence of the large development, in Ferindonald and Strathpeffer, of the ichthyolitic members of the Old Red, existing chiefly as fetid bituminous breccias and dark-colored sandstones: [Pg 366] the boulder-clay of the locality forms the dressings, not of red, but of blackish-gray rocks; and, as almost everywhere else in Scotland, its trail lies to the east of the strata, from which it was detached in the character of an impalpable mud by the age-protracted grindings of the denuding agent. It abounds in masses of bituminous breccia, some of which, of great size, seem to have been drifted direct from the valley of Strathpeffer, and are identical in structure and composition with the rock in which the mineral springs of the Strath have their rise, and to which they owe their peculiar qualities.

After walking on for about eight miles, through noble woods and a lovely country, I struck from off the high road at the pretty little village of Evanton, and pursued the course of the river Auldgrande, first through intermingled fields and patches of copsewood, and then through a thick fir wood, to where the bed of the stream contracts from a boulder-strewed bottom of ample breadth, to a gloomy fissure, so deep and dark, that in many places the water cannot be seen, and so narrow, that the trees which shoot out from the opposite sides interlace their branches atop. Large banks of the gray boulder-clay, laid open by the river, and charged with fragments of dingy sandstone and dark-colored breccia, testify, along the lower reaches of the stream, to the near neighborhood of the ichthyolitic member of the Old Red; but where the banks contract, we find only its lowest member, the Great Conglomerate. This last is by far the most picturesque member of the system,—abrupt and bold of outline in its hills, and mural in its precipices. And nowhere does it exhibit a wilder or more characteristic beauty than at the tall narrow portal of the Auldgrande, where the river,—after wailing for miles in a pent-up channel, narrow as one of the lanes of old Edinburgh, and hemmed in by walls quite as perpendicular, and nearly [Pg 367] twice as lofty,—suddenly expands, first into a deep brown pool, and then into a broad tumbling stream, that, as if permanently affected in temper by the strict severity of the discipline to which its early life had been subjected, frets and chafes in all its after course, till it loses itself in the sea. The banks, ere we reach the opening of the chasm, have become steep, and wild, and densely wooded; and there stand out on either hand, giant crags, that plant their iron feet in the stream; here girdled with belts of rank succulent shrubs, that love the damp shade and the frequent drizzle of the spray; and there hollow and bare, with their round pebbles sticking out from the partially decomposed surface, like the piled-up skulls in the great underground cemetery of the Parisians. Massy trees, with their green fantastic roots rising high over the scanty soil, and forming many a labyrinthine recess for the frog, the toad, and the newt, stretch forth their gnarled arms athwart the stream. In front of the opening, with but a black deep pool between, there lies a midway bank of huge stones. Of these, not a few of the more angular masses still bear, though sorely worn by the torrent, the mark of the blasting iron, and were evidently tumbled into the chasm from the fields above. But in the chasm there was no rest for them, and so the arrowy rush of the water in the confined channel swept them down till they dropped where they now lie, just where the widening bottom first served to dissipate the force of the current. And over the sullen pool in front we may see the stern pillars of the portal rising from eighty to a hundred feet in height, and scarce twelve feet apart, like the massive obelisks of some Egyptian temple; while, in gloomy vista within, projection starts out beyond projection, like column beyond column in some narrow avenue of approach to Luxor or Carnac. The precipices are green, with some moss or byssus, that like the miner, chooses a [Pg 368] subterranean habitat,—for here the rays of the sun never fall; the dead, mossy water beneath, from which the cliffs rise so abruptly, bears the hue of molten pitch; the trees, fast anchored in the rock, shoot out their branches across the opening, to form a thick tangled roof, at the height of a hundred and fifty feet overhead; while from the recesses within, where the eye fails to penetrate, there issues a combination of the strangest and wildest sounds ever yet produced by water: there is the deafening rush of the torrent, blent as if with the clang of hammers, the roar of vast bellows, and the confused gabble of a thousand voices. The sun, hastening to its setting, shone red, yet mellow, through the foliage of the wooded banks on the west, where, high above, they first curve from the sloping level of the fields, to bend over the stream; or fell more direct on the jutting cliffs and bosky dingles opposite, burnishing them as if with gold and fire; but all was coldly-hued at the bottom, where the torrent foamed gray and chill under the brown shadow of the banks; and where the narrow portal opened an untrodden way into the mysterious recesses beyond, the shadow deepened almost into blackness. The scene lacked but a ghost to render it perfect. An apparition walking from within like the genius in one of Goldsmith's essays "along the surface of the water," would have completed it at once.

Laying hold of an overhanging branch, I warped myself upwards from the bed of the stream along the face of a precipice, and, reaching its sloping top, forced my way to the wood above, over a steep bank covered with tangled underwood, and a slim succulent herbage, that sickened for want of the sun. The yellow light was streaming through many a shaggy vista, as, threading my way along the narrow ravine as near the steep edge as the brokenness of the ground permitted, I reached a huge mass of [Pg 369] travelled rock, that had been dropped in the old boulder period within a yard's length of the brink. It is composed of a characteristic granitic gneiss of a pale flesh-color, streaked with black, that, in the hand specimen, can scarce be distinguished from a true granite, but which, viewed in the mass, presents, in the arrangement of its intensely dark mica, evident marks of stratification, and which is remarkable, among other things, for furnishing almost all the very large boulders of this part of the country. Unlike many of the granitic gneisses, it is a fine solid stone, and would cut well. When I had last the pleasure of spending a few hours with the late Mr. William Laidlaw, the trusted friend of Sir Walter Scott, he intimated to me his intention,—pointing to a boulder of this species of gneiss,—of having it cut into two oblong pedestals, with which he purposed flanking the entrance to the mansion-house of the chief of the Rosses,—the gentleman whose property he at that time superintended. It was, he said, both in appearance and history, the most remarkable stone on the lands of Balnagown; and so he was desirous that it should be exhibited at Balnagown Castle to the best advantage. But as he fell shortly after into infirm health, and resigned his situation, I know not that he ever carried his purpose into effect. The boulder here, beside the chasm, measures about twelve feet in length and breadth, by from five to six in height, and contains from eight to nine hundred cubic feet of stone. On its upper table-like surface I found a few patches of moss and lichen, and a slim reddening tuft of the *Vaccinium myrtillus,* still bearing, late as was the season, its half-dozen blaeberries. This pretty little plant occurs in great profusion along the steep edges of the Auldgrande, where its delicate bushes, springing up amid long heath and ling, and crimsoned by the autumnal tinge, gave a peculiar warmth [Pg 370] and richness this evening to those bosky spots under the brown trees, or in immediate contact with the dark chasm on which the sunlight fell most strongly; and on all the more perilous projections, I found the dark berries still shrivelling on their stems. Thirty years earlier I would scarce have left them there; and the more perilous the crag on which they had grown, the more deliciously would they have eaten. But every period of life has its own playthings; and I was now chiefly engaged with the deep chasm and the huge boulder. Chasm and boulder had come to have greatly more of interest to me than the delicate berries, or than even that sovereign dispeller of ennui and low spirits, an adventurous scramble among the cliffs.

In what state did the chasm exist when the huge boulder,—detached, mayhap, at the close of a severe frost, from some island of the archipelago that is now the northern Highlands of Scotland,—was suffered to drop beside it, from some vast ice-floe drifting eastwards on the tide? In all probability merely as a fault in the Conglomerate, similar to many of those faults which in the Coal Measures of the southern districts we find occupied by continuous dikes of trap. But in this northern region, where the trap-rocks are unknown, it must have been filled up with the boulder-clay, or with some still more ancient accumulation of debris. And when the land had risen, and the streams, swollen into rivers, flowed along the hollows which they now occupy, the loose rubbish would in the lapse of ages gradually wash downwards to the sea, as the stones thrown from the fields above were washed downwards in a later time; and thus the deep fissure would ultimately be cleared out. The boulder-stones lie thickly in this neighborhood, and over the eastern half of Ross-shire, and the Black Isle generally; though for the [Pg 371] last century they have been gradually disappearing from the more cultivated tracts on which there were fences or farm-steadings to be built, or where they obstructed the course of the plough. We found them occurring in every conceivable situation,—high on hill-sides, where the shepherd crouches beside them for shelter in a shower,—deep in the open sea, where they entangle the nets of the fisherman,—on inland moors, where in some remote age they were painfully rolled together, to form the Druidical circle or Picts'-house,—or on the margin of the coast, where they had been piled over one another at a later time, as protecting bulwarks against the encroachments of the waves. They lie strewed more sparingly over extended plains, or on exposed heights, than in hollows sheltered from the west by high land, where the current, when it dashed high on the hill-sides, must have been diverted from its easterly course, and revolved in whirling eddies. On the top of the fine bluff hill of Fyrish, which I so admired to-day, each time I caught a glimpse of its purple front through the woods, and which shows how noble a mountain the Old Red Sandstone may produce, the boulders lie but sparsely. I especially marked, however, when last on its summit, a ponderous traveller of a vividly green hornblende, resting on a bed of pale yellow sandstone, fully a thousand feet over the present high-water level. But towards the east, in what a seaman would term the *bight* of the hill, the boulders have accumulated in vast numbers. They lie so closely piled along the course of the river Alness, about half a mile above the village, that it is with difficulty the waters, when in flood, can force their passage through. For here, apparently, when the tide swept along the hill-side, many an ice-floe, detained in the shelter by the revolving eddy, dashed together in rude collision, and shook their stony burdens to the bottom. [Pg 372] Immediately to the east of the low promontory on which the town of Cromarty is built there is another extensive accumulation of boulders, some of them of great size. They occupy exactly the place to which I have oftener than once seen the drift-ice of the upper part of the Cromarty Frith, set loose by a thaw, and then carried seawards by the retreating tide, forced back by a violent storm from, the east, and the fragments ground against each other into powder. And here, I doubt not, of old, when the sea stood greatly higher than now, and the ice-floes were immensely larger and more numerous than those formed, in the existing circumstances, in the upper shallows of the Frith, would the fierce north-east have charged home with similar effect, and the broken masses have divested themselves of their boulders.

The Highland chieftain of one of our old Gaelic traditions conversed with a boulder-stone, and told to it the story which he had sworn never to tell to man. I too, after a sort, have conversed with boulder-stones, not, however, to tell them any story of mine, but to urge them to tell theirs to me. But, lacking the fine ear of Hans Anderson, the Danish poet, who can hear flowers and butterflies talk, and understand the language of birds, I have as yet succeeded in extracting from them no such articulate reply

"As Memnon's image, long renowned of old
By fabling Nilus, to the quivering touch
Of Titan's ray, with each repulsive string
Consenting, sounded through the warbling air."

And yet, who can doubt that, were they a little more communicative, their stories of movement in the past, with the additional circumstances connected with the places which they have occupied ever since they gave [Pg 373] over travelling, would be exceedingly curious ones? Among the boulder group to the east of Cromarty, the most ponderous individual stands so exactly on the low-water line of our great Lammas tides, that though its shoreward edge may be reached dry-shod from four to six times every twelvemonth, no one has ever succeeded in walking dry shod round it. I have seen a strong breeze from the west, prolonged for a few days, prevent its drying, when the Lammas stream was at its point of lowest ebb, by from a foot to eighteen inches,—an indication, apparently, that to that height the waters of the Atlantic may be heaped up against our shores by the impulsion of the wind. And the recurrence, during at least the last century, of certain ebbs each season, which, when no disturbing atmospheric phenomena interfere with their operation, are sure to lay it dry, demonstrate, that during that period no change, even the most minute, has taken place on our coasts, in the relative levels of sea and shore. The waves have considerably encroached, during even the last half-century, on the shores immediately opposite; but it must have been, as the stone shows, simply by the attrition of the waves, and the consequent lowering of the beach,—not through any rise in the ocean, or any depression of the land.

The huge boulder here has been known for ages as the *Clach Malloch*, or accursed stone, from the circumstance, says tradition, that a boat was once wrecked upon it during a storm, and the boatmen drowned. Though little more than seven feet in height, by about twelve in length, and some eight or nine in breadth, its situation on the extreme line of ebb imparts a peculiar character to the various productions, animal and vegetable, which we find adhering to it. They occur in zones, just as on lofty hills the botanist finds his agricultural, moorland, and alpine [Pg 374] zones rising in succession as he ascends, the one over the other. At its base, where the tide rarely falls, we find two varieties of *Lobularia digitata*, dead man's hand, the orange colored and the pale, with a species of sertularia; and the characteristic vegetable is the rough-stemmed tangle, or cuvy. In the zone immediately above the lowest, these productions disappear; the characteristic animal, if animal it be, is a flat yellow sponge,—the *Halichondria papillaris*,—remarkable chiefly for its sharp siliceous spicula and its strong phosphoric smell; and the characteristic vegetable is the smooth-stemmed tangle, or queener. In yet another zone we find the common limpet and the vesicular kelp-weed; and the small gray balanus and serrated kelp-weed form the productions of the top. We may see exactly the same zones occurring in broad belts along the shore,—each zone indicative of a certain overlying depth of water; but it seems curious enough to find them all existing in succession on one boulder. Of the boulder and its story, however, more in my next.

[Pg 375]

CHAPTER VIII.

Imaginary Autobiography of the *Clach Malloch* Boulder—Its Creation—Its long night of unsummed Centuries—Laid open to light on a desert Island—Surrounded by an Arctic Vegetation—Undermined by the rising Sea—Locked up and floated off on an Ice-field—At rest on the Sea-bottom—Another Night of unsummed Years—The Boulder raised again above the waves by the rising of the Land—Beholds an altered Country—Pine Forests and Mammals—Another Period of Ages passes—The Boulder again floated off by an Ice-berg—Finally at rest on the Shore of Cromarty Bay—Time and Occasion of naming it—Strange Phenomena accounted for by Earthquakes—How the Boulder of Petty Bay was moved—The Boulder of Auldgrande—The old Highland Paupers—The little Parsi Girl—Her Letter to her Papa—But one Human Nature on Earth—Journey resumed—Conon Burying Ground—An aged Couple—Gossip.

The natural, and, if I may so speak, topographical, history of the *Clach Malloch,*—including, of course, its zoölogy and botany, with notes of those atmospheric effects on the tides, and of that stability for ages of the existing sea-level, which it indicates,—would of itself form one very interesting chapter: its geological history would furnish another. It would probably tell, if it once fairly broke silence and became autobiographical, first of a feverish dream of intense molten heat and overpowering pressure; and then of a busy time, in which the free molecules, as at once the materials and the artisans of the mass, began to build, each according to its nature, under the superintendence of a curious chemistry,—here forming sheets of black mica, there rhombs of a dark-green hornblende and a flesh-colored feldspar, yonder amorphous masses of a translucent quartz. It would add further, that at length, when the [Pg 376] slow process was over, and the entire space had been occupied to the full by plate, molecule, and crystal, the red fiery twilight of the dream deepened into more than midnight gloom, and a chill unconscious night descended on the sleeper. The vast Palæozoic period passes by,—the scarce less protracted Secondary ages come to a close,—the Eocene, Miocene, Pliocene epochs are ushered in and terminate,—races begin and end,—families and orders are born and die; but the dead, or those whose deep slumber admits not of dreams, take no note of time; and so it would tell how its long night of unsummed centuries seemed, like the long night of the grave, compressed into a moment.

The marble silence is suddenly broken by the rush of an avalanche, that tears away the superincumbent masses, rolling them into the sea; and the ponderous block, laid open to the light, finds itself on the bleak shore of a desert island of the northern Scottish archipelago, with a wintry scene of snow-covered peaks behind, and an ice-mottled ocean before. The winter passes, the cold severe spring comes on, and day after day the field-ice goes floating by,—now gray in shadow, now bright in the sun. At length vegetation, long repressed, bursts forth, but in no profuse luxuriance. A few dwarf birches unfold their leaves amid the rocks; a few sub-arctic willows hang out their catkins beside the swampy runnels; the golden potentilla opens its bright flowers on slopes where the evergreen *Empetrum nigrum* slowly ripens its glossy crow-berries; and from where the sea-spray dashes at full tide along the beach, to where the snow gleams at midsummer on the mountain-summits, the thin short sward is dotted by the minute cruciform stars of the scurvy-grass, and the crimson blossoms of the sea-pink. Not a few of the plants of our existing sea-shores and of our loftier hill-tops are still identical [Pg 377] in species; but wide zones of rich herbage, with many a fertile field and many a stately tree, intervene between the bare marine belts and the bleak insulated eminences; and thus the alpine, notwithstanding its identity with the littoral flora, has been long divorced from it; but in this early time the divorce had not yet taken place, nor for ages thereafter; and the same plants that sprang around the sea-margin rose also along the middle slopes to the mountain-summits. The landscape is treeless and bare, and a hoary lichen whitens the moors, and waves, as the years pass by, in pale tufts, from the disinterred stone, now covered with weather-stains, green and gray, and standing out in bold and yet bolder relief from the steep hill-side as the pulverizing frosts and washing rains bear away the lesser masses from around it. The sea is slowly rising, and the land, in proportion, narrowing its flatter margins, and yielding up its wider valleys to the tide; the low green island of one century forms the half-tide skerry, darkened with algæ, of another, and in yet a third exists but as a deep-sea rock. As its summit disappears, groups of hills, detached from the land, become islands, skerries, deep-sea rocks, in turn. At length the waves at full wash within a few yards of the granitic block. And now, yielding to the undermining influences, just as a blinding snow-shower is darkening the heavens, it comes thundering down the steep into the sea, where it lies immediately beneath the high-water line, surrounded by a wide float of pulverized ice, broken by the waves. A keen frost sets in; the half-fluid mass around is bound up for many acres into a solid raft, that clasps fast in its rigid embrace the rocky fragment; a stream-tide, heightened by a strong gale from the west, rises high on the beach; the consolidated ice-field moves, floats, is detached from the shore, creeps slowly outwards into the offing, bearing atop the boulder; and, [Pg 378] finally, caught by the easterly current, it drifts away into the open ocean. And then, far from its original bed in the rock, amid the jerkings of a cockling sea, the mass breaks through the supporting float, and settles far beneath, amid the green and silent twilight of the bottom, where its mosses and lichens yield their place to stony encrustations of deep purple, and to miniature thickets of arboraceous zoöphites.

The many-colored Acalephæ float by; the many-armed Sepiadæ shoot over; while shells that love the profounder depths,—the black Modiola and delicate Anomia,—anchor along the sides of the mass; and where thickets of the deep-sea tangle spread out their long, streamer-like fronds to the tide, the strong Cyprina and many-ribbed Astarte shelter by scores amid the reticulations of the short woody stems and thick-set roots. A sudden darkness comes on, like that which fell upon Sinbad when the gigantic roc descended upon him; the sea-surface is fully sixty fathoms over head; but even at this great depth an enormous iceberg grates heavily against the bottom, crushing into fragments in its course, Cyprina, Modiola, Astarte, with many a hapless mollusc besides; and furrows into deep grooves the very rocks on which they lie. It passes away; and, after many an unsummed year has also passed, there comes another change. The period of depression and of the boulder-clay is over. The water has shallowed as the sea-line gradually sank, or the land was propelled upwards by some elevatory process from below; and each time the tide falls, the huge boulder now raises over the waters its broad forehead, already hung round with flowing tresses of brown sea-weed, and looks at the adjacent coast. The country has strangely altered its features: it exists no longer as a broken archipelago, scantily covered by a semi-arctic vegetation, but as a continuous land, still whitened, where the [Pg 379] great valleys open to the sea, by the pale gleam of local glaciers, and snow-streaked on its loftier hill-tops. But vast forests of dark pine sweep along its hill-sides or selvage its shores; and the sheltered hollows are enlivened by the lighter green of the oak, the ash, and the elm. Human foot has not yet imprinted its sward; but its brute inhabitants have become numerous. The cream-colored coat of the wild bull,—a speck of white relieved against a ground of dingy green,—may be seen far amid the pines, and the long howl of the wolf heard from the nearer thickets. The gigantic elk raises himself from his lair, and tosses his ponderous horns at the sound; while the beaver, in some sequestered dell traversed by a streamlet, plunges alarmed into his deep coffer-dam, and, rising through the submerged opening of his cell, shelters safely within, beyond reach of pursuit. The great transverse valleys of the country, from its eastern to its western coasts, are still occupied by the sea,—they exist as broad ocean-sounds; and many of the detached hills rise around its shores as islands. The northern Sutor forms a bluff high island, for the plains of Easter Ross are still submerged; and the Black Isle is in reality what in later times it is merely in name,—a sea-encircled district, holding a midway place between where the Sound of the great Caledonian Valley and the Sounds of the Valleys of the Conon and Carron open into the German Ocean. Though the climate has greatly softened, it is still, as the local glaciers testify, ungenial and severe. Winter protracts his stay through the later months of spring; and still, as of old, vast floats of ice, detached from the glaciers, or formed in the lakes and shallower estuaries of the interior, come drifting down the Sounds every season, and disappear in the open sea, or lie stranded along the shores.

Ages have again passed: the huge boulder, from the further sinking of the waters, lies dry throughout the [Pg 380] neaps, and is covered only at the height of each stream-tide; there is a float of ice stranded on the beach, which consolidates around it during the neap, and is floated off by the stream; and the boulder, borne in its midst, as of old, again sets out a voyaging. It has reached the narrow opening of the Sutors, swept downwards by the strong ebb current, when a violent storm from the north-east sets in; and, constrained by antagonist forces,—the sweep of the tide on the one hand, and the roll of the waves on the other,—the ice-raft deflects into the little bay that lies to the east of the promontory now occupied by the town of Cromarty. And there it tosses, with a hundred more jostling in rude collision; and at length bursting apart, the *Clach Malloch*, its journeyings forever over, settles on its final resting-place. In a period long posterior it saw the ultimate elevation of the land. Who shall dare say how much more it witnessed, or decide that it did not form the centre of a rich forest vegetation, and that the ivy did not cling round it, and the wild rose shed its petals over it, when the Dingwall, Moray, and Dornoch Friths existed as sub-aërial valleys, traversed by streams that now enter the sea far apart, but then gathered themselves into one vast river, that, after it had received the tributary waters of the Shin and the Conon, the Ness and the Beauly, the Helmsdale, the Brora, the Findhorn, and the Spey, rolled on through the flat secondary formations of the outer Moray Frith,—Lias, and Oölite, and Greensand, and Chalk,—to fall into a gulf of the Northern Ocean which intervened between the coasts of Scotland and Norway, but closed nearly opposite the mouth of the Tyne, leaving a broad level plain to connect the coasts of England with those of the Continent! Be this as it may, the present sea-coast became at length the common boundary of land and sea. And the boulder continued to [Pg 381] exist for centuries still later as a nameless stone, on which the tall gray heron rested moveless and ghost-like in the evenings, and the seal at mid-day basked lazily in the sun. And then there came a night of fierce tempest, in which the agonizing cry of drowning men was heard along the shore. When the morning broke, there lay strewed around a few bloated corpses, and the fragments of a broken wreck; and amid wild execrations and loud sorrow the boulder received its name. Such is the probable history, briefly told, because touched at merely a few detached points, of the huge *Clach Malloch*. The incident of the second voyage here is of course altogether imaginary, in relation to at least this special boulder; but it is to second voyages only that all our positive evidence testifies in the history of its class. The boulders of the St. Lawrence, so well described by Sir Charles Lyell, voyage by thousands every year; [18] and there are few of my northern readers who have not heard of the short trip taken nearly half a century ago by the boulder of Petty Bay, in the neighborhood of Culloden.

[Pg 382]

A Highland minister of the last century, in describing, for Sir John Sinclair's Statistical Account, a large sepulchral cairn in his parish, attributed its formation to an *earthquake*! Earthquakes, in these latter times, are introduced, like the heathen gods of old, to bring authors out of difficulties. I do not think, however,—and I have the authority of the old critic for at least half the opinion,—that either gods or earthquakes should be resorted to by poets or geologists, without special occasion: they ought never to be called in except as a last resort, when there is no way of getting on without them. And I am afraid there have been few more gratuitous invocations of the earthquake than on a certain occasion, some five years ago, when it was employed by the inmate of a north-country manse, at once to account for the removal of the boulder-stone of Petty Bay, and to annihilate at a blow the geology of the Free Church editor of the *Witness*. I had briefly stated in one of my papers, in referring to this curious incident, that the boulder of the bay had been "borne nearly three hundred yards outwards into the sea by an enclasping mass of ice, in the course of a single tide." "Not at all," said the northern clergyman; "the cause assigned is wholly insufficient to produce such an effect. All the ice ever formed in the bay would be insufficient to remove such a boulder a distance, not of three hundred, but even of *three* yards." The removal of the stone "*is referrible to an* EARTHQUAKE!" The country, it would seem, took a sudden lurch, and the stone tumbled off. It fell athwart the flat surface of the bay, as a soup tureen sometimes falls athwart the table of a storm-beset steamer, vastly to the discomfort of the passengers, and again caught the ground as the land righted. Ingenious, certainly! It does appear a little wonderful, however, that in a shock so tremendous nothing should have fallen [Pg 383] off except the stone. In an earthquake on an equally great scale, in the present unsettled state of society, endowed clergymen would, I am afraid, be in some danger of falling out of their charges.

The boulder beside the Auldgrande has not only, like the *Clach Malloch*, a geologic history of its own, but, what some may deem of perhaps equal authority, a *mythologic* history also. The inaccessible chasm, impervious to the sun, and ever resounding the wild howl of the tortured water, was too remarkable an object to have escaped the notice of the old imaginative Celts; and they have married it, as was their wont, to a set of stories quite as wild as itself. And the boulder, occupying a nearly central position in its course, just where the dell is deepest, and narrowest, and blackest, and where the stream bellows far underground in its wildest combination of tones, marks out the spot where the more extraordinary incidents have happened, and the stranger sights have been seen. Immediately beside the stone there is what seems to be the beginning of a path leading down to the water; but it stops abruptly at a tree,—the last in the descent,—and the green and dewy rock sinks beyond for more than a hundred feet, perpendicular as a wall. It was at the abrupt termination of this path that a Highlander once saw a beautiful child smiling and stretching out its little hand to him, as it hung half in air by a slender twig. But he well knew that it was no child, but an evil spirit, and that if he gave it the assistance which it seemed to crave, he would be pulled headlong into the chasm, and never heard of more. And the boulder still bears, it is said, on its side,—though I failed this evening to detect the mark,—the stamp, strangely impressed, of the household keys of Balconie. [19]

[Pg 384]

The sun had now got as low upon the hill, and the ravine had grown as dark, as when, so long before, the lady of Balconie took her last walk along the sides of the Auldgrande; and I struck up for the little alpine bridge of a few undressed logs, which has been here thrown across the chasm, at the height of a hundred and thirty feet over the water. As I pressed through the thick underwood, I startled a strange-looking apparition in one of the open spaces beside the gulf, where, as shown by the profusion of plants of *vaccinium*, the blaeberries had greatly abounded in their season. It was that of an extremely old woman, cadaverously pale and miserable looking, with dotage glistening in her inexpressive, rheum-distilling eyes, and attired in a blue cloak, that had been homely when at its best, and was now exceedingly tattered. She had been poking with her crutch among the bushes, as if looking for berries; but my approach had alarmed her; and she stood muttering in Gaelic what seemed, from the tones and repetition, to be a few deprecatory sentences. I addressed her in English, and inquired what could have brought to a place so wild and lonely, one so feeble and helpless. "Poor object!" she muttered in reply,—"poor object!—very hungry;" but her scanty English could carry her no further. I slipped into her hand a small piece of silver, for which she overwhelmed me with thanks and blessings; and, bringing her to one of the broader avenues, traversed by a road which leads out of the wood, I saw her fairly entered upon the path in the right direction, and then, retracing my steps crossed the log-bridge. The old woman,—little, I should suppose from her appearance, under ninety,—was I doubt not, one of our ill-provided Highland paupers, that starve under a law which, while it has dried up the genial streams of voluntary charity in the country and presses hard upon the means of the humbler classes, alleviates little, if at all, [Pg 385] the sufferings of the extreme poor. Amid present suffering and privation there had apparently mingled in her dotage some dream of early enjoyment,—a dream of the days when she had plucked berries, a little herd-girl, on the banks of the Auldgrande; and the vision seemed to have sent her out, far advanced in her second childhood, to poke among the bushes with her crutch.

My old friend the minister of Alness,—uninstalled at the time in his new dwelling,—was residing in a house scarce half a mile from the chasm, to which he had removed from the parish manse at the Disruption; and, availing myself of an invitation of long standing, I climbed the acclivity on which it stands, to pass the night with him. I found, however, that with part of his family, he had gone to spend a few weeks beside the mineral springs of Strathpeffer, in the hope of recruiting a constitution greatly weakened by excessive labor, and that the entire household at home consisted of but two of the young ladies his daughters, and their ward, the little Buchubai Hormazdji.

And who, asks the reader, is this Buchubai Hormazdji? A little Parsi girl, in her eighth year, the daughter of a Christian convert from the ancient faith of Zoroaster, who now labors in the Free Church Mission at Bombay. Buchubai, his only child, was on his conversion, forcibly taken from him by his relatives, but restored again by a British court of law; and he had secured her safety by sending her to Europe, a voyage of many thousand miles, with a lady, the wife of one of our Indian missionaries, to whom she had become attached, as her second but true mamma, and with whose sisters I now found her. The little girl, sadly in want of a companion this evening, was content, for lack of a better, to accept of me as a playfellow; and she showed me all her rich eastern dresses, and all her toys, and a very fine emerald, set in the oriental fashion, which, [Pg 386] when she was in full costume, sparkled from her embroidered tiara. I found her exceedingly like little girls at home, save that she seemed more than ordinarily observant and intelligent,—a consequence mayhap, of that early development, physical and mental, which characterizes her race. She submitted to me, too, when I had got very much into her confidence, a letter she had written to her papa from Strathpeffer, which was to be sent him by the next Indian mail. And as it may serve to show that the style of little girls whose fathers were fire-worshippers for three thousand years and more differs in no perceptible quality from the style of little girls whose fathers in considerably less than three thousand were Pagans, Papists, and Protestants by turns, besides passing through the various intermediate forms of belief, I must, after pledging the reader to strict secrecy, submit it to his perusal:—

"My dearest Papa,—I hope you are quite well. I am visiting mamma at present at Strathpeffer. She is much better now than when she was travelling. Mamma's sisters give their love to you, and mamma, and Mr. and Mrs. F. also. They all ask you to pray for them, and they will pray also. There are a great many at water here for sick people to drink out of. The smell of the water is not at all nice. I sometimes drink it. Give my dearest love to Narsion Skishadre, and tell her that I will write to her.—Dearest papa," etc.

It was a simple thought, which required no reach of mind whatever to grasp,—and yet an hour spent with little Buchubai made it tell upon me more powerfully than ever before,—that there is in reality but one human nature on the face of the earth. Had I simply read of Buchubai Hormazdji corresponding with her father Hormazdji Pestonji, and sending her dear love to her old companion Narsion Skishadre, the names so specifically different from those [Pg 387] which we ourselves employ in designating our country folk, would probably have led me, through a false association, to regard the parties to which they attach as scarcely less specifically different from our country folk themselves. I suspect we are misled by associations of this kind when we descant on the peculiarities of race as interposing insurmountable barriers to the progress of improvement, physical or mental. We overlook, amid the diversities of form, color, and language, the specific identity of the human family. The Celt, for instance, wants, it is said, those powers of sustained application which so remarkably distinguish the Saxon; and so we agree on the expediency of getting rid of our poor Highlanders by expatriation as soon as possible, and of converting their country into sheep-walks and hunting-parks. It would be surely well to have philosophy enough to remember what, simply through the exercise of a wise faith, the Christian missionary never forgets, that the peculiarities of race are not specific and ineradicable, but mere induced habits and idiosyncracies engrafted on the stock of a common nature by accident of circumstance or development; and that, as they have been wrought into the original tissue through the protracted operation of one set of causes, the operation of another and different set, wisely and perseveringly directed, could scarce fail to unravel and work them out again. They form no part of the inherent design of man's nature, but have merely stuck to it in its transmissive passage downwards and require to be brushed off. There was a time, some four thousand years ago, when Celt and Saxon were represented by but one man and his wife, with their children and their children's wives; and some sixteen or seventeen centuries earlier all the varieties of the species,—Caucasian and Negro, Mongolian and Malay,—lay close packed up in the world's single family. In short, Buchubai's amusing prattle proved [Pg 388] to me this evening no bad commentary on St. Paul's sublime enunciation to the Athenians, that God has "made of one blood all nations of men to dwell on all the face of the earth." I was amused to find that the little girl, who listened intently as I described to the young ladies all I had seen and knew of the Auldgrande, had never before heard of a ghost, and could form no conception of one now. The ladies explained, described, defined; carefully guarding all they said, however, by stern disclaimers against the ghost theory altogether, but apparently to little purpose. At length Buchubai exclaimed, that she now knew what they meant, and that she herself had seen a great many ghosts in India. On explanation, however, her ghosts, though quite frightful enough, turned out to be not at all spiritual: they were things of common occurrence in the land she had come from,—exposed bodies of the dead.

Next morning—as the white clouds and thin mist-streaks of the preceding day had fairly foretold—was close and wet; and the long trail of vapor which rises from the chasm of the Auldgrande in such weather, and is known to the people of the neighborhood as the "smoke of the lady's baking," hung, snake-like, over the river. About two o'clock the rain ceased, hesitatingly and doubtfully, however, as if it did not quite know its own mind; and there arose no breeze to shake the dank grass, or to dissipate the thin mist-wreath that continued to float over the river under a sky of deep gray. But the ladies, with Buchubai, impatient to join their friends at Strathpeffer, determined on journeying notwithstanding; and, availing myself of their company and their vehicle, I travelled on with them to Dingwall, where we parted. I had purposed exploring the gray dingy sandstones and fetid breccias developed along the shores on the northern side of the bay, about two [Pg 389] miles from the town, and on the sloping acclivities between the mansion-houses of Tulloch and Fowlis; but the day was still unfavorable, and the sections seemed untemptingly indifferent; besides, I could entertain no doubt that the dingy beds here are identical in place with those of Cadboll on the coast of Easter Ross, which they closely resemble, and which alternate with the lower ichthyolitic beds of the Old Red Sandstone; and so, for the present at least, I gave up my intention of exploring them.

In the evening, the sun, far gone down towards its place of setting, burst forth in great beauty; and, under the influence of a kindly breeze from the west, just strong enough to shake the wet leaves, the sky flung off its thick mantle of gray. I sauntered out along the high-road, in the direction of my old haunts at Conon-side, with, however, no intention of walking so far. But the reaches of the river, a little in flood, shone temptingly through the dank foliage, and the cottages under the Conon woods glittered clear on their sweeping hill-side, "looking cheerily out" into the landscape; and so I wandered on and on, over the bridge, and along the river, and through the pleasure grounds of Conon-house, till I found myself in the old solitary burying-ground beside the Conon, which, when last in this part of the country, I was prevented from visiting by the swollen waters. The rich yellow light streamed through the interstices of the tall hedge of forest-trees that encircles the eminence, once an island, and fell in fantastic patches on the gray tombstone and the graves. The ruinous little chapel in the corner, whose walls a quarter of a century before I had distinctly traced, had sunk into a green mound; and there remained over the sward but the arch-stone of a Gothic window, with a portion of the moulded transom attached, to indicate the character and style of the vanished building. The old dial-stone, with the wasted gnomon, [Pg 390] has also disappeared; and the few bright-colored *throch-stanes*, raw from the chisel, that had been added of late years to the group of older standing, did not quite make up for what time in the same period had withdrawn. One of the newer inscriptions, however, recorded a curious fact. When I had resided in this part of the country so long before, there was an aged couple in the neighborhood, who had lived together, it was said, as man and wife, for more than sixty years: and now, here was their tombstone and epitaph. They had lived on long after my departure; and when, as the seasons passed, men and women whose births and baptisms had taken place since their wedding-day were falling around them well stricken in years, death seemed to have forgotten *them*; and when he came at last, their united ages made up well nigh two centuries. The wife had seen her ninety-sixth and the husband his hundred and second birthday. It does not transcend the skill of the actuary to say how many thousand women must die under ninety-six for every one that reaches it, and how many tens of thousands of men must die under a hundred and two for every man who attains to an age so extraordinary; but he would require to get beyond his tables in order to reckon up the chances against the woman destined to attain to ninety-six being courted and married in early life by the man born to attain to a hundred and two.

After enjoying a magnificent sunset on the banks of the Conon, just where the scenery, exquisite throughout, is most delightful, I returned through the woods, and spent half an hour by the way in the cottage of a kindly-hearted woman, now considerably advanced in years, whom I had known, when she was in middle life, as the wife of one of the Conon-side hinds, and who not unfrequently, when I was toiling at the mallet in the burning sun, hot and thirsty, and rather loosely knit for my work, had brought [Pg 391] me—all she had to offer at the time—a draught of fresh whey. At first she seemed to have wholly forgotten both her kindness and the object of it. She well remembered my master, and another Cromarty man who had been grievously injured, when undermining an old building, by the sudden fall of the erection; but she could bethink her of no third Cromarty man whatever. "Eh, sirs!" she at length exclaimed, "I daresay ye'll be just the sma' prentice laddie. Weel, what will young folk no come out o'? They were amaist a' stout big men at the wark except yoursel'; an' you're now stouter and bigger than maist o' them. Eh, sirs!—an' are ye still a mason?" "No; I have not wrought as a mason for the last fourteen years; but I have to work hard enough for all that." "Weel, weel, it's our appointed lot; an' if we have but health an' strength, an' the wark to do, why should we repine?" Once fairly entered on our talk together, we gossipped on till the night fell, giving and receiving information regarding our old acquaintances of a quarter of a century before; of whom we found that no inconsiderable proportion had already sunk in the stream in which eventually we must all disappear. And then, taking leave of the kindly old woman, I walked on in the dark to Dingwall, where I spent the night. I could fain have called by the way on my old friend and brother-workman, Mr. Urquhart,—of a very numerous party of mechanics employed at Conon-side in the year 1821 the only individual now resident in this part of the country; but the lateness of the hour forbade. Next morning I returned by the Conon road, as far as the noble old bridge which strides across the stream at the village, and which has done so much to banish the water-wraith from the fords; and then striking off to the right, I crossed, by a path comparatively little frequented, the insulated group of hills which separates the valley of [Pg 392] the Conon from that of the Peffer. The day was mild and pleasant, and the atmosphere clear; but the higher hills again exhibited their ominous belts of vapor, and there had been a slight frost during the night,—at this autumnal season the almost certain precursor of rain.

[Pg 393]

CHAPTER IX.

The Great Conglomerate—Its Undulatory and Rectilinear Members—Knock Farril and its Vitrified Fort—The old Highlanders an observant race—The Vein of Silver—Summit of Knock Farril—Mode of accounting for the Luxuriance of Herbage in the ancient Scottish Fortalices—The green Graves of Culloden—Theories respecting the Vitrification of the Hill-forts—Combined Theories of Williams and Mackenzie probably give the correct account—The Author's Explanation—Transformations of Fused Rocks—Strathpeffer—The Spa—Permanent Odoriferous Qualities of an ancient Sea-bottom converted into Rock—Mineral Springs of the Spa—Infusion of the powdered rock a substitute—Belemnite Water—The lively young Lady's Comments—A befogged Country seen from a hill-top—Ben-Wevis—Journey to Evanton—A Geologist's Night-mare—The Route Home—Ruins of Craighouse—Incompatibility of Tea and Ghosts—End of the Tour.

I was once more on the Great Conglomerate,—here, as elsewhere, a picturesque, boldly-featured deposit, traversed by narrow, mural-sided valleys, and tempested by bluff abrupt eminences. Its hills are greatly less confluent than those of most of the other sedimentary formations of Scotland; and their insulated summits, recommended by their steep sides and limited areas to the old savage Vaubans of the Highlands, furnished, ere the historic eras began, sites for not a few of the ancient hill-forts of the country. The vitrified fort of Craig Phadrig, of the Ord Hill of Kessock, and of Knock Farril,—two of the number, the first and last, being the most celebrated erections of their kind in the north of Scotland,—were all formed on hills of the

Great Conglomerate. The Conglomerate exists here as a sort of miniature Highlands, set down at the northern side of a large angular bay of Palæozoic rock, which indents the *true* Highlands of the country, and which exhibits in its central [Pg 394] area a prolongation of the long moory ridge of the Black Isle, formed, as I have already had occasion to remark, of an *upper* deposit of the same lower division of the Old Red,—a deposit as noticeable for affecting a confluent, rectilinear character in its elevations, as the Conglomerate is remarkable for exhibiting a detached and undulatory one. Exactly the same features are presented by the same deposits in the neighborhood of Inverness; the *undulatory* Conglomerate composing, to the north and west of the town, the picturesque wavy ridge comprising the twin-eminences of Munlochy Bay, the Ord Hill of Kessock, Craig Phadrig, and the fir-covered hill beyond in the line of the Great Valley; while on the south and east the *rectilinear* ichthyolitic member of the system, with the arenaceous beds that lie over it, form the continuous straight-lined ridge which runs on from beyond the moor of the Leys to beyond the moor of Culloden. There is a pretty little loch in this dwarf Highlands of the Brahan district, into which the old Celtic prophet Kenneth Ore, when, like Prospero, he relinquished his art, buried "deep beyond plummet sound" the magic stone in which he was wont to see the distant and the future. And with the loch it contains a narrow, hermit-like dell, bearing but a single row of fields, and these of small size, along its flat bottom, and whose steep gray sides of rustic Conglomerate resemble Cyclopean walls. It, besides, includes among its hills the steep hill of Knock Farril, which, rising bluff and bold immediately over the southern slopes of Strathpeffer, adds so greatly to the beauty of the valley, and bears atop perhaps the finest specimen of the vitrified fort in Scotland; and the bold frontage of cliff presented by the group to the west, over the pleasure grounds of Brahan, is, though on no very large scale, one of the most characteristic of the Conglomerate formation which can be seen anywhere. [Pg 395] It is formed of exactly such cliffs as the landscape gardener would make if he could,—cliffs with their rude prominent pebbles breaking the light over every square foot of surface, and furnishing footing, by their innumerable projections, to many a green tuft of moss, and many a sweet little flower. Some of the masses, too, that have rolled down from the precipices among the Brahan woods far below, and stand up, like the ruins of cottages, amid the trees, are of singular beauty,—worth all the imitation-ruins ever erected, and obnoxious to none of the disparaging associations which the mere show and make-believe of the artificial are sure always to awaken.

Whatever exhibited an aspect in any degree extraordinary was sure to attract the notice of the old Highlanders,—an acutely observant race, however slightly developed their reflective powers; and the great natural objects which excited their attention we always find associated with some traditionary story. It is said that in the Conglomerate cliffs above Brahan, a retainer of the Mackenzie, one of the smiths of the tribe, discovered a rich vein of silver, which he wrought by stealth, until he had filled one of the apartments of his cottage with bars and ingots. But the treasure, it is added, was betrayed by his own unfortunate vanity, to his chief, who hanged him in order to serve himself his heir; and no one since his death has proved ingenious enough to convert the rude rock into silver. Years had, I found, wrought their changes amid the miniature Highlands of the Conglomerate. The sapplings of the straggling wood on the banks of Loch Ousy,—the pleasant little lake, or lochan rather, of this upland region,—that I remembered having seen scarce taller than myself, had shot into vigorous treehood; and the steep slopes of Knock Farril, which I had left covered with their dark screen of pine, were now thickly mottled [Pg 396] over with half-decayed stumps, and bore that peculiarly barren aspect which tracts cleared of their wood so frequently assume in their transition state, when the plants that flourished in the shade have died out in consequence of the exposure, and plants that love the open air and the unbroken sunshine have not yet sprung up in their place. I found the southern acclivities of the hill covered with scattered masses of vitrified stone, that had fallen from the fortalice atop; and would recommend to the collector in quest of a characteristic specimen, that instead of laboring, to the general detriment of the pile, in detaching one from the walls above, he should set himself to seek one here. The blocks, uninjured by the hammer, exhibit, in most cases, the angular character of the original fragments better than those forcibly detached from the mass, and preserve in fine keeping those hollower interstices which were but partially filled with the molten matter, and which, when shattered by a blow, break through and lose their character.

One may spend an hour very agreeably on the green summit of Knock Farril. And at almost all seasons of the year a green summit it is,—greener considerably than any other hill-top in this part of the country. The more succulent grasses spring up rich and strong within the walls, here and there roughened by tufts of nettles, tall and rank, and somewhat perilous of approach,—witnesses, say the botanists, that man had once a dwelling in the immediate neighborhood. The green luxuriance which characterizes so many of the more ancient fortalices of Scotland seems satisfactorily accounted for by Dr. Fleming, in his "Zoölogy of the Bass." 'The summits and sides of those hills which were occupied by our ancestors as *hill-forts*," says the naturalist, "usually exhibit a far richer herbage than corresponding heights in the neighborhood [Pg 397] with the mineral soil derived from the same source. It is to be kept in view, that these positions of strength were at the same time occupied as *hill-folds*, into which, during the threatened or actual invasion of the district by a hostile tribe, the cattle were driven, especially during the night, as to places of safety, and sent out to pasture in the neighborhood during the day. And the droppings of these collected herds would, as takes place in analogous cases at present, speedily improve the soil to such an extent as to induce a permanent fertility." The further instance adduced by the Doctor, in showing through what protracted periods causes transitory in themselves may remain palpably influential in their effects, is curiously suggestive of the old metaphysical idea, that as every effect has its cause, "recurring from cause to cause up to the abyss of eternity, so every cause has also its effects, linked forward in succession to the end of time." On the bleak moor of Culloden the graves of the slain still exist as patches of green sward, surrounded by a brown groundwork of stunted heather. The animal matter,—once the nerves, muscles, and sinews of brave men,—which originated the change, must have been wholly dissipated ages ago. But the effect once produced has so decidedly maintained itself, that it remains not less distinctly stamped upon the heath in the present day than it could have been in the middle of the last century, only a few years after the battle had been stricken.

The vitrification of the rampart which on every side incloses the grassy area has been more variously, but less satisfactorily, accounted for than the green luxuriance within. It was held by Pennant to be an effect of volcanic fire, and that the walls of this and all our other vitrified strongholds are simply the crater-rims of extinct volcanoes,—a hypothesis wholly as untenable in reference to the [Pg 398] hill-forts as to the lime-kilns of the country: the vitrified forts are as little volcanic as the vitrified kilns. Williams, the author of the "Mineral Kingdom," and one of our earlier British geologists, after deciding, on data which his peculiar pursuits enabled him to collect and weigh, that they are *not* volcanic, broached the theory, still prevalent, as their name testifies, that they are artificial structures, in which vitrescency was designedly induced, in order to cement into solid masses accumulations of loose materials. Lord Woodhouselee advocated an opposite view. Resting on the fact that the vitrification is but of partial occurrence, be held that it had been produced, not of design by the builders of the forts, but in the process of their demolition by a besieging enemy, who, finding, as he premised, a large portion of the ramparts composed of wood, had succeeded in setting them on fire. This hypothesis, however, seems quite as untenable as that of Pennant. Fires not unfrequently occur in cities, among crowded groups of houses, where walls of stone are surrounded by a much greater profusion of dry woodwork than could possibly have entered into the composition of the ramparts of a hill-fort; but who ever saw, after a city fire, masses of wall from eight to ten feet in thickness fused throughout? The sandstone columns of the aisles of the Old Greyfriars in Edinburgh, surrounded by the woodwork of the galleries, the flooring, the seating, and the roof, were wasted, during the fire which destroyed the pile, into mere skeletons of their former selves; but though originally not more than three feet in diameter, they exhibited no marks of vitrescency. And it does not seem in the least probable that the stonework of the Knock Farril rampart could, if surrounded by wood at all, have been surrounded by an amount equally great, in proportion to its mass, as that which enveloped the aisle-columns of the Old Greyfriars.

[Pg 399]

The late Sir George Mackenzie of Coul adopted yet a fourth view. He held that the vitrification is simply an effect of the ancient beacon-fires kindled to warn the country of an invading enemy. But how account, on this hypothesis, for ramparts continuous, as in the case of Knock Farril, all round the hill? A powerful fire long kept up might well fuse a heap of loose stones into a solid mass; the bonfire lighted on the summit of Arthur Seat in 1842, to welcome the Queen on her first visit to Scotland, particularly fused numerous detached fragments of basalt, and imparted, in some spots to the depth of about half an inch, a vesicular structure to the solid rock beneath. But no fire, however powerful, could have constructed a rampart running without break for several hundred feet round an insulated hill-top. "To be satisfied," said Sir George, "of the reason why the signal-fires should be kindled on or beside a heap of stones, we have only to imagine a gale of wind to have arisen when a fire was kindled on the bare ground. The fuel would be blown about and dispersed, to the great annoyance of those who attended. The plan for obviating the inconvenience thus occasioned which would occur most naturally and readily would be to raise a heap of stones, on either side of which the fire might be placed to windward; and to account for the vitrification appearing all round the area, it is only necessary to allow the inhabitants of the country to have had a system of signals. A fire at one end might denote something different from a fire at the other, or in some intermediate part. On some occasions two or more fires might be necessary, and sometimes a fire along the whole line. It cannot be doubted," he adds, "that the rampart was originally formed with as much regularity as the nature of the materials would allow, both in order to render it more durable, and to make it serve the purposes of defence." This, I am [Pg 400] afraid, is still very unsatisfactory. A fire lighted along the entire line of a wall inclosing nearly an acre of area could not be other than a very attenuated, wire-drawn line of fire indeed, and could never possess strength enough to melt the ponderous mass of rampart beneath, as if it had been formed of wax or resin. A thousand loads of wood piled in a ring round the summit of Knock Farril, and set at once into a blaze, would wholly fail to affect the broad rampart below; and long ere even a thousand, or half a thousand, loads could have been cut down, collected, and fired, an invading enemy would have found time enough to moor his fleet and land his forces, and possess himself of the lower country. Again, the unbroken continuity of the vitrified line militates against the signal-system theory. Fire trod so closely upon the heels of fire, that the vitrescency induced by the one fire impinged on and mingled with the vitrescency induced by the others beside it. There is no other mode of accounting for the continuity of the fusion; and how could definite meanings possibly be attached to the various parts of a line so minutely graduated, that the centre of the fire kindled on any one graduation could be scarce ten feet apart from the centre of the fire kindled on any of its two neighboring graduations? Even by day, the exact compartment which a fire occupied could not be distinguished, at the distance of half a mile, from its neighboring compartments, and not at all by night, at any distance, from even the compartments farthest removed from it. Who, for instance, at the distance of a dozen miles or so, could tell whether the flame that shone out in the darkness, when all other objects around it were invisible, was kindled on the east or west end of an eminence little more than a hundred yards in length? Nay, who could determine,—for such is the requirement of the hypothesis,—whether it rose from a [Pg 401] compartment of the summit a hundred feet distant from its west or east end, or from a compartment merely ninety or a hundred and ten feet distant from it? The supposed signal system, added to the mere beacon hypothesis, is palpably untenable.

The theory of Williams, however, which is, I am inclined to think, the true one in the main, seems capable of being considerably modified and improved by the hypothesis of Sir George. The hill-fort,—palpably the most primitive form of fortalice or stronghold originated in a mountainous country,—seems to constitute man's first essay towards neutralizing, by the art of fortification, the advantages of superior force on the side of an assailing enemy. It was found, on the discovery of New Zealand, that the savage inhabitants had already learned to erect exactly such hill-forts amid the fastnesses of that country as those which were erected two thousand years earlier by the Scottish aborigines amid the fastnesses of our own. Nothing seems more probable, therefore, than that the forts of eminences such as Craig Phadrig and Knock Farril, originally mere inclosures of loose, uncemented stones, may belong to a period not less ancient than that of the first barbarous wars of Scotland, when, though tribe battled with tribe in fierce warfare, like the red men of the West with their brethren ere the European had landed on their shores, navigation was yet in so immature a state in Northern Europe as to secure to them an exemption from foreign invasion. In an after age, however, when the roving Vikings had become formidable, many of the eminences originally selected, from *their inaccessibility*, as sites for hill-forts, would come to be chosen, from *their prominence in the landscape*, as stations for beacon-fires. And of course the previously erected ramparts, higher always than the inclosed areas, would furnish on such hills [Pg 402] the conspicuous points from which the fires could be best seen. Let us suppose, then, that the rampart-crested eminence of Knock Farril, seen on every side for many miles, has become in the age of northern invasion one of the beacon-posts of the district, and that large fires, abundantly supplied with fuel by the woods of a forest-covered country, and blown at times into intense heat by the strong winds so frequent in that upper stratum of air into which the summit penetrates, have been kindled some six or eight times on some prominent point of the rampart, raised, mayhap, many centuries before. At first the heat has failed to tell on the stubborn quartz and feldspar which forms the preponderating material of the gneisses, granites, quartz rocks, and coarse conglomerate sandstones on which it has been brought to operate; but each fire throws down into the interstices a considerable amount of the fixed salt of the wood, till at length the heap has become charged with a strong flux; and then one powerful fire more, fanned to a white heat by a keen, dry breeze, reduces the whole into a semi-fluid mass. The same effects have been produced on the materials of the rampart by the beacon-fires and the alkali, that were produced, according to Pliny, by the fires and the soda of the Phœnician merchants storm-bound on the sands of the river Belus. But the state of civilization in Scotland at the time is not such as to permit of the discovery being followed up by similar results. The semi-savage guardians of the beacon wonder at the *accident*, as they well may; but those happy accidents in which the higher order of discoveries originate occur in only the ages of cultivated minds; and so they do not acquire from it the art of manufacturing glass. It could not fail being perceived, however, by intellects at all human, that the consolidation which the fires of one week, or month, or year, as the case [Pg 403] happened, had effected on one portion of the wall, might be produced by the fires of another week, or month, or year, on another portion of it; that, in short, a loose incoherent rampart, easy of demolition, might be converted, through the newly-discovered process, into a rampart as solid and indestructible as the rock on which it rested. And so, in course of time, simply by shifting the beacon-fires, and bringing them to bear in succession on every part of the wall, Knock Farril, with many a similar eminence in the country, comes to exhibit its completely vitrified fort where there had been but a loosely-piled hill-fort before. It in no degree militates against this compound theory,—borrowed in part from Williams and in part from Sir George,—that there are detached vitrified masses to be found on eminences evidently never occupied by hill-forts; or that there are hill-forts on other eminences only partially fused, or hill-forts on many of the less commanding sites that bear about them no marks of fire at all. Nothing can be more probable than that in the first class of cases we have eminences that had been selected as beacon-stations, which had not previously been occupied by hill-forts; and in the last, eminences that had been occupied by hill-forts which, from their want of prominence in the general landscape, had not been selected as beacon-stations. And in the intermediate class of cases we have probably ramparts that were only partially vitrified, because some want of fuel in the neighborhood had starved the customary fires, or because fires had to be less frequently kindled upon them than on the more important stations; or, finally, because these hill-forts, from some disadvantage of situation, were no longer used as places of strength, and so the beacon-keepers had no motive to attempt consolidating them throughout by the piecemeal application of the vitrifying agent. But the old Highland [Pg 404] mode of accounting for the present appearance of Knock Farril and its vitrified remains is perhaps, after all, quite as good in its way as any of the modes suggested by the philosophers. [20]

I spent some time, agreeably enough, beside the rude rampart of Knock Farril, in marking the various appearances exhibited by the fused and semi-fused materials of which it is composed,—the granites, gneisses, mica-schists, hornblendes, clay-slates, and red sandstones of the locality. One piece of rock, containing much lime, I found resolved into a yellow opaque substance, not unlike the coarse earthenware used in the making of ginger-beer bottles; but though it had been so completely molten that it had dropped into a hollow beneath in long viscid trails, it did not contain a single air-vesicle; while another specimen, apparently a piece of fused mica-schist, was so filled with air-cells, that the dividing partitions were scarcely the tenth of a line in thickness. I found bits of schistose gneiss resolved into green glass; the Old Red Sandstone basis of the Conglomerate, which forms the hill, into a semi-metallic scoria, like that of an iron-smelter's furnace; mica into a gray, waxy-looking stone, that scratched glass; and pure white quartz into porcellanic trails of white, that ran in one instance along the face of a darker-colored rock below, like streaks of cream along the sides of a burnt china jug. In one mass of pale large-grained granite I found that the feldspar, though it had acquired a vitreous gloss on the surface, still retained its peculiar rhomboidal cleavage; while the less stubborn quartz around it had become scarce less vesicular and light than a piece of pumice. On some of the other masses there was impressed, as if by a seal, the stamp [Pg 405] of pieces of charcoal; and so sharply was the impression retained, that I could detect on the vitreous surface the mark of the yearly growths, and even of the medullary rays, of the wood. In breaking open some of the others, I detected fragments of the charcoal itself, which, hermetically locked up in the rock, had retained all its original carbon. These last reminded me of specimens not unfrequent among the trap-rocks of the Carboniferous and Oölitic systems. From an intrusive overlying wacke in the neighborhood of Linlithgow I have derived for my collection pieces of carbonized wood in so complete a state of keeping, that under the microscope they exhibit unbroken all the characteristic reticulations of the coniferæ of the Coal Measures.

I descended the hill, and, after joining my friends at Strathpeffer,—Buchubai Hormazdji among the rest,—visited the Spa, in the company of my old friend the minister of Alness. The thorough identity of the powerful effluvium that fills the pump-room with that of a muddy sea-bottom laid bare in warm weather by the tide, is to the dweller on the sea-coast very striking. It *is* identity,—not mere resemblance. In most cases the organic substances undergo great changes in the bowels of the earth. The animal matter of the Caithness ichthyolites exists, for instance, as a hard, black, insoluble bitumen, which I have used oftener than once as sealing-wax; the vegetable mould of the Coal Measures has been converted into a fire-clay, so altered in the organic pabulum, animal and vegetable, whence it derived its fertility, that, even when laid open for years to the meliorating effects of the weather and the visits of the winged seeds, it will not be found bearing a single spike or leaf of green. But here, in smell, at least, that ancient mud, swum over by the Diplopterus and the Diplacanthus, and in which the Coccosteus and Pterichthys burrowed, has undergone no change. The soft ooze has become solid rock, but [Pg 406] its odoriferous qualities have remained unaltered. I next visited an excavation a few hundred yards on the upper side of the pump-room, in which the gray fetid breccia of the Strath has been quarried for dyke-building, and examined the rock with some degree of care, without, however, detecting in it a single plate or scale. Lying over that Conglomerate member of the system which, rising high in the Knock Farril range, forms the southern boundary of the valley, it occupies the place of the lower ichthyolitic bed, so rich in organisms in various other parts of the country; but here the bed, after it had been deposited in thin horizontal laminæ, and had hardened into stone, seems to have been broken up, by some violent movement, into minute sharp-edged fragments, that, without wear or attrition, were again consolidated into the breccia which it now forms. And its ichthyolites, if not previously absorbed, were probably destroyed in the convulsion. Detached scales and spines, however, if carefully sought for in the various openings of the valley, might still be found in the original laminæ of the fragments. They must have been amazingly abundant in it once; for so largely saturated is the rock with the organic matter into which they have been resolved, that, when struck by the hammer, the impalpable dust set loose sensibly affects the organs of taste, and appeals very strongly to those of smell. It is through this saturated rock that the mineral springs take their course. Even the surface-waters of the valley, as they pass over it contract in a perceptible degree its peculiar taste and odor. With a little more time to spare, I would fain have made this breccia of the Old Red the subject of a few simple experiments. I would have ground it into powder, and tried upon it the effect both of cold and hot infusion. Portions of the water are sometimes carried in casks and bottles, for the use of invalids, to a considerable distance; but it is quite possible that [Pg 407] a little of the *rock*, to which the water owes its qualities, might, when treated in this way, have all the effects of a considerable quantity of the *spring*. It might be of some interest, too, to ascertain its qualities when crushed, as a soil, or its effect on other soils; whether, for instance, like the old sterile soils of the Carboniferous period, it has lost, through its rock-change, the fertilizing properties which it once possessed; or whether it still retains them, like some of the coprolitic beds of the Oölite and Greensand, and might not, in consequence, be employed as a manure. A course of such experiments could scarce fail to furnish with agreeable occupation some of the numerous annual visitants of the Spa, who have to linger long, with but little to engage them, waiting for what, if it once fairly leave a man, returns slowly, when it returns at all.

In mentioning at the dinner-table of my friend my scheme of infusing rock in order to produce Spa water, I referred to the circumstance that the Belemnite of our Liasic deposits, when ground into powder, imparts to boiling water a peculiar taste and smell, and that the infusion, taken in very small quantities, sensibly affects both palate and stomach. And I suggested that Belemnite water, deemed sovereign of old, when the Belemnite was regarded as a thunderbolt, in the cure of bewitched cattle, might be in reality medicinal, and that the ancient superstition might thus embody, as ancient superstitions not unfrequently do, a nucleus of fact. The charm, I said, might amount to no more than simply the administration of a medicine to sick cattle, that did harm in no case, and good at times. The lively comment of one of the young ladies on the remark amused us all. If an infusion of stone had cured, in the last age, cattle that were bewitched, the Strathpeffer water, she argued, which was, it seems, but an infusion of stone, might cure cattle that were sick now; and so, though [Pg 408] the biped patients of the Strath could scarce fail to decrease when they knew that its infused stone contained but the strainings of old mud, and the juices of dead unsalted fish, it was gratifying to think that the poor Spa might still continue to retain its patients, though of a lower order. The pump-room would be converted into a rustic, straw-thatched shed, to which long trains of sick cattle, affected by weak nerves and dyspepsia, would come streaming along the roads every morning and evening, to drink and gather strength.

The following morning was wet and lowering, and a flat ceiling of gray cloud stretched across the valley, from the summit of the Knock Farril ridge of hills on the one side, to the lower flanks of Ben-Wevis on the other. I had purposed ascending this latter mountain,—the giant of the north-eastern coast, and one of the loftiest of our second-class Scottish hills anywhere,—to ascertain the extreme upper line at which travelled boulders occur in this part of the country. But it was no morning for wading knee-deep through the trackless heather; and after waiting on, in the hope the weather might clear up, watching at a window the poorer invalids at the Spa, as they dragged themselves through the rain to the water, I lost patience, and sallied out, beplaided and umbrellaed, to see from the top of Knock Farril how the country looked in a fog. At first, however, I saw much fog, but little country; but as the day wore on, the flat mist-ceiling rose together, till it rested on but the distant hills, and the more prominent features of the landscape began to stand out amid the more general gray, like the stronger lines and masses in a half-finished drawing, boldly dashed off in the neutral tint of the artist. The portions of the prospect generically distinct are, notwithstanding its great extent and variety, but few; and the partial veil of haze, [Pg 409] by glazing down its distracting multiplicity of minor points, served to bring them out all the more distinctly. There is, first stretching far in a southern and eastern direction along the landscape, the rectilinear ridge of the Black Isle,—not quite the sort of line a painter would introduce into a composition, but true to geologic character. More in the foreground, in the same direction, there spreads a troubled cockling sea of the Great Conglomerate. Turning to the north and west, the deep valley of Strathpeffer, with its expanse of rich level fields, and in the midst its old baronial castle, surrounded by coeval trees of vast bulk, lies so immediately at the foot of the eminence, that I could hear in the calm the rush of the little stream, swollen to thrice its usual bulk by the rains of the night. Beyond rose the thick-set Ben-Wevis,—a true gneiss mountain, with breadth enough of shoulders, and amplitude enough of base, to serve a mountain thrice as tall, but which, like all its cogeners of this ancient formation, was arrested in its second stage of growth, so that many of the slimmer granitic and porphyritic hills of the country look down upon it, as Agamemnon, according to Homer, looked down upon Ulysses.

"Broad is his breast, his shoulders larger spread,
Though great Atrides overtops his head."

All around, as if topling, wave-like, over the outer edges of the comparatively flat area of Palæozoic rock which composes the middle ground of the landscape, rose a multitude of primary hill-peaks, barely discernible in the haze; while the long withdrawing Dingwall Frith, stretching on towards the open sea for full twenty miles, and flanked on either side by ridges of sandstone, but guarded at the opening by two squat granitic columns, completed the [Pg 410] prospect, by adding to its last great feature. All was gloomy and chill; and as I turned me down the descent, the thick wetting drizzle again came on; and the mist-wreaths, after creeping upwards along the hill-side, began again to creep down. When I had first visited the valley, more than a quarter of a century before, it was on a hot breathless day of early summer, in which, though the trees in fresh leaf seemed drooping in the sunshine, and the succulent luxuriance of the fields lay aslant, half-prostrated by the fierce heat, the rich blue of Ben-Wevis, far above, was thickly streaked with snow, on which it was luxury even to look. It gave one iced fancies, wherewithal to slake, amid the bright glow of summer, the thirst in the mind. The recollection came strongly upon me, as the fog from the hill-top closed dark behind, like that sung by the old blind Englishman, which

"O'er the marish glides,
And gathers ground fast at the lab'rer's heel,
Homeward returning."

But the contrast had nothing sad in it; and it was pleasant to feel that it had not. I had resigned many a baseless hope and many an idle desire since I had spent a vacant day amid the sunshine, now gazing on the broad placid features of the snow-streaked mountain; and now sauntering under the tall ancient woods, or along the heath-covered slopes of the valley; but in relation to never-tiring, inexhaustible nature, the heart was no fresher at that time than it was now. I had grown no older in my feelings or in my capacity of enjoyment; and what then was there to regret?

I rode down the Strath in an omnibus which plies between the Spa and Dingwall, and then walked on to the village of Evanton, which I reached about an hour after [Pg 411] nightfall, somewhat in the circumstances of the "damp stranger," who gave Beau Brummel the cold. There were, however, no Beau Brummels in the quiet village inn in which I passed the night, and so the effects of the damp were wholly confined to myself. I was soundly pummelled during the night by a frightful female, who first assumed the appearance of the miserable pauper woman whom I had seen beside the Auldgrande, and then became the Lady of Balconie; and, though sufficiently indignant, and much inclined to resist, I could stir neither hand nor foot, but lay passively on my back, jambed fast behind the huge gneiss boulder and the edge of the gulf. And yet, by a strange duality of perception, I was conscious all the while that, having got wet on the previous day, I was now suffering from an attack of nightmare: and held that it would be no very serious matter even should the lady tumble me into the gulf, seeing that all would be well again when I awoke in the morning. Dreams of this character, in which consciousness bears reference at once to the fictitious events of the vision and the real circumstances of the sleeper, must occupy, I am inclined to think, very little time,—single moments, mayhap, poised midway between the sleeping and waking state. Next day (Sunday) I attended the Free Church in the parish, where I found a numerous and attentive congregation,—descendants, in large part, of the old devout Munroes of Ferindonald,—and heard a good solid discourse. And on the following morning I crossed the sea at what is known as the Fowlis Ferry, to explore, on my homeward route, the rocks laid bare along the shore in the upper reaches of the Frith.

I found but little by the way: black patches of bitumen in the sandstone of one of the beds, with a bed of stratified clay, inclosing nodules, in which, however, I succeeded in detecting nothing organic; and a few fragments of clay-slate [Pg 412] locked up in the Red Sandstone, sharp and unworn at their edges, as if derived from no great distance, though there be now no clay-slate in the eastern half of Ross; but though the rocks here belong evidently to the ichthyolitic member of the Old Red, not a single fish, not a "nibble" even, repaid the patient search of half a day. I, however, passed some time agreeably enough among the ruins of Craighouse. When I had last seen, many years before, this old castle, [21] the upper stories were accessible; but they were now no longer so. Time, and the little herdboys who occasionally shelter in its vaults, had been busy in the interval; and, by breaking off a few projecting corners by which the climber had held, and by effacing a few notches into which he had thrust his toe-points, they had rendered what had been merely difficult impracticable. I remarked that the huge kitchen chimney of the building,—a deep hollow recess which stretches across the entire gable, and in which, it is said, two thrashers once plied the flail for a whole winter,—bore less of the stain of recent smoke than it used to exhibit twenty years before; and inferred that there would be fewer wraith-lights seen from the castle at nights than in those days of *evil spirits* and illicit stills, when the cottars in the neighborhood sent more smuggled whiskey to market than any equal number of the inhabitants of almost any other district in the north. It has been long alleged that there existed a close connection between the more ghostly spirits of the country and its distilled ones. "How do you account," said a north country minister of the last age (the late Rev. Mr. M'Bean of Alves) to a sagacious old elder of his Session, "for the almost total disappearance of the ghosts and fairies that used to be so common in your young days?" "Tak my word for 't, minister," replied the shrewd old man, "it's a' owing to [Pg 413] the *tea*; when the *tea* cam in, the ghaists an' fairies gaed out. Weel do I mind when at a' our neeborly meetings,—bridals, christenings, lyke-wakes, an' the like,—we entertained ane anither wi' rich nappy ale; an' whan the verra dowiest o' us used to get warm i' the face, an' a little confused in the head, an' weel fit to see amaist onything whan on the muirs on our way hame. But the tea has put out the nappy; an' I have remarked, that by losing the nappy we lost baith ghaists an' fairies."

Quitting the ruin, I walked on along the shore, tracing the sandstone as I went, as it rises from lower to higher beds; and where it ceases to crop out at the surface, and gravel and the red boulder-clays take the place of rock, I struck up the hill, and, traversing the parishes of Resolis and Cromarty, got home early in the evening. I had seen and done scarcely half what I had intended seeing or doing: alas, that in reference to every walk which I have yet attempted to tread, this special statement should be so invariably true to fact!—alas, that all my full purposes, should be coupled with but half realizations! But I had at least the satisfaction, that though I had accomplished little, I had enjoyed much; and it is something, though not all, nor nearly all, that, since time is passing, it should pass happily. In my next chapter I shall enter on my tour to Orkney. It dates one year earlier (1846) than the tour with which I have already occupied so many chapters; but I have thus inverted the order of *time*, by placing it last, that I may be able so to preserve the order of *space* as to render the tract travelled over in my narrative continuous from Edinburgh to the northern extremity of Pomona.

[Pg 414]

CHAPTER X.

Recovered Health—Journey to the Orkneys—Aboard the Steamer at Wick—Mr. Bremner—Masonry of the Harbor of Wick—The greatest Blunders result from good Rules misapplied—Mr. Bremner's Theory about sea-washed Masonry—Singular Fracture of the Rock near Wick—The Author's mode of accounting for it—"Simple but not obvious" Thinking—Mr. Bremner's mode of making stone Erections under Water—His exploits in raising foundered Vessels—Aspect of the Orkneys—- The ungracious Schoolmaster—In the Frith of Kirkwall—Cathedral of St. Magnus—Appearance of Kirkwall—Its "perished suppers"—Its ancient Palaces—Blunder of the Scotch Aristocracy—The patronate Wedge—Breaking Ground in Orkney—Minute gregarious Coccosteus—True Position of the Coccosteus' Eyes—Ruins of one of Cromwell's Forts—Antiquities of Orkney—The Cathedral—Its Sculptures—The Mysterious Cell—Prospect from the Tower—Its Chimes—Ruins of Castle Patrick.

A twelvemonth had gone by since a lingering indisposition, which bore heavily on the springs of life, compelled me to postpone a long-projected journey to the Orkneys, and led me to visit, instead, rich level England, with its well-kept roads and smooth railways, along which the enfeebled invalid can travel far without fatigue. I had now got greatly stronger; and, if not quite up to my old thirty miles per day, nor altogether so bold a cragsman as I had been only a few years before, I was at least vigorous enough to enjoy a middling long walk, and to breast a tolerably steep hill. And so I resolved on at least glancing over, if not exploring, the fossiliferous deposits of the Orkneys, trusting that an eye somewhat practised in the formations mainly developed in these islands might enable me to make some amends for seeing comparatively little, by seeing well. I took coach at Invergordon for Wick early in the morning of Friday; and, after a weary ride, in [Pg 415] a bleak gusty day, that sent the dust of the road whirling about the ears of the screly-tossed "outsides," with whom I had taken my chance, I alighted in Wick, at the inn-door, a little after six o'clock in the evening. The following morning was wet and dreary; and a tumbling sea, raised by the wind of the previous day and night, came rolling into the bay; but the waves bore with them no steamer; and when, some five hours after the expected time, she also came rolling in, her darkened and weather-beaten sides and rigging gave evidence that her passage from the south had been no holiday trip. Impatient, however, of looking out upon the sea for hours, from under dripping eaves, and through the dimmed panes of streaming windows, I got aboard with about half-a-dozen other passengers; and while the Wick goods were in the course of being transferred to two large boats alongside, we lay tossing in the open bay. The work of raising box and package was superintended by a tall elderly gentleman from the shore, peculiarly Scotch in his appearance,—the steam company's agent for this part of the country.

"That," said an acquaintance, pointing to the agent, "is a very extraordinary man,—in his own special walk, one of the most original-minded, and at the same time most thoroughly practical, you perhaps ever saw. That is Mr. Bremner of Wick, known now all over Britain for his success in raising foundered vessels, when every one else gives them up. In the lifting of vast weights, or the overcoming the *vis inertiæ* of the hugest bodies, nothing ever baffles Mr. Bremner. But come, I must introduce you to him. He takes an interest in your peculiar science, and is familiar with your geological writings."

I was accordingly introduced to Mr. Bremner, and passed, in his company the half-hour which we spent in the bay, in a way that made me wish the time doubled. I had been [Pg 416] struck by the peculiar style of masonry employed in the harbor of Wick, and by its rock-like strength. The gray ponderous stones of the flagstone series of which it is built, instead of being placed on their flatter beds, like common ashlar in a building, or horizontal strata in a quarry, are raised on end, like staves in a pail or barrel, so that at some little distance the work looks as if formed of upright piles or beams jambed fast together. I had learned that Mr. Bremner had been the builder, and adverted to the peculiarity of his style of building. "You have given a vertical tilt to your strata," I said: "most men would have preferred the horizontal position. It used to be regarded as one of the standing rules of my old profession, that the 'broad bed of a stone' is the best, and should be always laid 'below.'" "A good rule for the land," replied Mr. Bremner, "but no good rule for the sea. The greatest blunders are almost always perpetrated through the misapplication of good rules. On a coast like ours, where boulders of a ton weight are rolled about with every storm like pebbles, these stones, if placed on what a workman would term their best beds, would be scattered along the shore like sea-wrack, by the gales of a single winter. In setting aside the prejudice," continued Mr. Bremner, "that what is indisputably the best bed for a stone on dry land is also the best bed in the water on an exposed coast, I reasoned thus:—The surf that dashes along the beach in times of tempest, and that forms the enemy with which I have to contend, is not simply water, with an onward impetus communicated to it by the wind and tide, and a reäctive impetus in the opposite direction,—the effect of the backward rebound, and of its own weight, when raised by these propelling forces above its average level of surface. True, it is all this; but it is also something more. As its white breadth of foam indicates, it is a subtile mixture of [Pg 417] water and *air*, with a powerful *upward* action,—a consequence of the air struggling to effect its escape; and this upward action must be taken into account in our calculations, as certainly as the other and more generally recognized actions. In striking against a piece of building, this subtile mixture dashes through the interstices into the interior of the masonry, and, filling up all its cavities, has by its upward action, a tendency to *set the work afloat*. And the broader the beds of the stones, of course the more extensive are the surfaces which it has to act upon. One of these flat flags, ten feet by four, and a foot in thickness, would present to this upheaving force, if placed on end, a superficies of but *four* square feet; whereas, if placed on its broader base, it would present to it a superficies of *forty* square feet. Obviously, then, with regard to this aërial upheaving force, that acts upon the masonry in a direction in which no precautions are usually adopted to bind it fast,—for the existence of the force itself is not taken into account,—the greater bed of the stone must be just ten times over a worse bed than its lesser one; and on a tempestuous foam-encircled coast such as ours, this aërial upheaving force is in reality, though the builder may not know it, one of the most formidable forces with which he had to deal. And so, on these principles, I ventured to set my stones on end,—on what was deemed their *worst*, not their *best* beds,—wedging them all fast together, like staves in an anker; and there, to the scandal of all the old rules, are they fast wedged still, firm as a rock." It was no ordinary man that could have originated such reasonings on such a subject, or that could have thrown himself so boldly, and to such practical effect, on the conclusions to which they led.

Mr. Bremner adverted, in the course of our conversation, to a singular appearance among the rocks a little to [Pg 418] the east and south of the town of Wick, that had not, he said, attracted the notice it deserved. The solid rock had been fractured by some tremendous blow, dealt to it externally at a considerable height over the sea-level, and its detached masses scattered about like the stones of an ill-built harbor broken up by a storm. The force, whatever its nature, had been enormously great. Blocks of some thirty or forty tons weight had been torn from out the solid strata, and piled up in ruinous heaps, as if the compact precipice had been a piece of loose brickwork, or had been driven into each other, as if, instead of being composed of perhaps the hardest and toughest sedimentary rock in the country, they had been formed of sun-dried clay. "I brought," continued Mr. Bremner, "one of your itinerant geological lecturers to the spot, to get his opinion; but he could say nothing about the appearance: it was not in his books." "I suspect," I replied, "the phenomenon lies quite as much within your own province as within that of the geological lecturer. It is in all probability an illustration, on a large scale, of those floating forces with which you operate on your foundered vessels, joined to the forces, laterally exerted, by which you drag them towards the shore. When the sea stood higher, or the land lower, in the eras of the raised beaches, along what is now Caithness, the abrupt mural precipices by which your coast here is skirted must have secured a very considerable depth of water up to the very edge of the land;—your coast-line must have resembled the side of a mole or wharf: and in that glacial period to which the thick deposit of boulder-clay immediately over your harbor yonder belongs, icebergs of very considerable size must not unfrequently have brushed the brows of your precipices. An iceberg from eighty to a hundred feet in thickness, and perhaps half a square mile in area, could [Pg 419] not, in this old state of things, have come in contact with these cliffs without first catching the ground outside; and such an iceberg, propelled by a fierce storm from the north-east, could not fail to lend the cliff with which it came in collision a tremendous blow. You will find that your shattered precipice marks, in all probability, the scene of a collision of this character: some hard-headed iceberg must have set itself to run down the land, and got wrecked upon it for its pains." My theory, though made somewhat in the dark,—for I had no opportunity of seeing the broken precipice until after my return from Orkney,—seemed to satisfy Mr. Bremner; nor, on a careful survey of the phenomenon, the solution of which it attempted, did I find occasion to modify or give it up.

With just knowledge enough of Mr. Bremner's peculiar province to appreciate his views, I was much impressed by their broad and practical simplicity; and bethought me, as we conversed, that the character of the thinking, which, according to Addison, forms the staple of all writings of genius, and which he defines as "simple but not obvious," is a character which equally applies to *all* good thinking, whatever its special department. Power rarely resides in ingenious complexities: it seems to eschew in every walk the elaborately attenuated and razor-edged mode of thinking,—the thinking akin to that of the old metaphysical poets,—and to select the broad and massive style. Hercules, in all the representations of him which I have yet seen, is the *broad* Hercules. I was greatly struck by some of Mr. Bremner's views on deep-sea founding. He showed me how, by a series of simple, but certainly not obvious contrivances, which had a strong air of practicability about them, he could lay down his erection, course by course, inshore, in a floating caisson of peculiar construction, beginning a little beyond the low-ebb line, and warping out his [Pg 420] work piecemeal, as it sank, till it had reached its proper place, in, if necessary, from ten to twelve fathoms water, where, on a bottom previously prepared for it by the diving-bell, he had means to make it take the ground exactly at the required line. The difficulty and vast expense of building altogether by the bell would be obviated, he said, by the contrivance, and a solidity given to the work otherwise impossible in the circumstances: the stones could be laid in his floating caisson with a care as deliberate as on the land. Some of the anecdotes which he communicated to me on this occasion, connected with his numerous achievements in weighing up foundered vessels, or in floating off wrecked or stranded ones, were of singular interest; and I regretted that they should not be recorded in an autobiographical memoir. Not a few of them were humorously told, and curiously illustrative of that general ignorance regarding the "strength of materials" in which the scientific world has been too strangely suffered to lie, in this the world's most mechanical age; so that what ought to be questions of strict calculation are subjected to the guessings of a mere common sense, far from adequate, in many cases, to their proper resolution. "I once raised a vessel," said Mr. Bremner,—"a large collier, chock-full of coal,—which an English projector had actually engaged to raise with huge bags of India rubber, inflated with air. But the bags, of course taxed far beyond their strength, collapsed or burst; and so, when I succeeded in bringing the vessel up, through the employment of more adequate means, I got not only ship and cargo, but also a great deal of good India rubber to boot." Only a few months after I enjoyed the pleasure of this interview with the Brindley of Scotland, he was called south, to the achievement of his greatest feat in at least one special department,—a feat generally recognized and appreciated as the most herculean [Pg 421] of its kind ever performed,—the raising and warping off of the Great Britain steamer from her perilous bed in the sand of an exposed bay on the coast of Ireland. I was conscious of a feeling of sadness as, in parting with Mr. Bremner, I reflected, that a man so singularly gifted should have been suffered to reach a period of life very considerably advanced, in employments little suited to exert his extraordinary faculties, and which persons of the ordinary type could have performed as well. Napoleon,—himself possessed of great genius,—could have estimated more adequately than our British rulers the value of such a man. Had Mr. Bremner been born a Frenchman, he would not now be the mere agent of a steam company, in a third-rate seaport town.

The rain had ceased, but the evening was gloomy and chill; and the Orcades, which, on clearing the Caithness coast, came as fully in view as the haze permitted, were enveloped in an undress of cloud and spray, that showed off their flat low features to no advantage at all. The bold, picturesque Hebrides look well in any weather; but the level Orkney Islands, impressed everywhere, on at least their eastern coasts, by the comparatively tame character borne by the Old Red flagstones, when undisturbed by trap or the primary rocks, demand the full-dress auxiliaries of bright sun and clear sky, to render their charms patent. Then, however, in their sleek coats of emerald and purple, and surrounded by their blue sparkling sounds and seas, with here a long dark wall of rock, that casts its shadow over the breaking waves, and there a light fringe of sand and broken shells, they are, as I afterwards ascertained, not without their genuine beauties. But had they shared in the history of the neighboring Shetland group, that, according to some of the older historians, were suffered to lie uninhabited for centuries after their first discovery, I [Pg 422] would rather have been disposed to marvel this evening, not that they had been unappropriated so long, but that they had been appropriated at all. The late member for Orkney, not yet unseated by his Shetland opponent, was one of the passengers in the steamboat; and, with an elderly man, an ambitious schoolmaster, strongly marked by the peculiarities of the genuine dominie, who had introduced himself to him as a brother voyager, he was pacing the quarter-deck, evidently doing his best to exert, under an unintermittent hot-water *douche* of queries, the patient courtesy of a Member of Parliament on a visit to his constituency. At length, however, the troubler quitted him, and took his stand immediately beside me; and, too sanguinely concluding that I might take the same kind of liberty with the schoolmaster that the schoolmaster had taken with the Member, I addressed to him a simple query in turn. But I had mistaken my man; the schoolmaster permitted to unknown passengers in humble russet no such sort of familiarities as those permitted by the Member; and so I met with a prompt rebuff, that at once set me down. I was evidently a big, forward lad, who had taken a liberty with the master. It is, I suspect, scarce possible for a man, unless naturally very superior, to live among boys for some twenty or thirty years, exerting over them all the while a despotic authority, without contracting those peculiarities of character which the master-spirits,—our Scots, Lambs, and Goldsmiths,—have embalmed with such exquisite truth in our literature, and which have hitherto militated against the practical realization of those unexceptionable abstractions in behalf of the status and standing of the teacher of youth which have been originated by men less in the habit of looking about them than the poets. It is worth while remarking how invariably the strong common sense of the Scotch people [Pg 423] has run every scheme under water that, confounding the character of the "village schoolmaster" with that of the "village clergyman," would demand from the schoolmaster the clergyman's work.

We crossed the opening of the Pentland Frith, with its white surges and dark boiling eddies, and saw its twin lighthouses rising tall and ghostly amid the fog on our lee. We then skirted the shores of South Ronaldshay, of Burra, of Copinshay, and of Deerness; and, after doubling Moul Head, and threading the sound which separates Shapinshay from the Mainland, we entered the Frith of Kirkwall, and caught, amid the uncertain light of the closing evening, our earliest glimpse of the ancient Cathedral of St. Magnus. It seems at first sight as if standing solitary, a huge hermit-like erection, at the bottom of a low bay,—for its humbler companions do not make themselves visible until we have entered the harbor by a mile or two more, when we begin to find that it occupies, not an uninhabited tract of shore, but the middle of a gray straggling town, nearly a mile in length. We had just light enough to show us, on landing, that the main thoroughfare of the place, very narrow and very crooked, had been laid out, ere the country beyond had got highways, or the proprietors carts and carriages, with an exclusive eye to the necessities of the foot-passenger,—that many of the older houses presented, as is common in our northern towns, their gables to the street, and had narrow slips of closes running down along their fronts,—and that as we receded from the harbor, a goodly portion of their number bore about them an air of respectability, long maintained, but now apparently touched by decay. I saw, in advance of one of the buildings, several vigorous-looking planes, about forty feet in height, which, fenced by tall houses in front and rear, and flanked by the tortuosities of the street, had apparently forgotten that they were [Pg 424] in Orkney, and had grown quite as well as the planes of public thoroughfares grow elsewhere. After an abortive attempt or two made in other quarters, I was successful in procuring lodgings for a few days in the house of a respectable widow lady of the place, where I found comfort and quiet on very moderate terms. The cast of faded gentility which attached to so many of the older houses of Kirkwall,—remnants of a time when the wealthier Udallers of the Orkneys used to repair to their capital at the close of autumn, to while away in each other's society their dreary winters,—reminded me of the poet Malcolm's "Sketch of the Borough,"—a portrait for which Kirkwall is known to have sat,—and of the great revolution effected in its evening parties, when "tea and turn-out" yielded its place to "tea and turn-in." But the church-yard of the place, which I had seen, as I passed along, glimmering with all its tombstones in the uncertain light, was all that remained to represent those "great men of the burgh," who, according to the poet, used to "pop in on its card and dancing assemblies, about the eleventh hour, resplendent in top-boots and scarlet vests," or of its "suppression-of-vice sisterhood of moral old maids," who kept all their neighbors right by the terror of their tongues. I was somewhat in a mood, after my chill and hungry voyage, to recall with a hankering of regret the vision of its departed suppers, so luxuriously described in the "Sketch,"—suppers at which "large rounds of boiled beef smothered in cabbage, smoked geese, mutton hams, roasts of pork, and dishes of dog-fish and of Welsh rabbits melted in their own fat, were diluted by copious draughts of strong home-brewed ale, and etherealized by gigantic bowls of rum punch." But the past, which is not ours, who, alas, can recall! And, after discussing a juicy steak and a modest [Pg 425] cup of tea, I found I could regard with the indifferency of a philosopher, the perished suppers of Kirkwall.

I quitted my lodgings for church next morning about three-quarters of an hour ere the service commenced; and, finding the doors shut, sauntered up the hill that rises immediately over the town. The thick gloomy weather had passed with the night; and a still, bright, clear-eyed Sabbath looked cheerily down on green isle and blue sea. I was quite unprepared by any previous description, for the imposing assemblage of ancient buildings which Kirkwall presents full in the foreground, when viewed from the road which ascends along this hilly slope to the uplands. So thickly are they massed together, that, seen from one special point of view, they seem a portion of some magnificent city in ruins,—some such city though in a widely different style of architecture, as Palmyra or Baalbec. The Cathedral of St. Magnus rises on the right, the castle-palace of Earl Patrick Stuart on the left, the bishop's palace in the space between; and all three occupy sites so contiguous, that a distance of some two or three hundred yards abreast gives the proper angle for taking in the whole group at a glance. I know no such group elsewhere in Scotland. The church and palace of Linlithgow are in such close proximity, that, seen together, relieved against the blue gleam of their lake, they form one magnificent pile; but we have here a taller, and, notwithstanding its Saxon plainness, a nobler church, than that of the southern burgh, and at least one palace more. And the associations connected with the church, and at least one of the palaces ascend to a remoter and more picturesque antiquity. The castle-palace of Earl Patrick dates from but the time of James the Sixth; but in the palace of the bishop, old grim Haco died, after his defeat at Largs, "of grief," says Buchanan, "for the loss of his army, and of a valiant youth his relation;" and in the ancient Cathedral, [Pg 426] his body, previous to its removal to Norway, was interred for a winter. The church and palace belong to the obscure dawn of the national history, and were Norwegian for centuries before they were Scotch.

As I was coming down the hill at a snail's pace, I was overtaken by a countryman on his way to church. "Ye'll hae come," he said, addressing me, "wi' the great man last night?" "I came in the steamer," I replied, "with your Member, Mr. Dundas." "O, aye," rejoined the man; "but I'm no sure he'll be our Member next time. The Voluntaries yonder, ye see," jerking his head, as he spoke, in the direction of the United Secession chapel of the place, "are awfu' strong and unco radical; and the Free Kirk folk will soon be as bad as them. But I belong to the Establishment; and I side wi' Dundas." The aristocracy of Scotland committed, I am afraid, a sad blunder when they attempted strengthening their influence as a class by seizing hold of the Church patronages. They have fared somewhat like those sailors of Ulysses who, in seeking to appropriate their master's wealth, let out the winds upon themselves; and there is now, in consequence, a perilous voyage and an uncertain landing before them. It was the patronate wedge that struck from off the Scottish Establishment at least nine-tenths of the Dissenters of the kingdom,—its Secession bodies, its Relief body, and, finally, its Free Church denomination,—comprising in their aggregate amount a great and influential majority of the Scotch people. Our older Dissenters,—a circumstance inevitable to their position as such,—have been thrown into the movement party: the Free Church, in her present transition state, sits loose to all the various political sections of the country; but her natural tendency is towards the movement party also; and already, in consequence, do our Scottish aristocracy possess greatly less political influence in the kingdom of which they [Pg 427] own almost all the soil, than that wielded by their brethren the Irish and English aristocracy in their respective divisions of the empire. Were the representation of England and Ireland as liberal as that of Scotland, and as little influenced by the aristocracy, Conservatism, on the passing of the Reform Bill, might have taken leave of office for evermore. And yet neither the English nor Irish are naturally so Conservative as the Scotch. The patronate wedge, like that appropriated by Achan, has been disastrous to the people, for it has lost to them the great benefits of a religious Establishment, and very great these are; but it threatens, as in the case of the sons of Carmi of old, to work more serious evil to those by whom it was originally coveted,—"evil to themselves and all their house." As I approached the Free Church, a squat, sun-burned, carnal-minded "old wee wifie," who seemed passing towards the Secession place of worship, after looking wistfully at my gray maud, and concluding for certain that I could not be other than a Southland drover, came up to me, and asked, in a cautious whisper, "Will ye be wantin' a coo?" I replied in the negative; and the wee wifie, after casting a jealous glance at a group of grave-featured Free Church folk in our immediate neighborhood, who would scarce have tolerated Sabbath trading in a Seceder, tucked up her little blue cloak over her head, and hied away to the chapel.

In the Free Church pulpit I recognized an old friend, to whom I introduced myself at the close of the service, and by whom I was introduced, in turn, to several intelligent members of his session, to whose kindness I owed, on the following day, introductions to some of the less accessible curiosities of the place. I rose betimes on the morning of Monday, that I might have leisure enough before me to see them all, and broke my first ground in Orkney as a geologist in a quarry a few hundred yards to the south and east [Pg 428] of the town. It is strange enough how frequently the explorer in the Old Red finds himself restricted in a locality to well nigh a single organism,—an effect, probably, of some gregarious instinct in the ancient fishes of this formation, similar to that which characterizes so many of the fishes of the present time, or of some peculiarity in their constitution, which made each choose for itself a peculiar habitat. In this quarry, though abounding in broken remains, I found scarce a single fragment which did not belong to an exceedingly minute species of Coccosteus, of which my first specimen had been sent me a few years before by Mr. Robert Dick, from the neighborhood of Thurso, and which I at that time, judging from its general proportions, had set down as the young of the *Coccosteus cuspidatus*. Its apparent gregariousness, too, quite as marked at Thurso as in this quarry, had assisted, on the strength of an obvious enough analogy, in leading to the conclusion. There are several species of the existing fish, well known on our coasts, that, though solitary when fully grown, are gregarious when young. The coal-fish, which as the sillock of a few inches in length congregates by thousands, but as the colum-saw of from two and a half to three feet is a solitary fish, forms a familiar instance; and I had inferred that the Coccosteus, found solitary, in most instances, when at its full size, had, like the coal-fish, congregated in shoals when in a state of immaturity. But a more careful examination of the specimens leads me to conclude that this minute gregarious Coccosteus, so abundant in this locality that its fragments thickly speckle the strata for hundreds of yards together—(in one instance I found the dorsal plates of four individuals crowded into a piece of flag barely six inches square)—was in reality a distinct species. Though not more than one-fourth the size, measured linearly, of the *Coccosteus decipiens*, its plates exhibit [Pg 429] as many of those lines of increment which gave to the occipital buckler of the creature its tortoise-like appearance, and through which plates of the buckler species were at first mistaken for those of a Chelonian, as are exhibited by plates of the larger kinds, with an area ten times as great; its tubercles, too, some of them of microscopic size, are as numerous;—evidences, I think,—when we take into account that in the bulkier species the lines and tubercles increased in number with the growth of the plates, and that, once formed, they seem never to have been affected by the subsequent enlargement of the creature,—that this ichthyolite was not an *immature*, but really a *miniature* Coccosteus. We may see on the plates of the full-grown Coccosteus, as on the shells of bivalves, such as *Cardium echinatum*, or on those of spiral univalves, such as *Buccinum undatum*, the diminutive markings which they bore when the creature was young; and on the plates of this species we may detect a regular gradation of tubercles from the microscopic to the minute, as we may see on the plates of the larger kinds a regular gradation from the minute to the fall-sized. The average length of the dwarf Coccosteus of Thurso and Kirkwall, taken from the snout to the pointed termination of the dorsal plate, ranges from one and a-half to two inches; its entire length from head to tail probably from three to four. It was from one of Mr. Dick's specimens of this species that I first determined the true position of the eyes of the Coccosteus,—a position which some of my lately-found ichthyolites conclusively demonstrate, and which Agassiz, in his restoration, deceived by ill-preserved specimens, has fixed at a point considerably more lateral and posterior, and where eyes would have been of greatly less use to the animal. About a field's breadth below this quarry of the *Coccosteus minor*,—if I may take the liberty of extemporizing a name, until such time as some person [Pg 430] better qualified furnishes the creature with a more characteristic one,—there are the remains, consisting of fosse and rampart, with a single cannon lying red and honeycombed amid the ruins, of one of Cromwell's forts, built to protect the town against the assaults of an enemy from the sea. In the few and stormy years during which this ablest of British governors ruled over Scotland, he seems to have exercised a singularly vigilant eye. The claims on his protection of even the remote Kirkwall did not escape him.

The antiquities of the burgh next engaged me; and, as became its dignity and importance, I began with the Cathedral, a building imposing enough to rank among the most impressive of its class anywhere, but whose peculiar *setting* in this remote northern country, joined to the associations of its early history with the Scandinavian Rollos, Sigurds, Einars, and Hacos of our dingier chronicles, serve greatly to enhance its interest. It is a noble pile, built of a dark-tinted Old Red Sandstone,—a stone which, though by much too sombre for adequately developing the elegancies of the Grecian or Roman architecture, to which a light delicate tone of color seems indispensable, harmonizes well with the massier and less florid styles of the Gothic. The round arch of that ancient Norman school which was at one time so generally recognized as Saxon, prevails in the edifice, and marks out its older portions. A few of the arches present on their ringstones those characteristic toothed and zig-zag ornaments that are of not unfamiliar occurrence on the round squat doorways of the older parish churches of England; but by much the greater number exhibit merely a few rude mouldings, that bend over ponderous columns and massive capitals, unfretted by the tool of the carver. Though of colossal magnificence, the exterior of the edifice yields in effect, as in all true Gothic buildings,—for the Gothic is greatest in what the [Pg 431] Grecian is least,—to the sombre sublimity of the interior. The nave, flanked by the dim deep aisles, and by a double row of smooth-stemmed gigantic columns, supporting each a double tier of ponderous arches, and the transepts, with their three tiers of small Norman windows, and their bold semi-circular arcs, demurely gay with toothed or angular carvings, that speak of the days of Rolf and Torfeinar, are singularly fine,—far superior to aught else of the kind in Scotland; and a happy accident has added greatly to their effect. A rare Byssus,—the *Byssus aeruginosa* of Linnæus,—the *Leprasia aeruginosa* of modern botanists,—one of those gloomy vegetables of the damp cave and dark mine whose true habitat is rather under than upon the earth, has crept over arch, and column, and broad bare wall, and given to well nigh the entire interior of the building a close-fitted lining of dark velvety green, which, like the Attic rust of an ancient medal, forms an appropriate covering to the sculpturings which it enwraps without concealing, and harmonizes with at once the dim light and the antique architecture. Where the sun streamed upon it, high over head, through the narrow windows above, it reminded me of a pall of rich green velvet. It seems subject, on some of the lower mouldings and damper recesses, especially amid the tombs and in the aisles, to a decomposing mildew, which eats into it in fantastic map-like lines of mingled black and gray, so resembling Runic fret-work, that I had some difficulty in convincing myself that the tracery which it forms,—singularly appropriate to the architecture,—was not the effect of design. The choir and chancel of the edifice, which at the time of my visit were still employed as the parish church of Kirkwall, and had become a "world too wide" for the shrunken congregation, are more modern and ornate than the nave and transepts; and the round arch gives place, in at least their [Pg 432] windows, to the pointed one. But the unique consistency of the pile is scarce at all disturbed by this mixture of styles. It is truly wonderful how completely the forgotten architects of the darker ages contrived to avoid those gross offences against good taste and artistic feeling into which their successors of a greatly more enlightened time are continually falling. Instead of idly courting ornament for its own sake, they must have had as their proposed object the production of some definite effect, or the development of some special sentiment. It was perhaps well for them, too, that they were not so overladen as our modern architects with the *learning* of their profession. Extensive knowledge requires great judgment to guide it. If that high genius which can impart its own homogeneous character to very various materials be wanting, the more multifarious a man's ideas become, the more is he in danger of straining after a heterogeneous patch-work excellence, which is but excellence in its components, and deformity as a whole. Every new vista opened up to him on what has been produced in his art elsewhere presents to him merely a new avenue of error. His mind becomes a mere damaged kaleidoscope, full of little broken pieces of the fair and the exquisite, but devoid of that nicely reflective machinery which can alone cast the fragments into shapes of a chaste and harmonious beauty.

Judging from the sculptures of St. Magnus, the stone-cutter seems to have had but an indifferent command of his trade in Orkney, when there was a good deal known about it elsewhere. And yet the rudeness of his work here, much in keeping with the ponderous simplicity of the architecture, serves but to link on the pile to a more venerable antiquity, and speaks less of the inartificial than of the remote. I saw a grotesque hatchment high up among the arches, that, with the uncouth carvings below, [Pg 433] served to throw some light on the introduction into ecclesiastical edifices of those ludicrous sculptures that seem so incongruously foreign to the proper use and character of such places. The painter had set himself, with, I doubt not, fair moral intent, to exhibit a skeleton wrapped up in a winding-sheet; but, like the unlucky artist immortalized by Gifford, who proposed painting a lion, but produced merely a dog, his skill had failed in seconding his intentions, and, instead of achieving a Death in a shroud, he had achieved but a monkey grinning in a towel. His contemporaries, however, unlike those of Gifford's artist, do not seem to have found out the mistake, and so the betowelled monkey has come to hold a conspicuous place among the solemnities of the Cathedral. It does not seem difficult to conceive how unintentional ludicrosities of this nature, introduced into ecclesiastical erections in ages too little critical to distinguish between what the workman had purposed doing and what he had done, might come to be regarded, in a less earnest but more knowing age, as precedents for the introduction of the intentionally comic and grotesque. Innocent accidental monkeys in towels may have thus served to usher into serious neighborhoods monkeys in towels that were such with malice *prepense*.

I was shown an opening in the masonry, rather more than a man's height from the floor, that marked where a square narrow cell, formed in the thickness of the wall, had been laid open a few years before. And in the cell there was found depending from the middle of the roof a rusty iron chain, with a bit of barley-bread attached. What could the chain and bit of bread have meant? Had they dangled in the remote past over some northern Ugolino? or did they form in their dark narrow cell, without air-hole or outlet, merely some of the reserve terrors of the [Pg 434] Cathedral, efficient in bending to the authority of the Church the rebellious monk or refractory nun? Ere quitting the building, I scaled the great tower,—considerably less tall, it is said, than its predecessor, which was destroyed by lightning about two hundred years ago, but quite tall enough to command an extensive, and, though bare, not unimpressive prospect. Two arms of the sea, that cut so deeply into the mainland on its opposite sides as to narrow it into a flat neck little more than a mile and a half in breadth, stretch away in long vista, the one to the south, and the other to the north; and so immediately is the Cathedral perched on the isthmus between, as to be nearly equally conspicuous from both. It forms in each, to the inward-bound vessel, the terminal object in the landscape. There was not much to admire in the town immediately beneath, with its roofs of gray slate,—almost the only parts of it visible from this point of view,—and its bare treeless suburbs; nor yet in the tract of mingled hill and moor on either hand, into which the island expands from the narrow neck, like the two ends of a sand-glass; but the long withdrawing ocean-avenues between, that seemed approaching from south and north to kiss the feet of the proud Cathedral,—avenues here and there enlivened on their ground of deep blue by a sail, and fringed on the lee—for the wind blew freshly in the clear sunshine—with their border of dazzling white, were objects worth while climbing the tower to see. Ere my descent, my guide hammered out of the tower-bells, on my special behalf, somewhat, I daresay, to the astonishment of the burghers below, a set of chimes handed down entire, in all the notes, from the times of the monks, from which also the four fine bells of the Cathedral have descended as an heirloom to the burgh. The chimes would have delighted the heart of old Lisle Bowles, the poet of

[Pg 435] "Well-tun'd bell's enchanting harmony."

I could, however, have preferred listening to their music, though it seemed really very sweet, a few hundred yards further away; and the quiet clerical poet,—the restorer of the Sonnet in England, would, I doubt not, have been of the same mind. The oft-recurring tones of those bells that ring throughout his verse, and to which Byron wickedly proposed adding a *cap*, form but an ingredient of the poetry in which he describes them; and they are represented always as distant tones, that, while they mingle with the softer harmonies of nature, never overpower them.

"How sweet the tuneful bells responsive peal!

And, hark! with lessening cadence now they fall,
And now, along the white and level tide
They fling their melancholy music wide!
Bidding me many a tender thought recall
Of happy hours departed, and those years
When, from an antique tower, ere life's fair prime,
The mournful mazes of their mingling chime
First wak'd my wondering childhood into tears!"

From the Cathedral I passed to the mansion of Old Earl Patrick,—a stately ruin, in the more ornate castellated style of the sixteenth century. It stands in the middle of a dense thicket of what are *trying* to be trees, and have so far succeeded, that they conceal, on one of the sides, the lower story of the building, and rise over the *spring* of the large richly-decorated turrets. These last form so much nearer the base of the edifice than is common in our old castles, that they exhibit the appearance rather of hanging towers than of turrets,—of towers with their foundations cut away. The projecting windows, with their deep mouldings, square mullions, and cruciform shot-holes, are [Pg 436] rich specimens of their peculiar style; and, with the double-windowed turrets with which they range, they communicate a sort of *high-relief* effect to the entire erection, "the exterior proportions and ornaments of which," says Sir Walter Scott, in his Journal, "are very handsome." Though a roofless and broken ruin, with the rank grass waving on its walls, it is still a piece of very solid masonry, and must have been rather stiff working as a quarry. Some painstaking burgher had, I found, made a desperate attempt on one of the huge chimney lintels of the great hall of the erection,—an apartment which Sir Walter greatly admired, and in which he lays the scene in the "Pirate" between Cleveland and Jack Bunce, but the lintel, a curious example of what, in the exercise of a little Irish liberty, is sometimes termed a *rectilinear arch*, defied his utmost efforts; and, after half-picking out the keystone, he had to give it up in despair. The bishop's palace, of which a handsome old tower still remains tolerably entire, also served for a quarry in its day; and I was scarce sufficiently distressed to learn, that on almost the last occasion on which it had been wrought for this purpose, one of the two men engaged in the employment suffered a stone, which he had loosed out of the wall, to drop on the head of his companion, who stood watching for it below, and killed him on the spot.

[Pg 437]

CHAPTER XI.

The Bishop's Palace at Orkney—Haco the Norwegian—Icelandic Chronicle respecting his Expedition to Scotland—His Death—Removal of his Remains to Norway—Why Norwegian Invasion ceased—Straw-plaiting—The Lassies of Orkney—Orkney Type of Countenance—Celtic and Scandinavian—An accomplished Antiquary—Old Manuscripts—An old Tune-book—Manuscript Letter of Mary Queen of Scots—Letters of General Monck—The fearless Covenanter—Cave of the Rebels—Why the tragedy of "Gustavus Vasa" was prohibited—Quarry of Pickoquoy—Its Fossil Shells—Journey to Stromness—Scenery—Birth-place of Malcolm, the Poet—His History—One of his Poems—His Brother a Free Church Minister—New Scenery.

The "upper story" of the bishop's palace, in which grim old Haco died,—thanks to the economic burghers who converted the stately ruin into a quarry,—has wholly disappeared. Though the death of this last of the Norwegian invaders does not date more than ten years previous to the birth of the Bruce, it seems to belong, notwithstanding, to a different and greatly more ancient period of Scottish history; as if it came under the influence of a sort of aërial perspective, similar to that which makes a neighboring hill in a fog appear as remote as a distant mountain when the atmosphere is clearer. Our national wars with the English were rendered familiar to our country folk of the last age, and for centuries before by the old Scotch "*Makkaris,*" Barbour and Blind Harry, and in our own times by the glowing narratives of Sir Walter Scott,—magicians who, unlike those ancient sorcerers that used to darken the air with their incantations, possessed the rare power of dissipating [Pg 438] the mists and vapors of the historic atmosphere, and rendering it transparent. But we had no such chroniclers of the time, though only half an age further removed into the past,

"When Norse and Danish galleys plied
Their oars within the Frith of Clyde,
And floated Haco's banner trim
Above Norweyan warriors grim,
Savage of heart and large of limb."

And hence the thick haze in which it is enveloped. Curiously enough, however, this period, during which the wild Scot had to contend with the still wilder wanderers of Scandinavia in fierce combats that he was too little skilful to record, and which appears so obscure and remote to his descendants, presents a phase comparatively near, and an outline proportionally sharp and well-defined to the intelligent peasantry of Iceland. *Their* Barbours and Blind Harries came a few ages sooner than ours, and the fog, in consequence, rose earlier; and so, while Scotch antiquaries of no mean standing can say almost nothing about the expedition or death-bed of Haco, even the humbler Icelanders, taught from their Sagas in the long winter nights, can tell how, harassed by anxiety and fatigue, the monarch sickened, and recovered, and sickened again; and how, dying in the bishop's palace, his body was interred for a winter in the Cathedral, and then borne in spring to the burying-place of his ancestors in Norway. The only clear vista on the death of Haco which now exists is that presented by an Icelandic chronicler: to which, as it seems so little known even in Orkney that the burying-place of the monarch is still occasionally sought for in the Cathedral, I must introduce the reader. I quote from an extract containing the account of Haco's expedition against Scotland, which was translated from the original Icelandic by the Rev. [Pg 439] James Johnstone, chaplain to his Britannic Majesty's Envoy Extraordinary at the court of Denmark, and appeared in the "Edinburgh Magazine" for 1787.

"King Haco," says the chronicler, "now in the seven and fortieth year of his reign, had spent the summer in watchfulness and anxiety. Being often called to deliberate with his captains, he had enjoyed little rest; and when he arrived at Kirkwall, he was confined to his bed by his disorder. Having lain for some nights, the illness abated, and he was on foot for three days. On the first day he walked about in his apartments; on the second he attended at the bishop's chapel to hear mass; and on the third he went to Magnus Church, and walked round the shrine of St. Magnus, Earl of Orkney. He then ordered a bath to be prepared, and got himself shaved. Some nights after, he relapsed, and took again to his bed. During his sickness he ordered the Bible and Latin authors to be read to him. But finding his spirits were too much fatigued by reflecting on what he had heard, he desired Norwegian books might be read to him night and day: first the lives of saints; and, when they were ended, he made his attendants read the Chronicles of our Kings, from Holden the Black, and so of all the Norwegian monarchs in succession, one after the other. The king still found his disorder increasing. He therefore took into consideration the pay to be given to his troops, and commanded that a merk of fine silver should be given to each courtier, and half a merk to each of the masters of the lights, chamberlain, and other attendants on his person. He ordered all the silver-plate belonging to his table to be weighed, and to be distributed if his standard silver fell short.... King Haco received extreme unction on the night before the festival of St. Lucia. Thorgisl, Bishop of Stravanger, Gilbert, Bishop of Hainar, Henry, Bishop of Orkney, Albert Thorleif and [Pg 440] many other learned men, were present; and, before the unction, all present bade the king farewell with a kiss.... The festival of the Virgin St. Lucia happened on a Thursday; and on the Saturday after, the king's disorder increased to such a degree, that he lost the use of his speech; and at midnight Almighty God called King Haco out of this mortal life. This was matter of great grief to all those who attended, and to most of those who heard of the event. The following barons were present at the death of the king:—Briniolf Johnson, Erling Alfson, John Drottning, Ronald Urka, and some domestics who had been near the king's person during his illness. Immediately on the decease of the king, bishops and learned men were sent for to sing mass.... On Sunday the royal corpse was carried to the upper hall, and laid on a bier. The body was clothed in a rich garb, with a garland on its head, and dressed out as became a crowned monarch. The masters of the lights stood with tapers in their hands, and the whole hall was illuminated. All the people came to see the body, which appeared beautiful and animated; and the king's countenance was as fair and ruddy as while he was alive. It was some alleviation of the deep sorrow of the beholders to see the corpse of their departed sovereign so decorated. High mass was then sung for the deceased. The nobility kept watch by the body during the night. On Monday the remains of King Haco were carried to St. Magnus Church, where they lay in state that night. On Tuesday the royal corpse was put in a coffin, and buried in the choir of St. Magnus Church, near the steps leading to the shrine of St. Magnus, Earl of Orkney. The tomb was then closed, and a canopy was spread over it. It was also determined that watch should be kept over the king's grave all winter. At Christmas the bishop and Andrew Plytt furnished entertainments, as the king [Pg 441] had directed; and good presents were given to all the soldiers. King Haco had given orders that his remains should be carried east to Norway, and buried near his fathers and relatives. Towards the end of winter, therefore, that great vessel which he had in the west was launched, and soon got ready. On Ash Wednesday the corpse of King Haco was taken out of the ground: this happened the third of the nones of March. The courtiers followed the corpse to Skalpeid, where the ship lay, and which was chiefly under the direction of the Bishop Thorgisl and Andrew Plytt. They put to sea on the first Saturday in Lent; but, meeting with hard weather, they steered for Silavog. From this place they wrote letters to Prince Magnus, acquainting him with the news, and then sailed for Bergen. They arrived at Laxavog before the festival of St. Benedict. On that day Prince Magnus rowed out to meet the corpse. The ship was brought near to the king's palace, and the body was carried up to a summer-house. Next morning the corpse was removed to Christ's Church, and was attended by Prince Magnus, the two queens, the courtiers, and the town's people. The body was then interred in the choir of Christ's Church; and Prince Magnus addressed a long and gracious speech to those who attended the funeral procession. All the multitude present were much affected, and expressed great sorrow of mind."

So far the Icelandic chronicle. Each age has as certainly its own mode of telling its stories as of adjusting its dress or setting its cap; and the mode of this northern historian is somewhat prolix. I am not sure, however, whether I would not prefer the simple minuteness with which he dwells on every little circumstance, to that dissertative style of history characteristic of a more reflective age, that for series of facts substitutes bundles of theories. [Pg 442] Cowper well describes the historians of this latter school, and shows how, on selecting some little-known personage of a remote time as their hero,

> "They disentangle from the puzzled skein
> In which obscurity has wrapped them up,
> The threads of politic and shrewd design
> That ran through all his purposes, and charge
> His mind with meanings that he never had,
> Or, having, kept concealed."

I have seen it elaborately argued by a writer of this class, that those wasting incursions of the Northmen which must have been such terrible plagues to the southern and western countries of Europe, ceased in consequence of their conversion to Christianity; for that, under the humanizing influence of religion, they staid at home, and cultivated the arts of peace. But the hypothesis is, I fear, not very tenable. Christianity, in even a purer form than that in which it first found its way among the ancient Scandinavians, and when at least as generally recognized nationally as it ever was by the subjects of Haco, has failed to put down the trade of aggressive war. It did not prevent honest, obstinate George the Third from warring with the Americans or the French: it only led him to enjoin a day of thanksgiving when his troops had slaughtered a great many of the enemy, and to ordain a fast when the enemy had slaughtered, in turn, a great many of his troops. And Haco, who, though he preferred the lives of the saints, and even of his ancestors, who could not have been very great saints, to the Scriptures, seems, for a king, to have been a not undevout man in his way, and yet appears to have had as few compunctions visitings on the score of his Scottish war as George the Third on that of the French or the American one. Christianity, too, ere his invasion [Pg 443] of Scotland, had been for a considerable time established in his dominions, and ought, were the theory a true one, to have operated sooner. The Cathedral of St. Magnus, when he walked round the shrine of its patron saint, was at least a century old. The true secret of the cessation of Norwegian invasion seems to have been the consolidation, under vigorous princes, of the countries which had lain open to it,—a circumstance which, in the later attempts of the invaders, led to results similar to those which broke the heart of tough old Haco, in the bishop's palace at Kirkwall.

From the ruins I passed to the town, and spent a not uninstructive half-hour in sauntering along the streets in the quiet of the evening, acquainting myself with the general aspect of the people. I marked, as one of the peculiar features of the place, groups of tidily-dressed young women, engaged at the close-heads with their straw plait,—the prevailing manufacture of the town,—and enjoying at the same time the fresh air and an easy chat. The special contribution made by the lassies of Orkney to the dress of their female neighbors all over the empire, has led to much tasteful dressing among themselves. Orkney, on its gala, days, is a land of ladies. What seems to be the typical countenance of these islands unites an aquiline but not prominent nose to an oval face. In the ordinary Scotch and English countenance, when the nose is aquiline it is also prominent, and the face is thin and angular, as if the additional height of the central feature had been given it at the expense of the cheeks, and of lateral shavings from off the chin. The hard Duke-of-Wellington face is illustrative of this type. But in the aquiline type of Orkney the countenance is softer and fuller, and, in at least the female face, the general contour greatly more handsome. Dr. Kombst, in his ethnographic map of Britain and Ireland, [Pg 444] gives to the coast of Caithness and the Shetland Islands a purely Scandinavian people, but to the Orkneys a mixed race, which he designates the Scandinavian-Gaelic. I would be inclined, however,—preferring rather to found on those traits of person and character that are still patent, than on the unauthenticated statements of uncertain history,—to regard the people as essentially one from the northern extremity of Shetland to the Ord Hill of Caithness. Beyond the Ord Hill, and on to the northern shores of the Frith of Cromarty, we find, though unnoted on the map, a different race,—a race strongly marked by the Celtic lineaments, and speaking the Gaelic tongue. On the southern side of the Frith, and extending on to the Bay of Munlochy, the purely Scandinavian race again occurs. The sailors of the

Danish fleet which four years ago accompanied the Crown Prince in his expedition to the Faroe Islands were astonished when, on landing at Cromarty, they recognized in the people the familiar cast of countenance and feature that marked their country folk and relatives at home; and found that they were simply Scandinavians like themselves, who, having forgotten their Danish, spoke Scotch instead. Rather more than a mile to the west of the fishing village of Avoch there commences a Celtic district, which stretches on from Munlochy to the river Nairne; beyond which the Scandinavian and Teutonic-Scandinavian border that fringes the eastern coast of Scotland extends unbroken southwards through Moray, Banff, and Aberdeen, on to Forfar, Fife, the Lothians, and the Mearns. These two intercalated patches of Celtic people in the northern tract,—that extending from the Ord Hill to the Cromarty Frith, and that extending from the Bay of Munlochy to the Nairne,—still retaining, as they do, after the lapse of ages, a sharp distinctness of boundary in respect of language, character, [Pg 445] and personal appearance, are surely great curiosities. The writer of these chapters was born on the extreme edge of one of these patches, scarce a mile distant from a Gaelic-speaking population; and yet, though his humble ancestors were located on the spot for centuries, he can find trace among them of but one Celtic name; and their language was exclusively the Lowland Scotch. For many ages the two races, like oil and water, refused to mix.

I spent the evening very agreeably with one of the Free Church elders of the place, Mr. George Petrie, an accomplished antiquary; and found that his love of the antique, joined to an official connection with the county, had cast into his keeping a number of curious old papers of the sixteenth, seventeenth, and eighteenth centuries,—not in the least connected, some of them, with the legal and civic records of the place, but which had somehow stuck around these, in their course of transmission from one age to another, as a float of brushwood in a river occasionally brings down along with it, entangled in its folds, uprooted plants and aquatic weeds, that would otherwise have disappeared in the cataracts and eddies of the upper reaches of the stream. Dead as they seemed, spotted with mildew, and fretted by the moth, I found them curiously charged with what had once been intellect and emotion, hopes and fears, stern business and light amusement. I saw, among the other manuscripts, a thin slip of a book, filled with jottings, in the antique square-headed style of notation, of old Scotch tunes, apparently the work of some musical county-clerk of Orkney in the seventeenth century; but the paper, in a miserable state of decay, was blotted crimson and yellow with the rotting damps, and the ink so faded, that the notation of scarce any single piece in the collection seemed legible throughout. Less valuable and more modern, though curious from their [Pg 446] eccentricity, there lay, in company with the music, several pieces of verse, addressed by some Orcadian Claud Halcro of the last age, to some local patron, in a vein of compliment rich and stiff as a piece of ancient brocade. A peremptory letter, bearing the autograph signature of Mary Queen of Scots, to Torquil McLeod of Dunvegan, who had been on the eve, it would seem, of marrying a daughter of Donald of the Isles, gave the Skye chieftain, "to wit" that, as he was of the blood royal of Scotland, he could form no matrimonial alliance without the royal permission,—a permission which, in the case in point, was not to be granted. It served to show that the woman who so ill liked to be thwarted in her own amours could, in her character as the Queen, deal despotically enough with the love affairs of other people. Side by side with the letter of Mary there were several not less peremptory documents of the times of the Commonwealth, addressed to the Sheriff of Orkney and Shetland, in the name of his Highness the Lord Protector, and that bore the signature of George Monck. I found them to consist chiefly of dunning letters,—such letters as those duns write who have victorious armies at their back,—for large sums of money, the assessments laid on the Orkneys by Cromwell. Another series of letters, some ten or twelve years later in their date, form portions of the history of a worthy covenanting minister, the Rev. Alexander Smith of Colvine, banished to North Ronaldshay from the extreme south of Scotland, for the offence of preaching the gospel, and holding meetings for social worship in his own house; and, as if to demonstrate his incorrigibility, one of the series,—a letter under his own hand, addressed from his island prison to the Sheriff-Depute in Kirkwall,—showed him as determined and persevering in the offence as ever. It was written immediately after his arrival. "The poor inhabitants," [Pg 447] says the writer, "so many as I have yet seen, have received me with much joy. *I intend, if the Lord will, to preach Christ to them next*

Lord's day, without the least mixture of anything that may smell of sedition or rebellion. If I be farther troubled for yt, I resolve to suffer with meekness and patience." The Galloway minister must have been an honest man. Deeming preaching his true vocation,—a vocation from the exercise of which he dared not cease, lest he should render himself obnoxious to the woe referred to by the apostle,—he yet could not steal a march on even the Sheriff, whose professional duty it was to prevent him from doing *his;* and so he fairly warned him that he proposed breaking the law. The next set of papers in the collection dated after the Revolution, and were full charged with an enthusiastic Jacobitisin, which seems to have been a prevalent sentiment in Orkney from the death of Queen Anne, until the disastrous defeat at Culloden quenched in blood the hopes of the party. There is a deep cave still shown on the shores of Westray, within sight of the forlorn Patmos of the poor Covenanter, in which, when the sun got on the Whig side of the hedge, twelve gentlemen, who had been engaged in the rebellion of 1745, concealed themselves for a whole winter. So perseveringly were they sought after, that during the whole time they dared not once light a fire, nor attempt fishing from the rocks to supply themselves with food; and, though they escaped the search, they never, it is said, completely recovered the horrors of their term of dreary seclusion, but bore about with them, in broken constitutions, the effects of the hardships to which they had been subjected. They must have had full time and opportunity, during that miserable winter, for testing the justice of the policy that had sent poor Smith into exile, from his snug southern parish in the Presbytery of [Pg 448] Dumfries, to the remotest island of the Orkneys. The great lesson taught in Providence during the seventeenth and part of the eighteenth century to our Scottish country folk seems to have been the lesson of toleration; and as they were slow, stubborn scholars, the lash was very frequently and very severely applied. One of the Jacobite papers of Mr. Petrie's collection,—a triumphal poem on the victory of Gladsmuir,—which, if less poetical than the Ode of Hamilton of Bangour on the same subject, is in no degree less curious,—serves to throw very decided light on a passage in literary history which puzzled Dr. Johnson, and which scarce any one would think of going to Orkney to settle.

Johnson states, in his Life of the poet Thomson, that the "first operation" of the act passed in 1739 "for licensing plays" was the "prohibition of 'Gustavus Vasa,' a tragedy of Mr. Brook." "Why such a work should be obstructed," he adds, "it is hard to discover." We learn elsewhere,—from the compiler of the "Modern Universal History," if I remember aright,—that "so popular did the prohibitory order of the Lord Chamberlain render the play," that, "on its publication the same year, not less than a thousand pounds were the clear produce." It was not, however, until more than sixty years after, when both Johnson and Brook were in their graves, that it was deemed safe to license it for the stage. Now, the fact that a drama, in itself as little dangerous as "Cato" or "Douglas," should have been prohibited by the Government of the day, in the first instance, and should have brought the author, on its publication, so large a sum in the second, can be accounted for only by a reference to the keen partisanship of the period, and the peculiar circumstances of parties. The Jacobites, taught by the rebellion of 1715 at once the value of the Highlands and the incompetency [Pg 449] of the Chevalier St. George as a leader, had begun to fix their hopes on the Chevalier's son, Charles Edward, at that time a young but promising lad; and, with the tragedy of Brook before them, neither they, nor the English Government of the day could have failed to see the foreigner George the Second typified—unintentionally, surely, on the part of Brook, who was a "Prince of Wales" Whig—in the foreigner Christiern the Second, the Scotch Highlanders in the Mountaineers of Dalecarlia, and the young Prince in Gustavus. In the Jacobite manuscript of Mr. Petrie's collection, the parallelism is broadly traced, nor is it in the least probable, as the poem is a piece of sad mediocrity throughout, that it is a parallelism which was originated by its writer. It must have been that of his party; and led, I doubt not, five years before, to the prohibition of Brook's tragedy, and to the singular success which attended its publication. The passage in the manuscript suggestive of this view takes the form of an address to the victorious prince, and runs as follows:—

"Meanwhile, unguarded youth, thou stoodst alone;
The cruel Tyrant urged his Armie on;
But Truth and Goodness were the Best of Arms;
And, fearless Prince, Thou smil'd at Threatened harms.

Thus, Glorious Vasa worked in Swedish mines,—
Thus, Helpless, Saw his Enemy's Designs,—
Till, roused, his Hardy Highlanders arose,
And poured Destruction on their foreign foes."

I rose betimes next morning, and crossed the Peerie [little] Sea, a shallow prolongation of the Bay of Kirkwall, cut off from the main sea by an artificial mound, to the quarry of Pickoquoy, somewhat notable, only a few years ago, as the sole locality in which shells had been detected in the Old Red Sandstone of Scotland. But these have since been found in the neighborhood of [Pg 450] Thurso, by Mr. Robert Dick, associated with bones and plates of the Asterolepis, and by Mr. William Watt on the opposite side of the Mainland of Orkney, at Marwick Head. So far as has yet been ascertained, they are all of one species, and more nearly resemble a small Cyclas than any other shell. They are, however, more deeply sulcated in concentric lines, drawn, as if by a pair of compasses, from the umbone, and somewhat resembling those of the genus Astarte, than any species of Cyclas with which I am acquainted. In all the specimens I have yet seen, it appears to be rather a thick dark epidermis that survives, than the shell which it covered; nay, it seems not impossible that to its thick epidermis, originally an essentially different substance from that which composed the calcareous case, the shell may have owed its preservation as a fossil; while other shells, its contemporaries, from the circumstance of their having been unfurnished with any such covering, may have failed to leave any trace of their existence behind them. It seems at least difficult to conceive of a sea inhabited by many genera of fishes, each divided into several species, and yet furnished with but one species of shell. I found the quarry of Pickoquoy,—a deep excavation only a few yards beyond the high-water mark, and some two or three yards under the high-water level,—deserted by the quarrymen, and filled to the brim by the overflowing of a small stream. I succeeded, however, in detecting its shells *in situ*. They seem restricted chiefly to a single stratum, scarcely half an inch in thickness, and lie, not thinly scattered over the platform which they occupy, but impinging on each other, like all the gregarious shells, in thickly-set groups and clusters. There occur among them occasional scales of Dipteri; and on some of the fragments of rock long exposed around the quarry-mouth to the weather I found them assuming a [Pg 451] pale nacreous gloss,—an effect, it is not improbable, of their still retaining, attached to the epidermis, a thin film of the original shell. The world's history must be vastly more voluminous now, and greatly more varied in its contents, than when the stratum which they occupy formed the upper layer of a muddy sea-bottom, and they opened their valves by myriads, to prey on the organic atoms which formed their food, or shut them again, startled by the shadow of the Dipterus, as he descended from the upper depths of the water to prey upon them in turn. The palate of this ancient ganoid is furnished with a curious dental apparatus, formed apparenly, like that of the recent wolf-fish, for the purpose of crushing shells.

About mid-day I set out by the mail-gig for Stromness. For the first few miles the road winds through a bare solitary valley, overlooked by ungainly heath-covered hills of no great altitude, though quite tall enough to prevent the traveller from seeing anything but themselves. As he passes on, the valley opens in front on an arm of the sea, over which the range of hills on the right abruptly terminates, while that on the left deflects into a line nearly parallel to the shore, leaving a comparatively level strip of moory land, rather more than a mile in breadth, between the steeper acclivities and the beach. A tall naked house rises between the road and the sea. Two low islands immediately behind it, only a few acres in extent,—one of them bearing a small ruin on its apex,—give a little variety to the central point in the prospect which the naked house forms; but the arm of the sea, bordered, at the time I passed, by a broad brown selvage of sea-weed, is as tame and flat as a Dutch lake; the background beyond, a long monotonous ridge, is bare and treeless; and in front lies the brown moory plain, bordered by the dull line of hills and darkened by scattered stacks of peat.

[Pg 452]

The scene is not at all such a one as a poet would, for its own sake, delight to fancy; and yet, in the recollection of at least one very pleasing poet, its hills, and islands, and blue arm of the sea, its brown moory plain, and tall naked house rising in the midst, must have been surrounded by a sunlit atmosphere of love and desire, bright enough to impart to even its tamest features a glow of exquisiteness and beauty. Malcolm the poet was born, and spent his years of boyhood and early youth, in the tall naked house; and the surrounding landscape is that to which he refers in his "Tales of Flood and Field," as rising in imagination before him, bright in the red gleam of the setting sun, when, on the steep slopes of the Pyrenees, the "silent stars of night were twinkling high over his head," and the "tents of the soldiery glimmering pale through the gloom." The tall house is the manse of the parish of Frith and Stennis; and the poet was the son of the Rev. John Malcolm, its minister. Here, when yet a mere lad, dreaming, in the quiet obscurity of an Orkney parish, far removed from the seat of war and the literary circles, of poetic celebrity and military renown, he addressed a letter to the Duke of Kent, the father of our Sovereign Lady the reigning Monarch, expressing an ardent wish to obtain a commission in the army then engaged in the Peninsula. The letter was such as to excite the interest of his Royal Highness, who replied to it by return of post, requesting the writer to proceed forthwith to London; for which he immediately set out, and was received by the Duke with courtesy and kindness. He was instructed by him to take ship for Spain, in which he arrived as volunteer; and, joining the army, engaged at the time in the siege of St. Sebastian, under General Graham, he was promoted shortly after, through the influence of his generous patron, to a lieutenancy in the 42d Highlanders. He served in that [Pg 453] distinguished regiment on to the closing campaign of the Pyrenees; but received at the battle of Toulouse a wound so severe as to render him ever after incapable of active bodily exertion; and so he had to retire from the army on half-pay, and a pension honorably earned. The history of his career as a soldier he has told with singular interest, in one of the earlier volumes of "Constable's Miscellany;" and his poems abound in snatches of description painfully true, drawn from his experience of the military life,—of scenes of stern misery and grim desolation, of injuries received, and of sufferings inflicted,—that must have contrasted sadly in his mind, in their character as gross realities, with the dreamy visions of conquest and glory in which he had indulged at an earlier time. The ruin of St. Sebastian, complete enough, and attended with circumstances of the horrible extreme enough, to appal men long acquainted with the trade of war, must have powerfully impressed an imaginative susceptible lad, fresh from the domesticities of a rural manse, in whose quiet neighborhood the voice of battle had not been heard for centuries, and surrounded by a simple people, remarkable for the respect which they bear to human life. In all probability, the power evinced in his description of the siege, and of the utter desolation in which it terminated, is in part owing to the fresh impressibility of his mind at the time. Such, at least, was my feeling regarding it, as I caught myself muttering some of its more graphic passages, and saw, from the degree of alarm evinced by the boy who drove the mail-gig, that the sounds were not quite lost in the rattle of that somewhat rickety vehicle, and that he had come to entertain serious doubts respecting the sanity of his passenger:—

"Sebastian, when I saw thee last,
[Pg 454] It was in Desolation's day,
As through thy voiceless streets I passed,
Thy piles in heaps of rubbish lay;
The roofless fragments of each wall
Bore many a dent of shell and ball;
With blood were all thy gateways red,
And thou,—a city of the dead!

With fire and sword thy walks were swept:
Exploded mines thy streets had heaped
In hills of rubbish; they had been
Traversed by gabion and fascine,
With cannon lowering in the rear
In dark array,—a deadly tier,—
Whose thunder-clouds, with fiery breath,
Sent far around their iron death;
The bursting shell, in fragments flung
Athwart the skies, at midnight sung,
Or, on its airy pathway sent,
Its meteors sweep the firmament.
Thy castle, towering o'er the shore,
Keeled on its rock amidst the roar
Of thousand thunders, for it stood
In circle of a fiery flood;

And crumbling masses fiercely sent
From its high frowning battlement,
Smote by the shot and whistling shell,
With groan and crash in ruin fell.
Through desert streets the mourner passed,
Midst-walls that spectral shadows east,
Like some fair spirit wailing o'er
The failed scenes it loved of yore;
No human voice was heard to bless
That place of waste and loneliness.

I saw at eve the night-bird fly,
And vulture dimly flitting by,
To revel o'er each morsel stolen
From the cold corse, all black and swoln
That on the shattered ramparts lay,
Of him who perished yesterday,—
Of him whose pestilential steam
Rose reeking on the morning beam,—
Whose fearful fragments, nearly gone,
Were blackening from the bleaching bone.

[Pg 455] The house-dog bounded o'er each scene
Where cisterns had so lately been:
Away in frantic haste he sprung,
And sought to cool his burning tongue.
He howled, and to his famished cry
The dreary echoes gave reply;
And owlet's dirge, through shadows dim,
Rolled back in sad response to him."

The father was succeeded in his parish by the brother of Malcolm,—a gentleman to whom, during my stay in Orkney, I took the liberty of introducing myself in his snug little Free Church manse at the head of the bay, and in whose possession I found the only portrait of the poet which exists. It is that of a handsome and interesting looking *young* man, though taken not many years before his death; for, like the greater number of his class, he did not live to be an old one, dying under forty. His brother the clergyman kindly accompanied me to two quarries in the neighborhood of his new domicil, which I found, like almost all the dry-stone fences of the district, speckled with scales, occipital plates, and gill-covers, of Osteolepides and Dipteri, but containing no entire ichthyolites. He had taken his side in the Church controversy, he told me, firmly, but quietly; and when the Disruption came, and he found it necessary to quit the old manse, which had been a home to his family for well nigh two generations, and in which both he and his brother had been born, he scarce knew what his people were to do, nor in what proportion he was to have followers among them. Somewhat to his surprise, however, they came out with him almost to a man; so that his successor in the parish church had sometimes, he understood, to preach to congregations scarcely exceeding half a dozen. I had learned elsewhere how thoroughly Mr. Malcolm was loved and respected by his parishioners; and that unconsciousness on his own part of the strength of their affection [Pg 456] and esteem, which his statement evinced, formed, I thought, a very pleasing trait, and one that harmonized well with the finely-toned unobtrusiveness and unconscious elegance which characterized the genius of his deceased brother. A little beyond the Free Church manse the road ascends between stone walls, abounding in fragments of ichthyolites, weathered blue by exposure to the sun and wind; and the top of the eminence forms the water-shed in this part of the Mainland, and introduces the traveller to a scene entirely new. The prospect is of considerable extent; and, what seems strange in Orkney, nowhere presents the traveller,—though it contains its large inland lake,—with a glimpse of the sea.

[Pg 457]

CHAPTER XII.

Hills of Orkney—Their Geologic Composition—Scene of Scott's "Pirate"—Stromness—Geology of the District—"Seeking beasts"—Conglomerate in contact with Granite—A palæozoic Hudson's Bay—Thickness of Conglomerate of Orkney—Oldest Vertebrate yet discovered in Orkney—Its Size—Figure of a characteristic plate of the Asterolepis—Peculiarity of Old Red Fishes—Length of the Asterolepis—A rich Ichthyolite Bed—Arrangement of the Layers—Queries as to the Cause of it—Minerals—An abandoned Mine—A lost Vessel—Kelp for Iodine—A dangerous Coast—Incidents of Shipwreck—Hospitality—Stromness Museum—Diplopterus mistaken for Dipterus—Their Resemblances and Differences—Visit to a remarkable Stack—Paring the Soil for Fuel, and consequent Barrenness—Description of the Stack—Wave-formed Caves—Height to which the Surf rises.

The Orkneys, like the mainland of Scotland, exhibit their higher hills and precipices on their western coasts: the Ward Hill of Hoy attains to an elevation of sixteen hundred feet; and there are some of the precipices which skirt the island of which it forms so conspicuous a feature, that rise sheer over the breakers from eight hundred to a thousand. Unlike, however, the arrangement on the mainland, it is the newer rocks that attain to the higher elevations; the heights of Hoy are composed of that arenaceous upper member of the Lower Old Red Sandstone,—the last formed of the Palæozoic deposits of Orkney,—which overlies the ichthyolitic flagstones and shales of Caithness at Dunnet Head, and the ichthyolitic nodular beds of Inverness, Ross, and Cromarty, at Culloden, Tarbet Ness, within the Northern Sutor, and along the bleak ridge of the Maolbuie. It is simply a tall upper story of the formation, erected along the western line of coast in the Orkneys, which the eastern line wholly wants. Its screen of hills forms a noble background [Pg 458] to the prospect which opens on the traveller as he ascends the eminence beyond the Free Church manse of Frith and Stennis. A large lake, bare and treeless, like all the other lakes and lochs of Orkney, but picturesque of outline, and divided into an upper and lower sheet of water by two low, long promontories, that jut out from opposite sides, and so nearly meet as to be connected by a threadlike line of road, half mound, half bridge, occupies the middle distance. There are moory hills and a few cottages in front; and on the promontories, conspicuous in the landscape, from the relief furnished by the blue ground of the surrounding waters, stand the tall stones of Stennis,—one group on the northern promontory, the other on the south. A gray old-fashioned house, of no very imposing appearance, rises between the road and the lake. It is the house of Stennis, or Turmister, in which Scott places some of the concluding scenes of the "Pirate," and from which he makes Cleveland and his fantastic admirer Jack Bunce witness the final engagement, in the bay of Stromness, between the Halcyon sloop of war and the savage Goffe. Nor does it matter anything that neither sea nor vessels can be seen from the house of Turmister: the fact which would be so fatal to a dishonest historian tells with no effect against the honest "*maker*," responsible for but the management of his tale.

I got on to Stromness; and finding, after making myself comfortable in my inn, that I had a fine bright evening still before me, longer by some three or four degrees of north latitude than the July evenings of Edinburgh, I set out, hammer in hand, to explore. Stromness is a long, narrow, irregular strip of a town, fairly thrust by a steep hill into the sea, on which it encroaches in a broken line of wharf-like bulwarks, along which, at high water, vessels of a hundred tons burden float so immediately beside the houses, that [Pg 459] their pennants on gala days wave over the chimney tops. The steep hill forms part of a granitic axis, about six miles in length by a mile in breadth, which forms the backbone of the district, and against which the Great Conglomerate and lower schists of the Old Red are upturned at a rather high angle. It is wrapped round in some places by a thin caul of the stratified primary rocks. Immediately over the town, on the brow of the eminence, where the granitic axis had been laid bare in digging a foundation for the Free Church manse, I saw numerous masses of schistose-gneiss, passing in some of the beds into a coarse-grained mica-schist, and a lustrous hornblendic slate, that had been quarried from over it, and which may be still seen built up into the garden-wall of the erection. I walked out towards the west, to examine the junction of the granite and the Great Conglomerate, where it is laid bare by the sea, little more than a quarter of a mile outside the town. There was a horde of noisy urchins a little beyond the inn, who, having seen me alight from the mail-gig, had determined in their own minds that I was engaged in the political canvass going forward at the time, but had not quite ascertained my side. They now divided into two parties; and when the one, as I passed, set up a "Hurra for Dundas," the other met them from the opposite side of the street, with a counter cry of "Anderson forever." Immediately after clearing the houses, I was accosted by a man from the country. "Ye'll be seeking beasts," he said:

"what price are cattle gi'en the noo?" "Yes, seeking *beasts*," I replied, "but very old ones: I have come to hammer your rocks for petrified fish." "I see, I see," said the man; "I took ye by ye'er gray plaid for a drover; but I ken something about the stane fish too; there's lots o' them in the quarries at Skaill."

I found the great Conglomerate in immediate contact with the granite, which is a ternary of the usual components, [Pg 460] somewhat intermediate in color between that of Peterhead and Aberdeen, and which at this point bears none of the caul of stratified primary rock by which it is overlaid on the brow of the hill. When the great Conglomerate, which is mainly composed of it here, was in the act of forming, this granite must have been one of the surface rocks of the locality, and in no respect a different stone from what it is now. The widely-spread Conglomerate base of the Old Red Sandstone, which presents, over an area of so many thousand square miles, such an identity of character, that specimens taken from the neighborhood of Lerwick, in Shetland, can scarce be distinguished from specimens detached from the hills which rise over the great Caledonian Valley, contains in various places, as under the Northern Sutor, for instance, and along the shores of Navity, fragments of rock which have not been detected *in situ* in the districts in which they occur as agglomerated pebbles. In general, however, we find it composed of the debris of those very granites and gneisses which, as in the case of the granitic axis here, were forced through it, and through the overlying deposits, by deep-seated convulsions, long posterior in date to its formation. It appears to have been formed in a vast oceanic basin of primary rock,—a Palæozoic Hudson's or Baffin's Bay,—partially surrounded, mayhap, by bare primary continents, swept by numerous streams, rapid and headlong, and charged with the broken debris of the inhospitable regions which they drained. The graptolite-bearing grauwacke of Banffshire seems to have been the only fossiliferous rock that occurred throughout the extent of this ancient northern basin. The Conglomerate of Orkney, like that of Moray and Ross, varies from fifty to a hundred yards in thickness. It is not overlaid in this section by the thick bed of coarse-grained sandstone, so well-marked a member of the formation at Cromarty, [Pg 461] Nigg, and Gamrie, and along the northern shores of the Beauly Frith; but at once passes into those gray bituminous flagstones so immensely developed in Caithness and the Orkneys. I traced the formation upwards this evening, walking along the edges of the upheaved strata, from where the Conglomerate leans against the granite, till where it merges into the gray flagstones, and then pursued these from older and lower to newer and higher layers, anxious to ascertain at what distance over the base the more ancient organisms of the system first appear, and what their character and kind. And little more than a hundred *yards* over the granite, and somewhat less than a hundred *feet* over the upper stratum of the great Conglomerate, I found what I sought,—a well-marked bone, perhaps the oldest vertebrate remain yet discovered in Orkney, embedded in a light grayish-colored layer of hard flag.

What, asks the reader, was the character of the ancient denizen of the Palæozoic basin of which it had formed a part? Was it a large or small fish, or of a high or low order? Not certainly of a low order, and by no means of a small size. The organism in the rock was a specimen of that curious nail-shaped bone of the Asterolepis which occurs as a central ridge in the single plate that occupies in this genus the wide curve of the under jaw, and as it was fully five inches in length from head to point, the plate to which it belonged must have measured ten inches across, and the frontal occipital buckler with which it was associated, one foot two inches in length (not including the three accessory plates at the nape), by ten inches in breadth. And if built, as it probably was, in the same massy proportions as its brother Cœlacanths the Holoptychius or Glyptolepis, the individual to which the nail-shaped bone belonged must have been, judging from the size of the corresponding parts in these ichthyolites, at least twice as large an animal [Pg 462] as the splendid Clashbennie Holoptychius of the Upper Old Red, now in the British Museum. The bulkiest icthyolites yet found in any of the divisions of the Old Red system are of the genus Asterolepis; and to this genus, and to evidently an individual of no inconsiderable size, this oldest of the organisms of the Orkney belonged. I was so interested in the fact, that before leaving this part of the country, I brought Dr. Garson, Stromness, and Mr. William Watt, jun., Skaill, both very intelligent palæontologists, to mark the place and character of the fossil, that they might be able to point it out to geological visitors in the future, or, if they preferred removing it to their town Museum, to indicate to them the stratum in which it had lain. For the present, I merely request the reader to mark, in the passing, that the most ancient organic remain yet found in the Old Red of this part of the country, nay, judging from its place, one of the most ancient yet found in Scotland,—so far as I know, absolutely the *most* ancient,—belonged to a ganoid as bulky as a large porpoise, and which, as shown by its teeth and jaws, possessed that peculiar organization which characterized the reptile fish of the Upper Devonian and Carboniferous periods. As there are, however, no calculations more doubtful or more to be suspected than those on which the size and bulk of the extinct animals are determined from some surviving fragment of their remains,—plate or bone,—I must attempt laying before the scientific reader at least a portion of the data on which I found.

This figure represents not inadequately one of the most characteristic plates of the Asterolepis. A very considerable fragment of what seems to be the same plate has been figured by Agassiz from a cast of one of the huge specimens [Pg 463] of Professor Asmus ("Old Red," Table 32, Fig. 13); but as no evidence regarding its true place had turned up at the time it was supposed by the naturalist to form part of the opercular covering of the animal. It belonged, however, to a different portion of the head. In almost all the fish that appear at our tables the space which occurs within the arched sweep of the lower jaws is mainly occupied by a complicated osseous mechanism, known to anatomists as the hyoid bone and branchiostegous rays; and which serves both to support the branchial arches and the branchiostegous membrane. Now, in the fish of the Old Red Sandstone, if we except some of the Acanthodians, we find no trace of this piece of mechanism: the arched space is covered over with dermal plates of bone, as a window is filled up with panes. Three plates, resembling very considerably the three divisions of a pointed Gothic window, furnished with a single central mullion, divided atop into two branches, occupied the space in the genera Osteolepis and Diplopterus; and two plates resembling the divisions of a pointed Gothic window, whose single central mullion does *not* branch atop, filled it up in the genera Holoptychius and Glyptolepis. In the genus Asterolepis this arch-shaped space was occupied, as I have said, by a single plate,—that represented in the wood-cut; and the nail-shaped bone rose on its internal surface along the centre,—the nail-head resting immediately beneath the centre of the arch, and the nail-point bordering on the isthmus below, at which the two shoulder-bones terminated. Now, in all the specimens which I have yet examined, the form and proportions of this plate are such that it can be very nearly inscribed in a semi-circle, of which the length of the nail is the radius. A nail five inches in length must have belonged to a plate ten inches in its longer diameter. I have ascertained further, that this longer diameter was equal to [Pg 464] the shorter diameter of the creature's frontal buckler, measured across about two thirds of its entire length from the nape; and that a transverse diameter of ten inches at this point was associated in the buckler with a longitudinal diameter of fourteen inches from the nape to the snout. Thus five inches along the nail represent fourteen inches along the occipital shield. The proportion, however, which the latter bore to the entire body in this genus has still to be determined. The corresponding frontal shield in the Coccosteus was equal to about one-fifth the creature's entire length, and in the Osteolepis and Diplopterus, to nearly one-seventh its length; while the length of the *Glyptolepis leptopterus*, a fish of the same family as the Asterolepis, was about five and a half times that of its occipital shield. If the Asterolepis was formed in the proportions of the Diplopterus, the ancient individual to which this nail-like bone belonged must have been about eight feet two inches in length; but if moulded, as it more probably was, in the proportions of the Glyptolepis, only six feet five inches. All the Cœlacanths, however, were exceedingly massive in proportion to their length; they were fish built in the square, muscular, thick-set, Dirk-Hatterick and Balfour-of-Burley style; and of the Russian specimens, some of the larger bones must have belonged to individuals of from twice to thrice the length of the Stromness one.

Passing upwards along the strata, step by step, as along a fallen stair, each stratum presenting a nearly perpendicular front, but losing, in the downward slant of the *tread*, as a carpenter would say, the height attained in the *rise*, I came, about a quarter of a mile farther to the west, and several hundred feet higher in the formation, upon a fissile dark-colored bed, largely charged with ichthyolites. The fish I found ranged in three layers,—the lower layer consisting almost exclusively of Dipterians, chiefly Osteolepides; [Pg 465] the middle layer, of Acanthodians, of the genera Cheiracanthus and Diplacanthus; and the upper layer, of Cephalaspides, mostly of one species, the *Coccosteus decipiens*. I found exactly the same arrangement in a bed considerably higher in the system, which occurs a full mile farther on,—the Dipterians at the bottom, the Acanthodians in the middle, and the Cephalaspides atop; and was informed by Mr. William Watt, a competent authority in the case, that the arrangement is comparatively a common one in the quarries of Orkney. How account for the phenomenon? How account for the three storeys, and the apportionment of the floors, like those of a great city, each to its own specific class of society? Why should the first floor be occupied by Osteolepides, the second by Cheiracanthi and their congeners, and the third by Coccostei? Was the arrangement an effect of normal differences in the constitutions of the several families, operated upon by some deleterious gas or mineral poison, which, though it eventually destroyed the whole, did not so simultaneously, but consecutively,—the families of weakest constitution first, and the strongest last? Or were they exterminated by some disease, that seized upon the families, not at once, but in succession? Or did they visit the locality serially, as the haddock now visits our coasts in spring, and the herring towards the close of summer; and were then killed off, whether by poison or disease, as they came? These are questions which may never be conclusively answered. It is well, however, to observe, as a curious geological fact, that peculiar arrangement of the fossils by which they are suggested, and to record the various instances in which it occurs. The minerals which I remarked among the schists here as most abundant are a kind of black ironstone, exceedingly tough and hard, occurring in detached masses, and a variety of bright pyrites disseminated among the [Pg 466] darker flagstones, either as irregularly-formed, brassy-looking concretions of small size, or spread out on their surfaces in thin leaf-like films, that resemble, in some of the specimens, the icy-foliage with which a severe frost encrusts a window-pane. Still further on I came upon a vein of galena; but a miner's excavation in the solid rock, a little above high-water mark, quite as dark and nearly as narrow as a fox-earth, showed me that it had been known long before, and, as the workings seemed to have been deserted for ages, known to but little purpose. The crystals of ore, small and thinly scattered, are embedded in a matrix of barytes, stromnite, and other kindred minerals, and the thickness of the entire vein is not very considerable. I have since learned, from the "Statistical Account of the Parish of Sandwick," that the workings of the mine penetrate into the rock for about a hundred yards, but that it has been long abandoned, "as a speculation which would not pay."

I observed scattered over the beach, in the neighborhood of the lead mine, considerable quantities of the hard chalk of England; and, judging there could be no deposits of the hard chalk in this neighborhood, I addressed myself on my way back, to a kelp-burner engaged in wrapping up his fire for the night with a thick covering of weed, to ascertain how it had come there. "Ah, master," he replied, "that chalk is all that remains of a fine large English vessel, that was knocked to pieces here a few years ago. She was ballasted with the chalk; and as it is a light sort of stone, the surf has washed it ashore from that low reef in the middle of the tideway where she struck and broke up. Most of the sailors, poor fellows, lie in the old churchyard, beside the broken ruin yonder. It is a deadly shore this to seafaring-men." I had understood that the kelp-trade was wholly at an end in Orkney; and, remarking that the [Pg 467] sea-weed which he employed was chiefly of one kind,—the long brown fronds of tang dried in the sun,—I inquired of him to what purpose the substance was now employed, seeing that barilla and the carbonate of soda had supplanted it in the manufacture of soap and glass, and why he was so particular in selecting his weed. "It's some valuable medicine," he said, "that's made of the kelp now: I forget its name; but it's used for bad sores and cancer; and we must be particular in our weed, for it's not every kind of weed that has the medicine in't. There's most of it, we're told, in the leaves of the tang." "Is the name of the drug," I asked, "iodine?" "Ay, that must be just it," he replied,—"iodine; but it doesn't make such a demand for kelp as the glass and the soap." I afterwards learned that the kelp-burner's character of this strip of coast, as peculiarly fatal to the mariner, was borne out by many a sad casualty, too largely charged with the wild and the horrible to be lightly forgotten. The respected Free Church clergyman of Stromness, Mr. Learmonth, informed me that, ere the Disruption, while yet minister of the parish, there were on one sad occasion eight dead bodies carried of a Sabbath morning to his manse door. Some of the incidents connected with these terrible shipwrecks, as related with much graphic effect by a boatman who carried me across the sound, on an exploratory ramble to the island of Hoy, struck me as of a character considerably beyond the reach of the mere dealer in fiction. The master of one hapless vessel, a young man, had brought his wife and only child with him on the voyage destined to terminate so mournfully; and when the vessel first struck, he had rushed down to the cabin to bring them both on deck, as their only chance of safety. He had, however, unthinkingly shut the cabin-door after him; a second tremendous blow, as not unfrequently [Pg 468] happens in such cases, so affected the framework of the sides and deck, that the door was jammed fast in its frame. And long ere it could be cut open,—for no human hand could unfasten it,—the vessel had filled to the beams, and neither the master nor his wife and child were ever seen more. In another ship, wrecked within a cable-length of the beach, the mate, a man of Herculean proportions, and a skilful swimmer, stripped and leaped overboard, not doubting his ability to reach the shore. But he had failed to remark what in such circumstances is too often forgotten, that the element on which he flung himself, beaten into foam against the shallows, was, according to Mr. Bremner's shrewd definition, not water, but a mixture of water and air, specifically lighter than the human body; and so at the shore, though so close at hand, he never arrived, disappearing almost at the vessel's side. "The ground was rough," said my informant, "and the sea ran mountains high; and I can scarce tell you how I shuddered on finding, long ere his corpse was thrown up, his two eyes detached from their sockets, staring from a wreath of sea-weed." There is in this last circumstance, horrible enough surely for the wildest German tale ever written, a unique singularity, which removes it beyond the reach of invention.

At my inn I found a pressing invitation awaiting me from the Free Church manse, which I was urged to make my home so long as I remained in that part of the country. A geologist, however, fairly possessed by the enthusiasm without which weak man can accomplish nothing,—whether he be a deer-stalker or mammoth-fancier, or angle for live salmon or dead Pterichthyes,—has a trick of forgetting the right times of dining and taking tea, and of throwing the burden of his bodily requirements on early extempore breakfasts and late suppers; and so reporting myself a man [Pg 469] of irregular habits and bad hours, whose movements could not in the least be depended upon, I had to decline the hospitality which would fain have adopted me as its guest, notwithstanding the badness of the character that, in common honesty, I had to certify as my own. Next morning I breakfasted at the manse, and was introduced by Mr. Learmonth to two gentlemen of the place, who had been kindly invited to meet with me, and who, from their acquaintance with the geology of the district enabled me to make the best use of my time, by cutting direct on those cliffs and quarries in the neighborhood in which organic remains had been detected, instead of wearily rediscovering them for myself. There is a small but interesting museum in Stromness, rich in the fossils of the locality; and I began the geologic business of the day by devoting an hour to the examination of its organisms, chiefly ichthyolites. I saw among them several good specimens of the genus Pterichthys, and of what is elsewhere one of the rarer genera of the Dipterians,—the Diplopterus. A well-marked individual of the latter genus had, I found, been misnamed Dipterus by some geological visitor who had recently come the way,—a mistake which, as in both ichthyolites the fins are similarly placed, occasionally occurs, but which may be easily avoided, when the specimens are in a tolerable state of preservation, by taking note of a few well-marked characteristics by which the genera are distinguished. In both Dipterus and Diplopterus the bright enamel of the scales was thickly punctulated by microscopic points,—the exterior terminations of funnel-shaped openings, that communicated between the surface and the cells of the middle table of the scale; but the form of the scales themselves was different,—that of the Dipterus being nearly circular, and that of the Diplopterus, save on the dorsal ridge, rhomboidal. [Pg 470] Again, the lateral line of the Diplopterus was a raised line, running as a ridge along the scales;

whereas that of the Dipterus was a depressed one, existing as a furrow. Their heads, too, were covered by an entirely dissimilar arrangement of plates. The rounded snout-plate of the Diplopterus was suddenly contracted to nearly one-half its breadth by two semicircular inflections, which formed the orbits of the eyes; full in the centre, a little above these, a minute, lozenge-shaped plate seemed as if inlaid in the larger one, the analogue, apparently, of the anterior frontal; and over all there expanded a broad plate, the superior frontal, half divided vertically by a line drawn downwards from the nape, which, however, stopped short in the middle; and fretted transversely by two small but deeply-indented rectangular marks, which, crossing from the central to two lateral plates, assumed the semblance of connecting pins. The snout of the Dipterus was less round; it bore no mark of the eye-orbits; and the frontal buckler, broader in proportion to its length than that of the Diplopterus, consisted of many more plates. I may here mention that the frontal buckler of Diplopterus has not yet been figured nor described; whereas that of Dipterus, though unknown as such, has been given to the world as the occipital covering of a supposed Cephalaspian,—the Polyphractus. Polyphractus is, however, in reality a synonym for Dipterus,—the one name being derived from a peculiarity of the animal's fins: the other, from the great number of its occipital plates. There is no science founded on mere observation that can be altogether free, in its earlier stages, from mistakes of this character,—mistakes to which the palæontologist, however skilful, is peculiarly liable. The teeth of the two genera were essentially different. Those of the Dipterus, exclusively palatal, were blunt and squat, and ranged in two [Pg 471] rectangular patches; [22] while those of the Diplopterus bristled along its jaws and were slender and sharp. Their tails, too, though both heterocercal, were diverse in their type. In each, an angular strip of gradually-diminishing scales,—a prolongation of the scaly coat which protected the body, and which covered here a prolongation of the vertebral column,—ran on to the extreme termination of the upper lobe; but there was in the Diplopterus a greatly larger development of fin on the superior or dorsal side of the scaly strip than on that of the Dipterus. If the caudal fin of the Osteolepis be divided longitudinally into six equal parts, it will be found that one of these occurs on the upper side of the vertebral prolongation, and five on the under; in the caudal fin of the Diplopterus so divided, rather more than *two* parts will be found to occur on the upper side, and rather less than four on the under; while in the caudal fin of the Dipterus the development seems to have been restricted to the under side exclusively; at least, in none of the many individuals which I have examined have I found any trace of caudal rays on the upper side. These are minute and somewhat trivial particulars; but the geologist may find them of use; and the non-geologist may be disposed to extend to them some little degree of tolerance, when he considers that they distinguished two largely developed genera of animals, to which the Author of all did not deem it unworthy his wisdom to impart, in the act of creation, certain marked points of resemblance, and other certain points of dissimilarity.

[Pg 472]

From the Museum, accompanied by one of the gentlemen to whom Mr. Learmonth had introduced me at breakfast, and who obligingly undertook to act as my guide on the occasion, I set out to visit a remarkable stack on the sea-coast, about four miles north and west of Stromness. We scaled together the steep granitic hill immediately over the town, and then cut on the stack, straight as the bird flies, across a trackless common, bare and stony, and miserably pared by the *flaughter* spade. The landed proprietors in this part of the mainland are very numerous, and their properties small; and there are vast breadths of undivided common that encircle their little estates, as the Atlantic encircles the Orkneys. But the state in which I found the unappropriated parts of the district had in no degree the effect of making me an opponent of appropriation or the landholders. Our country, had it been left as a whole to all its people, as the Communist desiderates, would ere now be of exceedingly little value to any portion of them. The soil of the Orkney commons has been so repeatedly pared off and carried away for fuel, that there are now wide tracts on which there is no more soil to pare, and which present, for the original covering of peaty mould, a continuous surface of pale boulder-clay, here and there mottled by detached tufts of scraggy heath, and here and there roughened by projections of the underlying rock. All is unredeemable barrenness. On the other hand, wherever a bit of private property appears, though in the immediate neighborhood of these ruined wastes, the surface is swarded over, and the soil is the better, not the worse, for the services which it has rendered to man in the past. Whatever the Chartist and the Leveller may think of the matter, it is, I find, virtually on behalf of the many that the soil has been appropriated by the few. After passing from off the tract of moor which overlies the granitic axis of the district, to a [Pg 473] tract equally moory which spreads over the gray flagstones, I marked, more especially in the hollows and ravines, where minute springs ooze from the rock, vast quantities of bog-iron embedded in the soil, and presenting greatly the appearance of the scoria of a smith's forge. The apparent scoria here is simply a reproduction of the iron of the underlying flagstones, transferred, through the agency of water, to that stratum of vegetable mould and boulder-clay which represents the recent period.

I found the stack which I had been brought to see forming the picturesque centre of a bold tract of rock scenery. It stands out from the land as a tall insulated tower, about two hundred feet in height, sorely worn at its base by the breakers that ceaselessly fret against its sides, but considerably broader atop, where it bears a flat cover of sward on the same level with the tops of the precipices which in the lapse of ages have receded from around it. Like the sward-crested hammock left by a party of laborers, to mark the depth to which they have cut in removing a bank or digging a pond, it remains to indicate how the attrition of the surf has told upon the iron-bound coast; demonstrating that lines of precipices hard as iron, and of giddy elevation, are in full retreat before the dogged perseverance of an assailant that, though baffled in each single attack, ever returns to the charge, and gains by an aggregation of infinitesimals,—the result of the whole. From the edge of a steep promontory that commands an inflection of the coast, and of the wall of rock which sweeps round it, I watched for a few seconds the sea,—greatly heightened at the time by the setting in of the flood-tide,—as it broke, surge after surge, against the base of the tall dark precipices; and marked how it accomplished its work of disintegration. The flagstone deposit here abounds in vertical cracks and flaws; and in the line of each of the [Pg 474] many fissures which these form the waves have opened up a cave; so that for hundreds of yards together the precipices seem as if founded on arch-divided piers, and remind one of those ancient prints or drawings of Old London Bridge in which a range of tall sombre buildings is represented as rising high over a line of arches; or of rows of lofty houses in those cities of southern Europe in which the dwellings fronting the streets are perforated beneath by lines of squat piazzas, and present above a dingy and windowless breadth of wall. In course of time the piers attenuate and give way; the undermined precipices topple down, parting from the solid mass behind in those vertical lines by which they are traversed at nearly right angles with their line of stratification; the perpendicular front which they had covered comes to be presented, in consequence, to the sea; its faults and cracks gradually widen into caves, as those of the fallen front had gradually widened at an earlier period; in the lapse of centuries, it too, resigning its place, topples over headlong, an undermined mass; the surge dashes white and furious where the dense rock had rested before; and thus, in its slow but irresistible march, the sea gains upon the land. In the peculiar disposition and character of the prevailing strata of Orkney, as certainly as in the power of the tides which sweep athwart its coasts, and the wide extent of sea which, stretching around it, gives the waves scope to gather bulk and momentum, may be found the secret of the extraordinary height to which the surf sometimes rises against its walls of rock. During the fiercer tempests, masses of foam shoot upwards against the precipices, like inverted cataracts, fully two hundred feet over the ordinary tide-level, and, washing away the looser soil from their summits, leaves in its place patches of slaty gravel, resembling that of a common sea-beach. Rocks less perpendicular, however [Pg 475] great the violence of the wind and sea, would fail to project upwards bodies of surf to a height so extraordinary. But the low angle at which the strata lie, and the rectangularity maintained in relation to their line of bed by the fissures which traverse them, give to the Orkney precipices,—remarkable for their perpendicularity and their mural aspect,—exactly the angle against which the waves, as broken masses of foam, beat up to their greatest possible altitude. On a tract of iron-bound coast that skirts the entrance of the Cromarty Frith I have seen the surf rise, during violent gales from the north-west especially, against one rectangular rock, known as the White Rock, fully an hundred feet; while against scarcely any of the other precipices, more sloping, though equally exposed, did it rise more than half that height.

[Pg 476]

CHAPTER XIII.

Detached Fossils—Remains of the Pterichthys—Terminal Bones of the Coccosteus, etc., preserved—Internal Skeleton of Coccosteus—The shipwrecked Sailor in the Cave—Bishop Grahame—His Character, as drawn by Baillie—His Successor—Ruins of the Bishop's Country-house—Sub-aërial Formation of Sandstone—Formation near New Kaye—Inference from such Formation—Tour resumed—Loch of Stennis—Waters of the Loch fresh, brackish, and salt—Vegetation varied accordingly—Change produced in the Flounder by fresh water—The Standing Stones, second only to Stonehenge—Their purpose—Their Appearance and Situation—Diameter of the Circle—What the Antiquaries say of it—Reference to it in the "Pirate"—Dr. Hibbert's Account.

We returned to Stromness along the edge of the cliffs gradually descending from higher to lower ranges of prepices, and ever and anon detecting ichthyolite beds in the weathered and partially decomposed strata. As the rock moulders into an incoherent clay, the fossils which it envelops become not unfrequently wholly detached from it, so that, on a smart blow dealt by the hammer, they leap out entire, resembling, from the degree of compression which they exhibit, those mimic fishes carved out of plates of ivory or of mother-of-pearl, which are used as counters in some of the games of China or the East Indies. The material of which they are composed, a brittle jet, though better suited than the stone to resist the disintegrating influences, is in most cases greatly too fragile for preservation. One may, however, acquire from the fragments a knowledge of certain minute points in the structure of the ancient animals to which they belonged, respecting which specimens of a more robust texture give no evidence. The plates of Coccosteus sometimes spring out as unbroken [Pg 477] as when they covered the living animal, and, if the necessary skill be not wanting, may be set up in their original order. And I possess specimens of the head of Dipterus in which the nearly circular gill-covers may be examined on both surfaces, interior and exterior, and in which the cranial portion shows not only the enamelled plates of the frontal buckler, but also the strange mechanism of the palatal teeth, with the intervening cavities that had lodged both the brain and the occipital part of the spine. The fossils on the top of the cliffs here are chiefly Dipterians of the two closely allied genera, Diplopterus and Osteolepis.

A little farther on, I found, on a hill-side in which extensive slate-quarries had once been wrought, the remains of Pterichthys existing as mere patches, from which the color had been discharged, but in which the almost human-like outline of both body and arms were still distinctly traceable; and farther on still, where the steep wall of cliffs sinks into a line of grassy banks, I saw in yet another quarry, ichthyolites of all the three great ganoid families so characteristic of the Old Red,—Cephalaspians, Dipterians, and Acanthodians,—ranged in the three-storied order to which I have already referred as so inexplicable. The specimens, however, though numerous, are not fine. They are resolved into a brittle bituminous coal, resembling hard pitch or black wax, which is always considerably less tenacious than the matrix in which they are inclosed; and so, when laid open by the hammer, they usually split through the middle of the plates and scales, instead of parting from the stone at their surfaces, and resemble, in consequence, those dark, shadow-like profiles taken in Indian ink by the limner, which exhibit a correct outline, but no details. We find, however, in some of the genera, portions of the animal preserved that are rarely [Pg 478] seen in a state of keeping equally perfect in the ichthyolites of Cromarty, Moray, or Banff,—those terminal bones of the Coccosteos, for instance, that were prolonged beyond the plates by which the head and upper parts of the body were covered. Wherever the ichthyolites are inclosed in nodules, as in the more southerly counties over which the deposit extends, the nodule terminates, in almost every case, with the massier portions of the organism; for the thinner parts, too inconsiderable to have served as attractive nuclei to the stony matter when the concretion was forming, were left outside its pale, and so have been lost; whereas, in the northern districts of the deposit, where the fossils, as in Caithness and Orkney, occur in flagstone, these slimmer parts, when the general state of keeping is tolerably good, lie spread out on the planes of the slabs, entire often in their minutest rays and articulations. The numerous Coccostei of this quarry exhibit, attached to their upper plates, their long vertebral columns, of many joints, that, depending from the broad dorsal shields of the ichthyolite, remind one of those skeleton fishes one sometimes sees on the shores of a fishing village, in which the bared backbone joints on, cord-like, to the broad plates of the skull. None of the other fishes of the Old Red Sandstone possessed an internal skeleton so decidedly osseous as that of the Coccosteus, and none of them presented externally so large an extent of naked skin,—provisions which probably went together. For about three-fifths of the entire length of the animal the surface was unprotected by dermal plates; and the muscles must have found the fulcrums on which they acted in the internal skeleton exclusively. And hence a necessity for greater strength in their interior framework than in that of fishes as strongly fenced round externally by scales or plates as the coleoptera by their elytrine, or the crustacea [Pg 479] by their shells. Even in the Coccosteus, however, the ossification was by no means complete; and the analogies of the skeleton seem to have allied it rather with the skeletons of the sturgeon family than with the skeletons of the sharks or rays. The processes of the vertebræ were greatly more solid in their substance than the vertebræ themselves,—a condition which in the sharks and rays is always reversed; and they frequently survive, each with its little sprig of bone, formed like the letter Y, that attached it to its centrum, projecting from it, in specimens from which the vertebral column itself has wholly disappeared. I found frequent traces, during my exploratory labors in Orkney, of the dorsal and ventral fins of this ichthyolite; but no trace whatever of the pectorals or of the caudal fin. There seem to have been no pectorals; and the tail, as I have always had occasion to remark, was apparently a mere point, unfurnished with rays.

In descending from the cliffs upon the quarries, my companion pointed to an angular notch in the rock-edge, apparently the upper termination of one of the numerous vertical cracks by which the precipices are traversed, and which in so many cases on the Orkney coast have been hollowed by the waves into long open coves or deep caverns. It was up there, he said, that about twelve years ago the sole survivor of a ship's crew contrived to scramble, four days after his vessel had been dashed to fragments against the rocks below, and when it was judged that all on board had perished. The vessel was wrecked on a Wednesday. She had been marked, when in the offing, standing for the bay of Stromness; but the storm was violent, and the shore a lee one; and as it was seen from the beach that she could scarce weather the headland yonder, a number of people gathered along the cliffs, furnished with ropes, to render to the crew whatever assistance [Pg 480] might be possible in the circumstances. Human help, however, was to avail them nothing. Their vessel, a fine schooner, when within forty yards of the promontory, was seized broadside by an enormous wave, and dashed against the cliff, as one might dash a glass-phial against a stone-wall. One blow completed the work of destruction; she went rolling in entire from keel to mast-head, and returned, on the recoil of the broken surge, a mass of shapeless fragments, that continued to dance idly amid the foam, or were scattered along the beach. But of the poor men, whom the spectators had seen but a few seconds before running wildly about the deck, there remained not a trace; and the saddened spectators returned to their homes to say that all had perished. Four days after,—on the morning of the following Sabbath,—the sole survivor of the crew, saved, as if by miracle, climbed up the precipice, and presented himself to a group of astonished and terrified country people, who could scarce regard him as a creature of this world. The fissure, which at the top of the cliff forms but a mere angular inflection, is hollowed below into a low-roofed cave of profound depth, into the farther extremity of which the tide hardly ever penetrates. It is floored by a narrow strip of shingly beach; and on this bit of beach, far within the cave, the sailor found himself, half a minute after the vessel had struck and gone to pieces, washed in, he knew not how. Two pillows and a few dozen red herrings, which had been swept in along with him, served him for bed and board; a tin cover enabled him to catch enough of the fresh-water droppings of the roof to quench his thirst; several large fragments of wreck that had been jammed fast athwart the opening of the cave broke the violence of the wind and sea; and in that doleful prison, day after day, he saw the tides sink and rise, and lay, when the surf rolled high at the fall of [Pg 481] the tide, in utter darkness even at mid-day, as the waves outside rose to the roof, and inclosed him in a chamber as entirely cut off from the external atmosphere as that of a diving bell. He was oppressed in the darkness, every time the waves came rolling in and compressed his modicum of air, by a sensation of extreme heat,—an effect of the condensation; and then, in the interval of recession, and consequent expansion, by a sudden chill. At low ebb he had to work hard in clearing away the accumulations of stone and gravel which had been rolled in by the previous tide, and threatened to bury him up altogether. At length he succeeded, after many a fruitless attempt, in gaining an upper ledge that overhung his prison-mouth; and, by a path on which a goat would scarce have found footing, he scrambled to the top. His name was Johnstone; and the cave is still known as "Johnstone's Cave." Such was the narrative of my companion.

A little farther on, the undulating bank, into which the cliffs sink, projects into the sea as a flat green promontory, edged with hills of indurated sand, and topped by a picturesque ruin, that forms a pleasing object in the landscape. The ruin is that of a country residence of the bishops of Orkney during the disturbed and unhappy reign of Scotch Episcopacy, and bears on a flat tablet of weathered sandstone the initials of its founder, Bishop George Grahame, and the date of its erection, 1633. With a green cultivated oasis immediately around it, and a fine open sound, overlooked by the bold, picturesque cliffs of Hoy, in front, it must have been, for at least half the year, an agreeable, and, as its remains testify, a not uncomfortable habitation. But I greatly fear Scottish clergymen of the Establishment, whether Presbyterian or Episcopalian, when obnoxious, from their position or their tenets, to the great bulk of the Scottish people, have not been left, since [Pg 482] at least the Reformation, to enjoy either quiet or happy lives, however extrinsically favorable the circumstances in which they may have been placed. Bishop Grahame, only five years after the date of the erection, was tried before the famous General Assembly of 1638; and, being convicted of having "all the ordinar faults of a bishop," he was deposed, and ordered within a limited time "to give tokens of repentance, under paine of excommunication." "He was a curler on the ice on the Sabbath day," says Baillie,—"a setter of tacks to his sones and grandsones, to the prejudice of the Church; he oversaw adulterie; slighted charming; neglected preaching and doing of anie good; and held portions of ministers' stipends for building his cathedral." The concluding portion of his life, after his deposition, was spent in obscurity; nor did his successor in the bishoprick, subsequent to the reëstablishment of Episcopacy at the Restoration,—Bishop Honeyman,—close his days more happily. He was struck in the arm by the bullet which the zealot Mitchell had intended for Archbishop Sharp; and the shattered bone never healed; "for, though he lived some years after," says Burnet, "*they* were forced to lay open the wound every year, for an exfoliation;" and his life was eventually shortened by his sufferings. All seemed comfortable enough, and quite quiet enough, in the bishop's country-house to-day. There were two cows quietly chewing the cud in what apparently had been the dignitary's sitting-room, and patiently awaiting the services of a young woman who was approaching at some little distance with a pail. A large gray cat, that had been sunning herself in a sheltered corner of the court-yard, started up at our approach, and disappeared through a slit hole. The sun, now gone far down the sky, shone brightly on shattered gable-tops, and roofless, rough-edged walls, revealing many [Pg 483] a flaw and chasm in the yielding masonry; and their shadows fell with picturesque effect on the loose litter, rude implements, and gapped dry-stone fence, of the neglected farm-yard which surrounds the building.

I have said that the flat promontory occupied by the ruin is edged by hills of indurated sand. Existing in some places as a continuous bed of a soft gritty sandstone, scooped wave-like a-top, and varying from five to eight feet in thickness, they form a curious example of a sub-aërial formation,—the sand of which they are composed having been all blown from the sea-beach, and consolidated by the action of moisture on a calcareous mixture of comminuted shells, which forms from twenty to twenty-five per cent. of their entire mass. I found that the sections of the bed laid open by the encroachments of the sea, were scarce less regularly stratified than those of a subaqueous deposit, and that it was hollowed, where most exposed to the weather, into a number of spherical cells, which gave to those parts of the surface where they lay thickest, somewhat the aspect of a rude Runic fret-work,—an appearance not uncommon in weathered sandstones. With more time to spare, I could fain have studied the deposit more carefully, in the hope of detecting a few peculiarities of structure sufficient to distinguish sub-aërially-formed from subaqueously-deposited beds of stone. Sandstones of sub-aërial formation are of no very unfrequent occurrence among the recent deposits. On the coast of Cornwall there are cliffs of considerable height that extend for several miles, and have attained a degree of solidity sufficient to serve the commoner purposes of the architect, which at one time existed as accumulations of blown sand. "It is around the promontory of New Kaye," says Dr. Paris, in an interesting memoir on the subject, "that the most extensive formation of sandstone [Pg 484] takes place. Here it may be seen in different stages of induration, from a state in which it is too friable to be detached from the rock upon which it reposes, to a hardness so considerable, that it requires a violent blow from a sledge-hammer to break it. Buildings are here constructed of it; the church of Cranstock is entirely built with it; and it is also employed for various articles of domestic and agricultural uses. The geologist who has previously examined the celebrated specimen from Guadaloupe will be struck with the great analogy which it bears to this formation." Now, as vast tracts of the earth's surface,—in some parts of the world, as in Northern Africa, millions of square miles together,—are at present overlaid by accumulations of sand, which have this tendency to consolidate and become lasting sub-aërial formations, destined to occupy a place among the future strata of the globe, it seems impossible but that also in the old geologic periods there must have been, as now, sand-wastes and sub-aërial formations. And as the representatives of these may still exist in some of our sandstone quarries, it might be well to be possessed of a knowledge of the peculiarities by which they are to be distinguished from deposits of subaqueous origin. In order that I might have an opportunity of studying these peculiarities where they are to be seen more extensively developed than elsewhere on the eastern coast of Scotland, I here formed the intention of spending a day, on my return south, among the sand-wastes of Moray,—a purpose which I afterwards carried into effect. But of that more anon.

On the following morning, availing myself of a kind invitation, through Dr. Garson, from his brother, a Free Church minister resident in an inland district of the Mainland, in convenient neighborhood with the northern coasts of the island, and with several quarries, I set out from [Pg 485] Stromness, taking in my way the Loch and Standing Stones of Stennis, which I had previously seen from but my seat in the mail-gig as I passed. Mr. Learmonth, who had to visit some of his people in this direction, accompanied me for several miles along the shores of the loch, and lightened the journey by his interesting snatches of local history, suggested by the various objects that lay along our road,—buildings, tumuli, ancient battle-fields, and standing stones. The loch itself, an expansive sheet of water fourteen miles in circumference, I contemplated with much interest, and longed for an opportunity of studying its natural history. Two promontories,—those occupied by the Standing Stones, shoot out from the opposite sides, and approach so near as to be connected by a rustic bridge. They divide the loch into two nearly equal parts, the lower of which gives access to the sea, and is salt in its nether reaches and brackish in its upper ones, while the higher is merely brackish in its nether reaches, and fresh enough in its upper ones to be potable. The shores of both were strewed, at the time I passed, by a line of wrack, consisting, for the first few miles, from where the lower loch opens to the sea, of only marine plants, then of marine plants mixed with those of fresh-water growth, and then, in the upper sheet of water, of lacustrine plants exclusively. And the fauna of the loch, like its flora, is, I was led to understand, of the same mixed character; the marine and fresh-water animals having each their own reaches, with certain debatable tracts between, in which each expatiates with more or less freedom, according to its nature and constitution,—some of the sea-fishes advancing far on the fresh water, and others, among the proper denizens of the lake, encroaching far on the salt. The common fresh-water eel strikes out, I was told, farthest into the sea-water; in which, indeed, reversing the habits of the salmon, it is known in various [Pg 486] places to deposit its spawn; it seeks, too, impatient of a low temperature, to escape from the cold of winter, by taking refuge in water brackish enough in a climate such as ours to resist the influence of frost. Of the marine fishes; on the other hand, I found that the flounder got greatly higher than any of the others, inhabiting reaches of the lake almost entirely fresh. A memoir on the Loch of Stennis and its productions, animal and vegetable, such as a Gilbert White of Selborne could produce, would be at once a very valuable and very curious document. By dividing it into reaches, in which the average saltness of the water was carefully ascertained, and its productions noted, with the various modifications which these underwent as they receded upwards or downwards from their proper habitat towards the line at which they could no longer exist, much information might be acquired, of a kind important to the naturalist, and not without its use to the geological student. I have had an opportunity elsewhere of observing a curious change which fresh-water induces on the flounder. In the brackish water of an estuary it becomes, without diminishing in general size, thicker and more fleshy than when in its legitimate habitat the sea; but the flesh loses in quality what it gains in quantity;—it is flabby and insipid, and the margin-fin lacks always its delicious strip of transparent fat. I fain wish that some intelligent resident on the shores of Stennis would set himself carefully to examine its productions, and that then, after registering his observations for a few years, he would favor the world with its natural history.

The Standing Stones,—second in Britain of their kind, to only those of Stonehenge,—occur in two groups; the smaller group (composed, however, of the taller stones) on the southern promontory; the larger on the northern one. Rude and shapeless, and bearing no other impress of the designing faculty than that they are stuck endwise in the [Pg 487] earth, and form, as a whole, regular figures on the sward, there is yet a sublime solemnity about them, unsurpassed in effect by any ruin I have yet seen, however grand in its design or imposing in its proportions. Their very rudeness, associated with their ponderous bulk and weight, adds to their impressiveness. When there is art and taste enough in a country to hew an ornate column, no one marvels that there should also be mechanical skill enough in it to set it up on end; but the men who tore from the quarry these vast slabs, some of them eighteen feet in height over the soil, and raised them where they now stand, must have been ignorant savages, unacquainted with machinery, and unfurnished, apparently, with a single tool. And what, when contemplating their handiwork, we have to subtract in idea from their minds, we add, by an involuntary process, to their bodies: we come to regard the feats which they have accomplished as performed by a power not mechanical, but gigantic. The consideration, too, that these remains,—eldest of the works of man in this country,—should have so long survived all definite tradition of the purposes which they were raised to serve, so that we now merely know regarding them that they were religious in their uses,—products of that ineradicable instinct of man's nature which leads him in so many various ways to attempt conciliating the Powers of another world,—serves greatly to heighten their effect. History at the time of their erection had no existence in these islands: the age, though it sought, through the medium of strange, unknown rites, to communicate with Heaven, was not knowing enough to communicate, through the medium of alphabet or symbol, with posterity. The appearance of the obelisks, too, harmonizes well with their great antiquity and the obscurity of their origin. For about a man's height from the ground they are covered thick by the shorter lichens,—chiefly the [Pg 488] gray-stone parmelia,—here and there embroidered by golden-hued patches of the yellow parmelia of the wall; but their heads and shoulders, raised beyond the reach alike of the herd-boy and of his herd, are covered by an extraordinary profusion of a flowing beard-like lichen of unusual length,—the lichen *calicarus* (or, according to modern botanists, *Ramalina scopulorum*), in which they look like an assemblage of ancient Druids, mysteriously stern and invincibly silent and shaggy as the bard of Gray, when

> "Loose his beard and hoary hair
> Streamed like a meteor on the troubled air."

The day was perhaps too sunny and clear for seeing the Standing Stones to the best possible advantage. They could not be better placed than on their flat promontories, surrounded by the broad plane of an extensive lake, in a waste, lonely, treeless country, that presents no bold, competing features to divert attention from them as the great central objects of the landscape; but the gray of the morning, or an atmosphere of fog and vapor, would have associated better with the mystic obscurity of their history, their shaggy forms, and their livid tints, than the glare of a cloudless sun, that brought out in hard, clear relief their rude outlines, and gave to each its sharp dark patch of shadow. Gray-colored objects, when tall and imposing, but of irregular form, are seen always to most advantage in an uncertain light,—in fog or frost-rime, or under a scowling sky, or, as Parnell well expresses it, "amid the living gleams of night."

They appeal, if I may so express myself, to the sentiment of the ghostly and the spectral, and demand at least a partial envelopment of the obscure. Burns, with the true tact of the genuine poet, develops the sentiment almost instinctively in an exquisite stanza in one of his less-known songs, "The Posey,"—

> [Pg 489] "The hawthorn I will pu', *wi' its locks o' siller gray,*
> Where, *like an aged man, it stands at break o' day.*"

Scott, too, in describing these very stones, chooses the early morning as the time in which to exhibit them, when they "stood in the gray light of the dawning, like the phantom forms of antediluvian giants, who, shrouded in the habiliments of the dead, come to revisit, by the pale light, the earth which they had plagued with their oppression, and polluted by their sins, till they brought down upon it the vengeance of long-suffering heaven." On another occasion, he introduces them as "glimmering, a grayish white, in the rising sun, and projecting far to the westward their long gigantic shadows." And Malcolm, in the exercise of a similar faculty with that of Burns and of Scott, surrounds them, in his description, with a somewhat similar atmosphere of partial dimness and obscurity:—

> "The hoary rocks, of giant size,
> That o'er the land in circles rise,
> Of which tradition may not tell,
> Fit circles for the wizard's spell,
> Seen far *amidst the scowling storm,*
> Seem each a tall and phantom form,
> *As hurrying vapors o'er them flee,*
> Frowning in grim security,
> While, like a dread voice from the past,
> Around them moans the autumnal blast."

There exist curious analogies between the earlier stages of society and the more immature periods of life,—between the savage and the child; and the huge circle of Stennis seems suggestive of one of these. It is considerably more than four hundred feet in diameter, and the stones which compose it, varying from three to fourteen feet in height, must have been originally from thirty-five to forty in number, though only sixteen now remain erect. A mound and [Pg 490] fosse, still distinctly traceable, run round the whole; and there are several mysterious-looking tumuli outside, bulky enough to remind one of the lesser morains of the geologist. But the circle, notwithstanding its imposing magnitude, is but a huge child's house, after all,—one of those circles of stones which children lay down on their village green, and then, in the exercise of that imaginative faculty which distinguishes between the young of the human animal and those of every other creature, convert, by a sort of conventionalism, into a church or dwelling-house, within which they seat themselves, and enact their imitations of their seniors, whether domestic or ecclesiastical. The circle of Stennis was a circle, say the antiquaries, devoted to the sun. The group of stones on the southern promontory of the lake formed but a half-circle, and it was a half-circle dedicated to the moon. To the circular sun the great rude children of an immature age of the world had laid down a circle of stones on the one promontory; to the moon, in her half-orbed state, they had laid down a half-circle on the other; and in propitiating these material deities, to whose standing in the old Scandinavian worship the names of our *Sun*day and *Mon*day still testify, they employed in their respective inclosures, in the exercise of a wild unregulated fancy, uncouth irrational rites, the extremeness of whose folly was in some measure concealed by the horrid exquisiteness of their cruelty. We are still in the nonage of the species, and see human society sowing its wild oats in a thousand various ways, very absurdly often, and often very wickedly; but matters seem to have been greatly worse when, in an age still more immature, the grimly-bearded, six-feet children of Orkney were laying down their stone-circles on the green. Sir Walter, in the parting scene between Cleveland and Minna Troil, which he describes as having taken place amid the lesser group of stones, refers to an immense slab "lying flat and prostrate in the middle of the others, supported by [Pg 491] short pillars, of which some relics are still visible," and which is regarded as the sacrificial stone of the erection. "It is a current belief," says Dr. Hibbert, in an elaborate paper in the "Transactions of the Scottish Antiquaries," that upon this stone a victim of royal birth was immolated. Halfdan the Long-legged, the son of Harold the Fair-haired, in punishment for the aggressions of Orkney, had made an unexpected descent upon its coasts, and acquired possession of the Jarldom. In the autumn succeeding Halfdan was retorted upon, and, after an inglorious contest, betook himself to a place of concealment, from which he was the following morning unlodged, and instantly doomed to the Asæ. Einar, the Jarl of Orkney, with his sword carved the captive's back into the form of an eagle, the spine being longitudinally divided, and the ribs being separated by a transverse cut as far as the loins. He then extracted the lungs, and dedicated them to Odin for a perpetuity of victory, singing a wild song,—'I am revenged for the slaughter of Rognvalld: this have the Nornæ decreed. In my fiording the pillar of the people has fallen. Build up the cairn, ye active youths, for victory is with us. From the stones of the sea-shore will I pay the Long-legged a hard seat.' There is certainly no trace to be detected, in this dark story, of a golden age of the world: the golden age is, I would fain hope, an age yet to come. There at least exists no evidence that it is an age gone by. It will be the full-grown *manly* age of the world when the race, as such, shall have attained to their years of discretion. They are at present in their froward boyhood, playing at the mischievous games of war, and diplomacy, and stock-gambling, and site-refusing, and it is not quite agreeable for quiet honest people to be living amongst them. But there would be nothing gained by going back to that more infantine state of society in which the Jarl Einar carved into a red eagle the back of Halfdan the Long-legged.

[Pg 492]

CHAPTER XIV.

On Horseback—A pared Moor—Small Landholders—Absorption of small holdings in England and Scotland—Division of Land favorable to Civil and Religious Rights—Favorable to social Elevation—An inland Parish—The Landsman and Lobster—Wild Flowers of Orkney—Law of Compensation illustrated by the Tobacco Plant—Poverty tends to Productiveness—Illustrated in Ireland—Profusion of Ichthyolites—Orkney a land of Defunct Fishes—Sandwick—A Collection of Coccostean Flags—A Quarry full of

Heads of Dipteri—The Bergil, or Striped Wrasse—Its Resemblance to the Dipterus—Poverty of the Flora of the Lower Old Red—No true Coniferous Wood in the Orkney Flagstones—Departure for Hoy—The intelligent Boatman—Story of the Orkney Fisherman.

While yet lingering amid the Standing Stones, I was joined by Mr. Garson, who had obligingly ridden a good many miles to meet me, and now insisted that I should mount and ride in turn, while he walked by my side, that I might be fresh, he said, for the exploratory ramble of the evening. I could have ventured more readily on taking the command of a vessel than of a horse, and with fewer fears of mutiny; but mount I did; and the horse, a discreet animal, finding he was to have matters very much his own way, got upon honor with me, and exerted himself to such purpose that we did not fall greatly more than a hundred yards behind Mr. Garson. We traversed in our journey a long dreary moor, so entirely ruined, like those which I had seen on the previous day, by belonging to everybody in general, as to be no longer of the slightest use to anybody in particular. The soil seems to have been naturally poor; but it must have taken a good deal of spoiling to render it the sterile, verdureless waste it is now; for even where it had been poorest, I found that in the island-like [Pg 493] appropriated patches by which it is studded, it at least bears, what it has long ceased to bear elsewhere, a continuous covering of green sward. But if disposed to quarrel with the commons of Orkney, I found in close neighborhood with them that with which I could have no quarrel,—numerous small properties farmed by the proprietors, and forming, in most instances, farms by no means very large. There are parishes in this part of the mainland divided among from sixty to eighty landowners.

A nearly similar state of things seems to have obtained in Scotland about the beginning of the eighteenth century, and for the greater part of the previous one. I am acquainted with old churchyards in the north of Scotland that contain the burying-grounds of from six to ten landed proprietors, whose lands are now merged into single properties. And, in reading the biographies of our old covenanting ministers, I have often remarked as curious, and as bearing in the same line, that no inconsiderable proportion of their number were able to retire, in times of persecution, to their own little estates. It was during the disastrous wars of the French Revolution,—wars from the effects of which Great Britain will, I fear, never fully recover,—that the smaller holdings were finally absorbed. About twenty years ere the war began, the lands of England were parcelled out among no fewer than two hundred and fifty thousand families; before the peace of 1815, they had fallen into the hands of thirty-two thousand. In less than half a century, that base of actual proprietorship on which the landed interest of any country must ever find its surest standing, had contracted in England to less than one-seventh its former extent. In Scotland the absorption of the great bulk of the lesser properties seems to have taken place somewhat earlier; but in it also the revolutionary war appears to have given them the final blow; [Pg 494] and the more extensive proprietors of the kingdom are assuredly all the less secure in consequence of their extinction. They were the smaller stones in the wall, that gave firmness in the setting to the larger, and jammed them fast within those safe limits determined by the line and plummet, which it is ever perilous to overhang. Very extensive territorial properties, wherever they exist, create almost necessarily—human nature being what it is—a species of despotism more oppressive than even that of great unrepresentative governments. It used to be remarked on the Continent, that there was always less liberty in petty principalities, where the eye of the ruler was ever on his subjects, than under the absolute monarchies. [23] And in a country such as ours, the accumulation of landed property in the hands of comparatively a few individuals has the effect often of bringing the territorial privileges of the [Pg 495] great landowner into a state of antagonism with the civil and religious rights of the people, that cannot be other than perilous to the landowner himself. In a district divided, like Orkney, among many owners, a whole country-side could not be shut up against its people by some ungenerous or intolerant proprietor,—greatly at his own risk and to his own hurt,—as in the case of Glen Tilt or the Grampians; nor, when met for purposes of public worship, could the population of a parish be chased from off its bare moors, at his instance, by the constable or the sheriff-officer, to worship God agreeably to their consciences amid the mire of a cross-road, or on the bare sea-beach uncovered by the ebb of the tide. The smaller properties of the country, too, served admirably as stepping-stones, by which the proprietors or their children, when possessed of energy and intellect, could mount to a higher walk of society. Here beside me, for instance, was my friend Mr. Garson, a useful and much-esteemed minister of religion in his native district; while his brother, a medical man of superior parts, was fast rising into extensive practice in the neighboring town. They had been prepared for their respective professions by a classical education; and yet the stepping-stone to positions in society at once so important and so respectable was simply one of the smaller holdings of Orkney, derived to them as the descendants of one of the old Scandinavian Udallers, and which fell short, I was informed, of a hundred a-year.

Mr. Garson's dwelling, to which I was welcomed with much hospitality by his mother and sisters, occupies the middle of an inclined hollow or basin, so entirely surrounded by low, moory hills, that at no point,—though the radius of the prospect averages from four to six miles,—does it command a view of the sea. I scarce expected being introduced in Orkney to a scene in which the traveller [Pg 496] could so thoroughly forget that he was on an island. Of the parish of Harray, which borders on Mr. Garson's property, no part touches the sea-coast; and the people of the parish are represented by their neighbors, who pride themselves upon their skill as sailors and boatmen, as a race of lubberly landsmen, unacquainted with nautical matters, and ignorant of the ocean and its productions. A Harray man is represented, in one of their stories, as entering into a compact of mutual forbearance with a lobster,—to him a monster of unknown powers and formidable proportions,—which he had at first attempted to capture, but which had shown fight, and had nearly captured him in turn. "Weel, weel, let a-be for let a-be," he is made to say; "if thou does na clutch me in thy grips, I'se no clutch thee in mine." It is to this primitive parish that David Vedder, the sailor-poet of Orkney, refers, in his "Orcadian Sketches," as "celebrated over the whole archipelago for the peculiarities of its inhabitants, their singular manners and habits, their uncouth appearance, and homely address. Being the most landward district in Pomona," he adds, "and consequently having little intercourse with strangers, it has become the stronghold of many ancient customs and superstitions, which modern innovation has pushed off from their pedestals in almost all the other parts of the island. The permanency of its population, too, is mightily in favor of 'old use and wont,' as it is almost entirely divided amongst a class of men yelept *pickie*, or petty lairds, each ploughing his own fields and reaping his own crops, much in the manner their great-great-grandfathers did in the days of Earl Patrick. And such is the respect which they entertain for their hereditary beliefs, that many of them are said still to cast a lingering look, not unmixed with reverence, on certain spots held sacred by their Scandinavian ancestors."

[Pg 497]

After an early dinner I set out for the barony of Birsay, in the northern extremity of the mainland, accompanied by Mr. Garson, and passed for several miles over a somewhat dreary country, bare, sterile, and brown, studded by cold, broad, treeless lakes, and thinly mottled by groups of gray, diminutive cottages, that do not look as if there was much of either plenty or comfort inside. But after surmounting the hills that form the northern side of the interior basin, I was sensible of a sudden improvement on the face of the country. Where the land slopes towards the sea, the shaggy heath gives place to a green luxuriant herbage; and the frequent patches of corn seem to rejoice in a more genial soil. The lower slopes of Orkney are singularly rich in wild flowers,—richer by many degrees than the fat loamy meadows of England. They resemble gaudy pieces of carpeting, as abundant in petals as in leaves: their luxuriant blow of red and white, blue and yellow, seems as if competing, in the extent of surface which it occupies, with their general ground of green. I have remarked a somewhat similar luxuriance of wild flowers in the more sheltered hollows of the bleak north-western coasts of Scotland. There is little that is rare to be found among these last, save that a few Alpine plants may be here and there recognized as occurring at a lower level than elsewhere in Britain; but the vast profusion of blossoms borne by species common to the greater part of the kingdom imparts to them an apparently novel character. We may detect, I am inclined to think, in this singular profusion, both in Orkney and the bleaker districts of the mainland of Scotland, the operation of a law not less influential in the animal than in the vegetable world, which, when hardship presses upon the life of the individual shrub or quadruped, so as to threaten its vitality, renders it fruitful in behalf of its species. I have seen the [Pg 498] principle strikingly exemplified in the common tobacco plant, when reared in a northern country in the open air. Year after year it continued to degenerate, and to exhibit a smaller leaf and a shorter stem, until the successors of what in the first year of trial had been vigorous plants of from three to four feet in height, had in the sixth or eighth become mere weeds of scarce as many inches. But while the more flourishing, and as yet undegenerate plant, had merely borne a-top a few florets, which produced a small quantity of exceedingly minute seeds, the stunted weed, its descendant, was so thickly covered over in its season with its pale yellow bells, as to present the appearance of a nosegay; and the seeds produced were not only bulkier in the mass, but also individually of much greater size. The tobacco had grown productive in proportion as it had degenerated and become poor. In the common scurvy grass, too, remarkable, with some other plants, as I have already had occasion to mention, for taking its place among both the productions of our Alpine heights and of our sea-shores, it will be found that in proportion as its habitat proves ungenial, and its stems and leaves become dwarfish and thin, its little white cruciform flowers increase, till, in localities where it barely exists, as if on the edge of extinction, we find the entire plant forming a dense bundle of seed-vessels, each charged to the full with seed. And in the gay meadows of Orkney, crowded with a vegetation that approaches its northern limit of production, we detect what seems to be the same principle, chronically operative; and hence, it would seem, their extraordinary gaiety. Their richly-blossoming plants are the poor productive *Irish* of the vegetable world; [24] for [Pg 499] Doubleday seems to be quite in the right in holding that the law extends to not only the inferior animals, but to our own species also. The lean, ill-fed sow and rabbit rear, it has been long known, a greatly more numerous progeny than the same animals when well cared for and fat; and every horse and cattle breeder knows, that to over-feed his animals proves a sure mode of rendering them sterile. The sheep, if tolerably well pastured, brings [Pg 500] forth only a single lamb at a birth; but if half-starved and lean, the chances are that it may bring forth two or three. And so it is also with the greatly higher human race. Place them in circumstances of degradation and hardship so extreme as almost to threaten their existence as individuals, and they increase, as if in behalf of the species, with a rapidity without precedent in circumstances of greater comfort. The aristocratic families of a country are continually running out; and it requires frequent creations to keep up the House of Lords; while our poor people seem increasing in some districts in almost the mathematical ratio. The county of Sutherland is already more populous than it was previous to the great clearings. In Skye, though fully two-thirds of the population emigrated early in the latter half of the last century, a single generation had scarce passed ere the gap was completely filled; and miserable Ireland, had the human family no other breeding-place or nursery, would of itself be sufficient in a very few ages to people the world.

We returned, taking in our way the cliffs of Marwick Head, in which I detected a few scattered plates and scales, and which, like nine-tenths of the rocks of Orkney, belong to the great flagstone division of the formation. I found the dry-stone fences on Mr. Garson's property still richer in detached fossil fragments than the cliffs; but there are few erections in the island that do not inclose in their walls portions of the organic. We find ichthyolite remains in the flagstones laid bare along the way-side,—in every heap of road-metal,—in the bottom of every stream,—in almost every cottage and fence. Orkney is a land of defunct fishes, and contains in its rocky folds more individuals of the waning ganoid family than are now to be found in all the existing seas, lakes, and rivers of the world. I enjoyed in a snug upper room a delectable [Pg 501] night's rest, after a day of prime exercise, prolonged till it just touched on toil, and again experienced, on looking out in the morning on the wide flat basin around, a feeling somewhat akin to wonder, that Orkney should possess a scene at once so extensive and so exclusively inland.

Towards mid-day I walked on to the parish manse of Sandwick, armed with a letter of introduction to its inmate, the Rev. Charles Clouston,—a gentleman whose descriptions of the Orkneys, in the very complete and tastefully written Guide-Book of the Messrs. Anderson of Inverness, and of his own parish in the "Statistical Account of Scotland," had, both from the high literary ability and the amount of scientific acquirement which they exhibit, rendered me desirous to see. I was politely received, though my visit must have been, as I afterwards ascertained, at a rather inconvenient time. It was now late in the week, and the coming Sabbath was that of the communion in the parish; but Mr. Clouston obligingly devoted to me at least an hour, and I found it a very profitable one. He showed me a collection of flags, with which he intended constructing a grotto, and which contained numerous specimens of Coccosteus, that he had exposed to the weather, to bring out the fine blue efflorescence,—a phosphate of iron which forms on the surface of the plates. They reminded me, from their peculiar style of coloring, and the grotesqueness of their forms, of the blue figuring on pieces of buff-colored china, and seemed to be chiefly of one species, very abundant in Orkney, the *Coccosteus decipiens.* We next walked out to see a quarry in the neighborhood of the manse, remarkable for containing in immense abundance the heads of Dipteri,—many of them in a good state of keeping, with all the multitudinous plates to which they owe their pseudo-name, Polyphractus, in their original places, and bearing unworn [Pg 502] and untarnished their minute carvings and delicate enamel, but existing in every case as mere detached heads. I found three of them lying in one little slaty fragment of two and a half inches by four, which I brought along with me. Mr. Clouston had never seen the curious arrangement of palatal plates and teeth which distinguishes the Dipterus; and, drawing his attention to it in an ill-preserved specimen which I found in the coping of his glebe-wall, I restored, in a rude pencil sketch, the two angular patches of teeth that radiate from the elegant dart-head in the centre of the palate, with the rhomboidal plate behind. "We have a fish, not uncommon on the rocky coasts of this part of the country," he said,—"the Bergil or Striped Wrasse (*Labras Balanus*),—which bears exactly such patches of angular teeth in its palate. They adhere strongly together; and, when found in our old Picts' houses, which occasionally happens, they have been regarded by some of our local antiquaries as artificial,—an opinion which I have had to correct, though it seems not improbable that, from their gem-like appearance, they may have been used in a rude age as ornaments. I think I can show you one disinterred here some years ago." It interested me to find, from Mr. Clouston's specimens that the palatal grinders of this recent fish of Orkney very nearly resemble those of its *Dipterus* of the Old Red Sandstone. The group is of nearly the same size in the modern as in the ancient fish, and presents the same angular form; but the individual teeth are more strongly set in the Bergil than in the Dipterus, and radiate less regularly from the inner rectangular point of the angle to its base outside. I could fain have procured an Orkney Bergil, in order to determine the general pattern of its palatal dentition with what is very peculiar in the more ancient fish,—the form of the lower jaw; and to ascertain farther, from [Pg 503] the contents of the stomach, the species of shell-fish or crustaceans on which it feeds; but, though by no means rare in Orkney, where it is occasionally used as food, I was unable, during my short stay, to possess myself of a specimen.

Mr. Clouston had, I found, chiefly directed his palæontological inquiries on the vegetable remains of the flagstones, as the department of the science in which, in relation to Orkney, most remained to be done; and his collection of these is the most considerable in the number of its specimens that I have yet seen. It, however, serves but to show how very extreme is the poverty of the flora of the Lower Old Red Sandstone. The numerous fishes of the period seem to have inhabited a sea little more various in its vegetation than in its molluscs. Among the specimens of Mr. Clouston's collection I could detect but two species of plants,—an imperfectly preserved vegetable, more nearly resembling a club-moss than aught I have seen, and a smooth-stemmed fucoid, existing as a mere coaly film on the stone, and distinguished chiefly from the other by its sharp-edged, well-defined outline, and from the circumstance that its stems continue to retain the same diameter for a considerable distance, and this, too, after throwing off at acute angles numerous branches, nearly equal in bulk to the parent trunk. In a specimen about two and a half feet in length, which I owe to the kindness of Mr. Dick of Thurso, there are stems continuous throughout, that, though they ramify into from six to eight branches in that space, are quite as thick atop as at bottom. They are the remains, in all probability, of a long flexible fucoid, like those fucoids of the intertropical seas that, streaming slantwise in the tide, rise not unfrequently to the surface in fifteen and twenty fathoms water. I saw among Mr. Clouston's specimens no such lignite as the fragment of [Pg 504] true coniferous wood which I had found at Cromarty a few years previous, and which, it would seem, is still unique among the fossils of the Old Red Sandstone. In the chart of the Pacific attached to the better editions of "Cook's Voyages," there are several entries along the track of the great navigator that indicate where, in mid-ocean, trees, or fragments of trees, had been picked up. The entries, however, are but few, though they belong to all the three voyages together: if I remember aright, there are only five entries in all,—two in the Northern and three in the Southern Pacific. The floating tree, at a great distance from land, is of rare occurrence in even the present scene of things, though the breadth of land be great, and trees numerous; and in the times of the Old Red Sandstone, when probably the breadth of land was *not* great, and trees *not* numerous, it seems to have been of rarer occurrence still. But it is at least something to know that in this early age of the world trees there were.

I walked on to Stromness, and on the following morning, that of Saturday, took boat for Hoy,—skirting, on my passage out, the eastern and southern shores of the intervening island of Græmsay, and, on the passage back again, its western and northern shores. The boatman, an intelligent man,—one of the teachers, as I afterwards ascertained, in the Free Church Sabbath-school,—lightened the way by his narratives of storm and wreck, and not a few interesting snatches of natural history. There is no member of the commoner professions with whom I better like to meet than with a sensible fisherman, who makes a right use of his eyes. The history of fishes is still very much what the history of almost all animals was little more than half a century ago,—a matter of mere external description, heavy often and dry, and of classification founded exclusively on anatomical details. We have still [Pg 505] a very great deal to learn regarding the character, habits and instincts of these denizens of the deep,—much, in short, respecting that faculty which is in them through which their natures are harmonized to the inexorable laws, and they continue to live wisely and securely, in consequence, within their own element, when man, with all his reasoning ability, is playing strange vagaries in his;—a species of knowledge this, by the way, which constitutes by far the most valuable part,—the *mental* department of natural history; and the notes of the intelligent fisherman, gleaned from actual observation, have frequently enabled me to fill portions of the wide hiatus in the history of fishes which it ought of right to occupy. In passing, as we toiled along the Græmsay coast, the ruins of a solitary cottage, the boatman furnished us with a few details of the history and character of its last inmate, an Orkney fisherman, that would have furnished admirable materials for one of the darker sketches of Crabbe. He was, he said, a resolute, unsocial man, not devoid of a dash of reckless humor, and remarkable for an extraordinary degree of bodily strength, which he continued to retain unbroken to an age considerably advanced, and which, as he rarely admitted of a companion in his voyages, enabled him to work his little skiff alone, in weather when even better equipped vessels had enough ado to keep the sea. He had been married in early life to a religiously-disposed woman, a member of some dissenting body; but, living with him in the little island of Græmsay, separated by the sea from any place of worship, he rarely permitted her to see the inside of a church. At one time, on the occasion of a communion Sabbath in the neighboring parish of Stromness, he seemed to yield to her entreaties, and got ready his yawl, apparently with the design of bringing her across the Sound to the town. They had, however, no [Pg 506] sooner quitted the shore than he sailed off to a green little Ogygia of a holm in the neighborhood, on which, reversing the old mythologic story of Calypso and Ulysses, he incarcerated the poor woman for the rest of the day till evening. I could see, from the broad grin with which the boatman greeted this part of the recital, that there was, unluckily, almost fun enough in the trick to neutralize the sense of its barbarity. The unsocial fisherman lived on, dreaded and disliked, and yet, when his skiff was seen boldly keeping the sea in the face of a freshening gale, when every other was making for port, or stretching out from the land as some stormy evening was falling, not a little admired also. At length, on a night of fearful tempest, the skiff was marked approaching the coast, full on an iron-bound promontory, where there could be no safe landing. The helm, from the steadiness of her course, seemed fast lashed, and, dimly discernible in the uncertain light, the solitary boatman could be seen sitting erect at the bows, as if looking out for the shore. But as his little bark came shooting inwards on the long roll of a wave, it was found that there was no speculation in his stony glance: the misanthropic fisherman was a cold and rigid corpse. He had died at sea, as English juries emphatically express themselves in such cases, under "the visitation of God."

[Pg 507]

CHAPTER XV.

Hoy—Unique Scenery—The Dwarfie Stone of Hoy—Sir Walter Scott's Account of it—Its Associations—Inscription of Names—George Buchanan's Consolation—The mythic Carbuncle of the Hill of Hoy—No Fossils at Hoy—Striking Profile of Sir Walter Scott on the Hill of Hoy—Sir Walter, and Shetland and Orkney—Originals of two Characters in "The Pirate"—Bessie Millie—Garden of Gow, the "Pirate"—Childhood's Scene of Byron's "Torquil"—The Author's Introduction to his Sister—A German Visitor—German and Scotch Sabbath-keeping habits contrasted—Mr. Watt's Specimens of Fossil Remains—The only new Organism found in Orkney—Back to Kirkwall—to Wick—Vedder's Ode to Orkney.

We landed at Hoy, on a rocky stretch of shore, composed of the gray flagstones of the district. They spread out here in front of the tall hills composed of the overlying sandstone, in a green undulating platform, resembling a somewhat uneven esplanade spread out in front of a steep rampart. With the upper deposit a new style of scenery commences, unique in these islands: the hills, bold and abrupt, rise from fourteen to sixteen hundred feet over the sea-level; and the valleys by which they are traversed,—no mere shallow inflections of the general surface, like most of the other valleys of Orkney,—are of profound depth, precipitous, imposing, and solitary. The sudden change from the soft, low, and comparatively tame, to the bold, stern, and high, serves admirably to show how much the character of a landscape may depend on the formation which composes it. A walk of somewhat less than two miles brought me into the depths of a brown, shaggy valley, so profoundly solitary, that it does not contain a single human habitation, nor, with one interesting exception, a single trace [Pg 508] of the hand of man. As the traveller approaches by a path somewhat elevated, in order to avoid the peaty bogs of the bottom, along the slopes of the northern side of the dell, he sees, amid the heath below, what at first seems to be a rhomboidal piece of pavement of pale Old Red Sandstone, bearing atop a few stunted tufts of vegetation. There are no neighboring objects of a known character by which to estimate its size; the precipitous hill-front behind is more than a thousand feet in height: the greatly taller Ward Hill of Hoy, which frowns over it on the opposite side, is at least five hundred feet higher; and, dwarfed by these giants, it seems a mere pavior's flag, mayhap some five or six feet square, by from eighteen inches to two feet in depth. It is only on approaching it within a few yards that we find it to be an enormous stone, nearly thirty feet in length by almost fifteen feet in breadth, and in some places, though it thins, wedge-like, towards one of the edges, more than six feet in thickness,—forming altogether such a mass as the quarrier would detach from the solid rock to form the architrave of some vast gateway, or the pediment of some colossal statue. A cave-like excavation, nearly three feet square, and rather more than seven feet in depth, opens on its gray and lichened side. The excavation is widened within, along the opposite walls, into two uncomfortably short beds, very much resembling those of the cabin of a small coasting vessel. One of the two is furnished with a protecting ledge and a pillow of stone, hewn out of the solid mass, while the other, which is some five or six inches shorter than its neighbor, and presents altogether more the appearance of a place of penance than of repose, lacks both cushion and ledge. An aperture, which seems to have been originally of a circular form, and about two and a half feet in diameter, but which some unlucky herd-boy, apparently in the want of better employment, has considerably mutilated and widened, [Pg 509] opens at the inner excavation of the extremity to the roof, as the hatch of a vessel opens from the hold to the deck; for it is by far too wide in proportion to the size of the apartment to be regarded as a chimney. A gray, rudely-hewn block of sandstone, which, though greatly too ponderous to be moved by any man of the ordinary strength, seems to have served the purpose of a door, lies prostrate beside the opening in front. And such is the famous Dwarfie Stone of Hoy, as firmly fixed in our literature by the genius of Sir Walter Scott, as in this wild valley by its ponderous weight and breadth of base, and regarding which—for it shares in the general obscurity of the other ancient remains of Orkney—the antiquary can do little more than repeat, somewhat incredulously, what tradition tells him, viz., that it was the work, many ages ago, of an ugly, malignant goblin, half-earth half-air,—the Elfin Trolld,—a personage, it is said, that even within the last century, used occasionally to be seen flitting about in its neighborhood.

I was fortunate in a fine breezy day, clear and sunshiny, save where the shadows of a few dense piled-up clouds swept dark athwart the landscape. In the secluded recesses of the valley all was hot, heavy and still; though now and then a fitful snatch of a breeze, the mere fragment of some broken gust that seemed to have lost its way, tossed for a moment the white cannach of the bogs, or raised spirally into the air, for a few yards, the light beards of some seeding thistle, and straightway let them down again. Suddenly, however, about noon, a shower broke thick and heavy against the dark sides and gray scalp of the Ward Hill, and came sweeping down the valley. I did what Norna of the Fitful Head had, according to the novelist, done before me in similar circumstances, crept for shelter into the larger bed of the cell, which, though rather scant, taken fairly lengthwise, [Pg 510] for a man of five feet eleven, I found, by stretching myself diagonally from corner to corner, no very uncomfortable lounging-place in a thunder-shower. Some provident herd-boy had spread it over, apparently months before, with a littering of heath and fern, which now formed a dry, springy conch; and as I lay wrapped up in my plaid, listening to the rain-drops as they pattered thick and heavy atop, or slanted through the broken hatchway to the vacant bed on the opposite side of the excavation, I called up the wild narrative of Norna, and felt all its poetry. The opening passage of the story is, however, not poetry, but good prose, in which the curious visitor might give expression to his own conjectures, if ingenious enough either to form or to express them so well. "With my eyes fixed on the smaller bed," the sorceress is made to say, "I wearied myself with conjectures regarding the origin and purpose of my singular place of refuge. Had it been really the work of that powerful Trolld to whom the poetry of the Scalds referred it? or was it the tomb of some Scandinavian chief, interred with his arms and his wealth, perhaps also with his immolated wife, that what he loved best in life might not in death be divided from him? or was it the abode of penance chosen by some devoted anchorite of later days? or the idle work of some wandering mechanic, whom chance, and whim, and leisure, had thrust upon such an undertaking?" What follows this sober passage is the work of the poet. "Sleep," continues Norna, "had gradually crept upon me among my lucubrations, when I was startled from my slumbers by a second clap of thunder, and when I awoke, I saw through the dim light which the upper aperture admitted, the unshapely and indistinct form of Trolld the dwarf, seated opposite to me on the lesser couch, which his square and misshapen bulk seemed absolutely to fill up. I was startled, but not affrighted; for the blood of the ancient race of [Pg 511] Lochlin was warm in my veins. He spoke, and his words were of Norse,—so old, that few save my father, or I myself could have comprehended their import,—such language as was spoken in these islands ere Olave planted his cross on the ruins of heathenism. His meaning was dark also, and obscure, like that which the pagan priests were wont to deliver, in the name of their idols, to the tribes that assembled at the *Helgafels*.... I answered him in nearly the same strain, for the spirit of the ancient Scalds of our race was upon me; and far from fearing the phantom with whom I sat cooped within so narrow a space, I felt the impulse of that high courage which thrust the ancient champions and Druidesses upon contests with the invisible world, when they thought that the earth no longer contained enemies worthy to be subdued by them.... The Demon scowled at me as if at once incensed and overawed; and then, coiling himself up in a thick and sulphurous vapor, he disappeared from his place. I did not till that moment feel the influence of fright, but then it seized me. I rushed into the open air, where the tempest had passed away, and all was pure and serene." Shall I dare confess, that I could fain have passed some stormy night all alone in this solitary cell, were it but to enjoy the luxury of listening, amid the darkness, to the clashing rain and the roar of the wind high among the cliffs, or to detect the brushing sound of hasty footsteps in the wild rustle of the heath, or the moan of unhappy spirits in the low roar of the distant sea. Or, mayhap,—again to borrow from the poet,—as midnight was passing into morning,

"To ponder o'er some mystic lay,
Till the wild tale had all its sway;
And in the bittern's distant shriek
I heard unearthly voices speak,
[Pg 512] Or thought the wizard priest was come
To claim again his ancient home!
And bade my busy fancy range
To frame him fitting shape and strange;
Till from the dream my brow I cleared,

And smiled to think that I had feared."

The Dwarfie Stone has been a good deal undervalued by some writers, such as the historian of Orkney, Mr. Barry; and, considered simply as a work of art or labor, it certainly does not stand high. When tracing, as I lay a-bed, the marks of the tool, which, in the harder portions of the stone, are still distinctly visible, I just thought how that, armed with pick and chisel, and working as I was once accustomed to work, I could complete such another excavation to order in some three weeks or a month. But then, I could not make my excavation a thousand years old, nor envelop its origin in the sun-gilt vapors of a poetic obscurity, nor connect it with the supernatural, through the influences of wild ancient traditions, nor yet encircle it with a classic halo, borrowed from the undying inventions of an exquisite literary genius. A half-worn pewter spoon, stamped on the back with the word *London,* which was found in a miserable hut on the banks of the Awatska by some British sailors, at once excited in their minds a thousand tender remembrances of their country. And it would, I suspect, be rather a poor criticism, and scarcely suited to grapple with the true phenomena of the case, that, wholly overlooking the magical influences of the associative faculty, would concentrate itself simply on either the-workmanship or the materials of the spoon. Nor is the Dwarfie Stone to be correctly estimated, independently of the suggestive principle, on the rules of the mere quarrier who sells stones by the cubic foot, or of the mere contractor for hewn work who dresses them by the square one.

The pillow I found lettered over with the names of visitors; [Pg 513] but the stone,—an exceedingly compact red sandstone,—had resisted the imperfect tools at the command of the traveller,—usually a nail or knife; and so there were but two of the names decipherable,—that of an "H. Ross, 1735," and that of a "P. Folster, 1830." The rain still pattered heavily overhead; and with my geological chisel and hammer I did, to beguile the time, what I very rarely do,—added my name to the others, in characters which, if both they and the Dwarfie Stone get but fair play, will be distinctly legible two centuries hence. In what state will the world then exist, or what sort of ideas will fill the head of the man who, when the rock has well-nigh yielded up its charge, will decipher the name for the last time, and inquire, mayhap, regarding the individual whom it now designates, as I did this morning, when I asked, "Who was this H. Ross, and who this P. Folster?" I remember when it would have saddened me to think that there would in all probability be as little response in the one case as in the other; but as men rise in years they become more indifferent than in early youth to "that life which wits inherit after death," and are content to labor on and be obscure. They learn, too, if I may judge from experience, to pursue science more exclusively for its own sake, with less, mayhap, of enthusiasm to carry them on, but with what is at least as strong to take its place as a moving force, that wind and bottom of formed habit through which what were at first acts of the will pass into easy half-instinctive promptings of the disposition. In order to acquaint myself with the fossiliferous deposits of Scotland, I have travelled, hammer in hand, during the last nine years, over fully ten thousand miles; nor has the work been in the least one of dry labor,—not more so than that of the angler, or grouse-shooter, or deer-stalker: it has occupied the mere leisure interstices of a somewhat busy life, and has served [Pg 514] to relieve its toils. I have succeeded, however, in accomplishing but little: besides, what is discovery to-day will be but rudimentary fact to the tyro-geologists of the future. But if much has not been done, I have at least the consolation of George Buchanan, when, according to Melvill, "fand sitting in his chair, teiching his young man that servit him in his chalmer to spell a, b, ab; e, b, eb. 'Better this,' quoth he, 'nor stelling sheipe.'"

The sun broke out in great beauty after the shower, glistening on a thousand minute runnels that came streaming down the precipices, and revealing, through the thin vapory haze, the horizontal lines of strata that bar the hill-sides, like courses of ashlar in a building. I failed, however, to detect, amid the general many-pointed glitter by which the blue gauze-like mist was bespangled, the light of the great carbuncle for which the Ward Hill has long been famous,—that wondrous gem, according to Sir Walter, "that, though it gleams ruddy as a furnace to them that view it from beneath, ever becomes invisible to him whose daring foot scales the precipices whence it darts its splendor." The Hill of Hoy is, however, not the only one in the kingdom that, according to tradition, bears a jewel in its forehead. The "great diamond" of the Northern Sutor was at one time scarce less famous than the carbuncle of the Ward Hill. "I have been oftener than once interrogated on the western coast of Scotland regarding the diamond rock of Cromarty; and have been told, by an old campaigner who fought under Abercrombie, that he has listened to the familiar story of its diamond amid the sand wastes of Egypt." But the diamond has long since disappeared; and we now see only the rock. Unlike the carbuncle of Hoy, it was never seen by day; though often, says the legend, the benighted boatmen has gazed, from amid the darkness, as he came rowing along [Pg 515] the shore, on its clear beacon-like flame, which, streaming from the precipice, threw a fiery strip across the water; and often have the mariners of other countries inquired whether the light which they saw so high among the cliffs, right over their mast, did not proceed from the shrine of some saint or the cell of some hermit. At length an ingenious ship-captain determined on marking its place, brought with him from England a few balls of chalk, and took aim at it in the night-time with one of his great guns. Ere he had fired, however, it vanished, as if suddenly withdrawn by some guardian hand; and its place in the rock front has ever since remained as undistinguishable, whether by night or by day, as the scaurs and clefts around it. The marvels of the present time abide examination more patiently. It seems difficult enough to conceive, for instance, that the upper deposit of the Lower Old Red in this locality, out of which the mountains of Hoy have been scooped, once overlaid the flag stones of all Orkney, and stretched on and away to Dunnet Head, Tarbet Ness, and the Black Isle; and yet such is the story, variously authenticated, to which their nearly horizontal strata, and their abrupt precipices lend their testimony. In no case has this superior deposit of the formation of the Coccosteus been known to furnish a single fossil; nor did it yield me on this occasion, among the Hills of Hoy, what it had denied me everywhere else on every former one. Sly search, however, was by no means either very prolonged or very careful.

I found I had still several hours of day-light before me; and these I spent, after my return on a rough tumbling sea to Stromness, in a second survey of the coast, westwards from the granitic axis of the island, to the bishop's palace, and the ichthyolitic quarry beyond. From this point of view the high terminal Hill of Hoy, [Pg 516] towards the west, presents what is really a striking profile of Sir Walter Scott, sculptured in the rock front by the storms of ages, on so immense a scale, that the Colossus of Rhodes, Pharos and all, would scarce have furnished materials enough to supply it with a nose. There are such asperities in the outline as one might expect in that of a rudely modelled bust, the work of a master, from which, in his fiery haste, he had not detached the superfluous clay; but these interfere in no degree with the fidelity, I had almost said spirit, of the likeness. It seems well, as it must have waited for thousands of years ere it became the portrait it now is, that the human profile, which it preceded so long, and without which it would have lacked the element of individual truth, should have been that of Sir Walter. Amid scenes so heightened in interest by his genius as those of Orkney, he is entitled to a monument. To the critical student of the philosophy and history of poetic invention it is not uninstructive to observe how completely the novelist has appropriated and brought within the compass of one fiction, in defiance of all those lower probabilities which the lawyer who pleaded before a jury court would be compelled to respect, almost every interesting scene and object in both the Shetland and Orkney islands. There was but little intercourse in those days between the two northern archipelagos. It is not yet thirty years since they communicated with each other, chiefly through the port of Leith, where their regular traders used to meet monthly; but it was necessary, for purposes of effect, that the dreary sublimities of Shetland should be wrought up into the same piece of rich tissue with the imposing antiquities of Orkney,—Sumburgh Head and Roost with the ancient Cathedral of St. Magnus and the earl's palace, and Fitful Head and the sand-enveloped kirk of St. Ringan with the Standing [Pg 517] Stones of Stennis and the Dwarfie Stone of Hoy; and so the little jury-court probabilities have been sacrificed without scruple, and that higher truth of character, and that exquisite portraiture of external nature, which give such reality to fiction, and make it sink into the mind more deeply than historic fact, have been substituted instead. But such,—considerably to the annoyance of the lesser critics,—has been ever the practice of the greater poets. The lesser critics are all critics of the jury-court cast; while all the great masters of fiction, with Shakspeare at their head, have been asserters of that higher truth which is not letter, but spirit, and contemners of the mere judicial probabilities. And so they have been continually fretting the little men with their extravagances, and they ever will. What were said to be the originals of two of Sir Walter's characters in the "Pirate" were living in the neighborhood of Stromness only a few years ago. An old woman who resided immediately over the town, in a little cottage, of which there now remains only the roofless walls, and of whom the sailors, weather-bound in the port, used occasionally to purchase a wind, furnished him with the first conception of his Norna of the Fitful Head; and an eccentric shopkeeper of the place, who to his dying day used to designate the "Pirate," with much bitterness, as a "lying book," and its author as a "wicked lying man," is said to have suggested the character of Bryce Snailsfoot the peddler. To the sorceress Sir Walter himself refers in one of his notes. "At the village of Stromness, on the Orkney main island, called Pomona, lived," he says, "in 1814, an aged dame called Bessie Millie, who helped out her subsistence by selling favorable winds to mariners. Her dwelling and appearance were not unbecoming her pretensions: her house, which was on the brow of the steep hill on which Stromness is founded, was only accessible by [Pg 518] a series of dirty and precipitous lanes, and, for exposure, might have been the abode of Æolus himself, in whose commodities the inhabitant dealt. She herself was, as she told us, nearly one hundred years old, withered and dried up like a mummy. A clay-colored kerchief, folded round her head, corresponded in color to her corpse-like complexion. Two light-blue eyes that gleamed with a lustre like that of insanity, an utterance of astonishing rapidity, a nose and chin that almost met together, and a ghastly expression of cunning, gave her the effect of Hecate. She remembered Gow the pirate, who had been a native of these islands, in which he closed his career. Such was Bessie Millie, to whom the mariners paid a sort of tribute, with a feeling betwixt jest and earnest."

On the opposite side of Stromness, where the arm of the sea, which forms the harbor, is about a quarter of a mile in width, there is, immediately over the shore, a small square patch of ground, apparently a *planticruive*, or garden, surrounded by a tall dry-stone fence. It is all that survives—for the old dwelling-house to which it was attached was pulled down several years ago—of the patrimony of Gow the "Pirate;" and is not a little interesting, as having formed the central nucleus round which,—like those bits of thread or wire on which the richly saturated fluids of the chemist solidify and crystallize,—the entire fiction of the novelist aggregated and condensed under the influence of forces operative only in minds of genius. A white, tall, old-fashioned house, conspicuous on the hill-side, looks out across the bay towards the square inclosure, which it directly fronts. And it is surely a curious coincidence, that while in one of these two erections, only a few hundred yards apart, one of the heroes of Scott saw the light, the other should have proved the scene of the childhood of one of the heroes of Byron,

[Pg 519] "Torquil, the nursling of the northern seas."

The reader will remember, that in Byron's poem of "The Island," one of the younger leaders of the mutineers is described as a native of these northern isles. He is drawn by the poet, amid the wild luxuriance of an island of the Pacific, as

"The blue-eyed northern child,
Of isles more known to man, but scarce less wild,—
The fair-haired offspring of the Orcades,
Where roars the Pentland with his whirling seas,—
Rocked in his cradle by the roaring wind,
The tempest-born in body and in mind,—
His young eyes, opening on the ocean foam,—
Had from that moment deemed the deep his home."

Judging from what I learned of his real history, which is well known in Stromness, I found reason to conclude that he had been a hapless young man, of a kindly, genial nature; and greatly "more sinned against than sinning," in the unfortunate affair of the mutiny with which his name is now associated, and for his presumed share in which, untried and unconvicted, he was cruelly left to perish in chains amid the horrors of a shipwreck. I had the honor of being introduced on the following day to his sister, a lady far advanced in life, but over whose erect form and handsome features the years seemed to have passed lightly, and whom I met at the Free Church of Stromness, to which, at the Disruption, she had followed her respected minister. It seemed a fact as curiously compounded as some of those pictures of the last age in which the thin unsubstantialities of allegory mingled with the tangibilities of the real and the material, that the sister of one of Byron's heroes should be an attached member of the Free Church.

[Pg 520]

On my return to the inn, I found in the public room a young German of some one or two and twenty, who, in making the tour of Scotland, had extended his journey into Orkney. My specimens, which had begun to accumulate in the room, on chimney-piece and window-sill, had attracted his notice, and led us into conversation. He spoke English well, but not fluently,—in the style of one who had been more accustomed to read than to converse in it; and he seemed at least as familiar with two of our great British authors,—Shakspeare and Sir Walter Scott,—as most of the better-informed British themselves. It was chiefly the descriptions of Sir Walter in the "Pirate" that had led him into Orkney. He had already visited the Cathedral of St. Magnus and the Stones of Stennis; and on the morrow he intended visiting the Dwarfie Stone; though I ventured to suggest that, as a broad sound lay between Stromness and Hoy, and as the morrow was the Sabbath, he might find some difficulty in doing that. His circle of acquirement was, I found, rather literary than scientific. It seemed, however, to be that of a really accomplished young man, greatly better founded in his scholarship than most of our young Scotchmen on quitting the national universities; and I felt, as we conversed together, chiefly on English literature and general politics, how much poorer a figure I would have cut in his country than he cut in mine. I found, on coming down from my room next morning to a rather late breakfast, that he had been out among the Stromness fishermen, and had returned somewhat chafed. Not a single boatman could he find in a populous seaport town that would undertake to carry him to the Dwarfie Stone on the Sabbath,—a fact, to their credit, which it is but simple justice to state. I saw him afterwards in the Free Church, listening attentively to a thoroughly earnest and excellent discourse, [Pg 521] by the Disruption minister of the parish, Mr. Learmonth; and in the course of the evening he dropped in for a short time to the Free Church Sabbath-school, where he took his seat beside one of the teachers, as if curious to ascertain more in detail the character of the instruction which had operated so influentially on the boatmen, and which he had seen telling from the pulpit with such evident effect. What would not his country now give,—now, while drifting loose from all its old moorings, full on the perils of a lee shore,—for the anchor of a faith equally steadfast! He was a Lutheran, he told me; but, as is too common in Germany, his actual beliefs appeared to be very considerably at variance with his hereditary creed. The creed was a tolerably sound one, but the living belief regarding it seemed to do little more than take cognizance of what he deemed the fact of its death.

I had carried with me a letter of introduction to Mr. William Watt, to whom I have already had occasion to refer as an intelligent geologist; but the letter I had no opportunity of delivering. Mr. Watt had learned, however, of my being in the neighborhood, and kindly walked into Stromness, some six or eight miles, on the morning of Monday, to meet with me, bringing me a few of his rarer specimens. One of the number,—a minute ichthyolite, about three inches in length,—I was at first disposed to set down as new, but I have since come to regard it as simply an imperfectly-preserved specimen of a Cromarty and Morayshire species,—the *Glyptolepis microlepidotus*; though its state of keeping is such as to render either conclusion an uncertainty. Another of the specimens was that of a fish, still comparatively rare, first figured in the first edition of my little volume on the "Old Red Sandstone," from the earliest found specimen, at a time while it was yet unfurnished with a name, but which has since had a place assigned to it in the [Pg 522] genus Diplacanthus, as the species longispinus. The scales, when examined by the glass, remind one, from their pectinated character, of shells covering the walls of a grotto,—a peculiarity to which, when showing my specimen to Agassiz, while it had yet no duplicate, I directed his attention, and which led him to extemporize for it, on the spot, the generic name Ostralepis, or shell-scale. On studying it more leisurely, however, in the process of assigning to it a place in his great work, where the reader may now find it figured (Table XIV., fig. 8), the naturalist found reason to rank it among the Diplacanthi. Mr. Watt's specimen exhibited the outline of the head more completely than mine; but the Orkney ichthyolites rarely present the microscopic minutiæ; and the shell-like aspect of the scales was shown in but one little patch, where they had left their impressions on the stone. His other specimens consisted of single plates of a variety of Coccosteus, undistinguishable in their form and proportions from those of the *Coccosteus decipiens*, but which exceeded by about one-third the average size of the corresponding parts in that species; and of a rib-like bone, that belonged apparently to what few of the ichthyolites of the Lower Old Red seem to have possessed,—an osseous internal skeleton. This last organism was the only one I saw in Orkney with which I had not been previously acquainted, or which I could regard as new, though possibly enough it may have formed part, not of an undiscovered genus, but of the known genus Asterolepis, of whose inner framework, judging from the Russian specimens at least, portions must have been bony. After parting from Mr. Watt, I travelled on to Kirkwall, which, after a leisurely journey, I reached late in the evening, and on the following morning took the steamer for Wick. I brought away with me, if not many rare specimens or many new geological facts, at least a few pleasing recollections of an interesting [Pg 523] country and a hospitable people. In the previous chapter I indulged in a brief quotation from Mr. David Vedder, the sailor-poet of Orkney, and I shall make no apology for availing myself in the present, of the vigorous, well-turned stanzas in which he portrays some of those peculiar features by which the land of his nativity may be best recognized and most characteristically remembered.

TO ORKNEY.

Land of the whirlpool,—torrent,—foam,
Where oceans meet in madd'ning shock;
The beetling cliff,—the shelving holm,—
The dark insidious rock.
Land of the bleak, the treeless moor,—
The sterile mountain, sered and riven,—
The shapeless cairn, the ruined tower,
Scathed by the bolts of heaven,—

The yawning gulf,—the treacherous sand,—
love thee still, MY NATIVE LAND.

Land of the dark, the Runic rhyme,—
The mystic ring,—the cavern hoar,—
The Scandinavian seer, sublime
In legendary lore.
Land of a thousand sea-kings' graves,—
Those tameless spirits of the past,
Fierce as their subject arctic waves,
Or hyperborean blast,—
Though polar billows round thee foam,
I love thee!—thou wert once my home.

With glowing heart and island lyre,
Ah! would some native bard arise
To sing, with all a poet's fire,
Thy stern sublimities,—
The roaring flood,—the rushing stream,—
The promontory wild and bare,—
The pyramid, where sea-birds scream,
Aloft in middle air,—
The Druid temple on the heath,
Old even beyond tradition's birth.

[Pg 524] Though I have roamed through verdant glades,
In cloudless climes, 'neath azure skies,
Or plucked from beauteous orient meads,
Flowers of celestial dies,—
Though I have laved in limpid streams,
That murmur over golden sands,
Or basked amid the fulgid beams
That flame o'er fairer lands,
Or stretched me in the sparry grot,—
My country! THOU wert ne'er forgot.

THE END.

FOOTNOTES:

[1] March 31, 1845.

[2] Professor Nicol of Aberdeen believes the Red Sandstones of the West Highlands are of Devonian age, and the quartzite and limestone of Lower Carboniferous.—*See Quarterly Journal of the Geological Society, February 1857.*—W.S.

[3] Sir R. Murchison considers these rocks Silurian. See "Quarterly Journal" of the Geological Society, Anniversary Address.

[4] Probably one of the Isastrea of Edwards.

[5] See a paper by the Rev. P.B. Brodie, on Lias Corals, "Edinburgh New Philosophic Journal," April, 1857.

[6] The verses here referred to are introduced into "My Schools and Schoolmasters," chapter tenth.

[7] For a description of this pond see "My Schools and Schoolmasters," chapter tenth.

[8] These remarks refer to the poem "On Seeing a Sun-Dial in a Churchyard," which was introduced here when these chapters were first published in the "Witness," but, having been afterwards inserted in the tenth chapter of "My Schools and Schoolmasters," is not here reproduced.

[9] Mr. Peach has discovered fossils in the Durness limestone, which rests above the quartzite rock of the west of Scotland, that covers the Red Sandstone long believed to be Old Red. The fossils are very obscure.—W.S.S.

[10] This second title hears reference to the extent of the author's geologic excursions in Scotland, during the nine years from 1840 to 1848 inclusive.

[11] Since the above was written, I have seen an interesting paper in "Hogg's Weekly Instructor," in which the Rev. Mr. Longmuir of Aberdeen describes a visit to the Lias clay at Blackpots. Mr. Longmuir seems to have given more time to his researches than I found it agreeable, in a very indifferent day to devote to mine; and his list

of fossils is considerably longer. Their evidence, however, runs in exactly the same tract with that of the shorter list. He had been told at Banff that the clay contained "petrified tangles;" and the first organism shown him by the workmen, on his arrival at the deposit, were some of the "tangles" in question. "These" he goes on to say, "we found, as may have already been anticipated, to be pieces of Belemnites, well known on the other side of the Frith as 'thunderbolts,' and esteemed of sovereign efficacy in the cure of bewitched cattle." Though still wide of the mark, there is here an evident descent from the supernatural to the physical, from the superstitious to the true. "Satisfied that we had a mass of Lias clay before us, we set vigorously to work, in order either to find additional characteristic fossils, or obtain data on which to form a conjecture as to the history of this out-of-the-way deposit; and our labor was not without its reward. We shall now present a brief account of the specimens we picked up. Observing a number of stones of different sizes, that had been thrown out, as they were struck, by the workman's shovel, we immediately commenced, and, like an inquisitor of old, knocked our victims on the head, that they might reveal their secrets; or, like a Roman haruspex, examined their interior,—not, however, to obtain a knowledge of the future, but only to take a peep into the past. 1. Here, then, we take up, not a regular Lias lime nodule, but what appears to have formed part of one; and the first blow has laid open part of a whorl of an Ammonite, which, when complete, must have measured three or four inches in diameter, and it is perfectly assimilated to the calcareous matrix. 2. Here is a mass of indurated clay; and a gentle blow has exposed part of two Ammonites, smaller than the former, but their shells are white and powdery like chalk. 3. Another fragment is laid open; and there, quite unmistakably, lie the umbo and greater portion of the *Plagiostoma concentricum*. 4. Another fragment of a granular gritty structure presents a considerable portion of the interior of one of the shells of a Pecten, but whether the attached fragment is part of one of its ears, or of the other valve turned backward, is not so easily determined. 5. Here is a piece of Belemnite in limestone, and the fracture in the fossil presents the usual glistening planes of cleavage. 6. Next we take up a piece of distinctly laminated Lias, with Ammonites as thick as they can lie on the pages of this black book of natural history. 7. Once more we strike, and we have the cast and part of the shell of another bivalve; but the valves have been jerked off each other, and have suffered a severe compound fracture; nevertheless we can have little hesitation in pronouncing it a species of *unio*. 8. Here is another piece of limestone, with its small fragment of another shell, of very delicate texture, with finely marked traverse striæ. We are unwilling to decide on such slight evidence, but feel inclined to refer it to some species of Plagiostoma. 9. Here is a piece of pyrites, not quite so large as the first, and so vegetable-like in its markings, that it might be mistaken for part of a branch of a tree. This is also characteristic of the Lias; for when the shales are deeply impregnated with bitumen and pyrites, they undergo a slow combustion when heaped up with faggots and set on fire; and in the cliffs of the Yorkshire coast, after rainy weather, they sometimes spontaneously ignite, and continue to burn for several months. 10. As we passed through the works, on our way to the clay, we observed a sort of reservoir, into which the clay, after being freed from its impurities, had been run in a liquid state; the water had evaporated, and the drying clay had cracked in every direction. Here we find its counterpart in this large mass of stone; only the clay here, mixed with a portion of lime is petrified, and the fissures filled up with carbonate of lime; thus forming the septaria, or cement stone. We have dressed a specimen of it for our guide, who has a friend that will polish it, when the dark Lias will be strikingly contrasted with the white lime, and form rather a pretty piece of natural mosaic. 11. Coming to a simple piece of machinery for removing fragments of shale and stone from the clay, we examined some of the bits so rejected, and found what we had no doubt were fish-scales. 12. We have yet to notice certain long slender bodies, outwardly brown, but inwardly nearly black, resembling whip-cord in size. Are we to regard these as specimens of a fucus, perhaps the *filum*, or allied to it, which is known in some places by the appropriate name of sea-laces? 13. Passing on to the office, we were shown a chop of wood that had been found in the clay, and was destined for the Banff Museum. It is about eighteen inches in length, and half as much in breadth; and although evidently water-worn, yet we could count between twenty-five and thirty concentric rings on one of its ends, which not only enabled us to form some conjecture of its age previous to its overthrow, but also justified us in referring it to the coniferæ of the *vorwelt*, or ancient world."

Mr. Longmuir makes the following shrewd remarks, in answering the question, "Whether have we here a mass of Lias clay, as originally deposited, or has it resulted from the breaking up of Lias-shale?" "The former alternative," says Mr. Longmuir, "we have heard, has been maintained; but we are inclined to adopt the latter, and that for the following reasons: 1. This clay, judging from other localities, is not *in situ,* but has every appearance of having been precipitated into a basin in the gneiss on which it rests, having apparently under it, although it is impossible to say to what extent, a bed of comminuted shells. 2. The fossils are all fragmentary and water-worn. This is especially the case with regard to the Belemnites, the pieces averaging from one to two inches in length, no workman having ever found a complete specimen, such as occurs in the Lias-shale at Cromarty, in which they may be found nine inches in length. 3. But perhaps the most satisfactory proof, and one that in itself may be deemed sufficient, is the frequent occurrence of pieces of Lias-shale, with their embedded Ammonites; which clearly show that the Lias had been broken up, tossed about in some violent agitation of the sea, and churned into clay, just as some denudating process of a similar nature swept away the chalk of Aberdeenshire, leaving on many of its hills and plains the water-worn flints, with the characteristic fossils of the Cretaceous formation."

[12] A description of Miss Bond and of her "Letters" here referred to, is given in the fifth chapter of "My Schools and Schoolmasters."

[13] The story here referred to is narrated in "Scenes and Legends of the North of Scotland," chap. xxv.

[14] *Scaur*, Scotice, a precipice of clay. There is no single English word that conveys exactly the same idea.

[15] Mr. Dick has since disinterred from out the boulder-clays of the Burn of Freswick, *Patella vulgata, Buccinum undatum, Fesus antiquus, Rostellaria, Pes pelicana,* a *Natica, Lutraria,* and *Balanus.*

[16] That similarity of condition in which the hazel and the harder cerealia thrive was noted by our north-country farmers of the old School, long ere it had been recorded by the botanist. Hence such remarks, familiarized into proverbs, as "A good *nut* year's a good *ait* year;" or, "As the *nut* fills the *ait* fills."

[17] For this story, see "Scenes and Legends of the North of Scotland," chap. xxv.

[18] "In the River St. Lawrence," says Sir Charles Lyell, "the loose ice accumulates on the shoals during the winter, at which season the water is low. The separate fragments of ice are readily frozen together in a climate where the temperature is sometimes thirty degrees below zero, and boulders become entangled with them; so that in the spring, when the river rises on the melting of the snow, the rocks are floated off, frequently conveying away the boulders to great distances. A single block of granite, fifteen feet long by ten feet both in width and height, and which could not contain less than fifteen hundred cubic feet of stone, was in this way moved down the river several hundred yards, during the late survey in 1837. Heavy anchors of ships, lying on the shore, have in like manner been closed in and removed. In October 1836, wooden stakes were driven several feet into the ground, at one point on the banks of the St. Lawrence, at high-water mark, and over them were piled many boulders as large as the united force of six men could roll. The year after, all the boulders had disappeared, and others had arrived, and the stakes had been drawn out and carried away by the ice."—'Elements,' first edition, p. 138.

[19] The story of the Lady of Balconie and her keys is narrated in "Scenes and Legends of the North of Scotland." chap. xi.

[20] This mode is described in a traditionary story regarding a gigantic tribe of *Fions*, narrated in "Scenes and Legends of the North of Scotland," chap. iv.

[21] See "My Schools and Schoolmasters," chap xi.

[22] I can entertain no doubt that the angular groups of palatal teeth figured by Agassiz and the Russian geologists as those of a supposed Placoid termed the Ctenodus, are in reality groups of the palatal teeth of Dipterus. In some of my specimens the frontal buckler of Polyphractus is connected with the gill-covers and scales of Dipterus, and bears in its palate what cannot he distinguished from the teeth of Ctenodus. The three genera resolve themselves into one.

[23] There is a very admirable remark to this effect in the "Travelling Memorandums" of the late Lord Gardenstone, which, as the

work has been long out of print, and is now scarce, may be new to many of my readers: "It is certain, and demonstrated by the experience of ages and nations," says his Lordship, in referring to the old principalities of France, "that the government of petty princes is less favorable to the security and interests of society than the government of monarchs, who possess great and extensive territories. The race of great monarchs cannot possibly preserve a safe and undisturbed state of government, without many delegations of power and office to men of approved abilities and practical knowledge, who are subject to complaint during their administration, and responsible when it is at an end; or yet without an established system of laws and regulations; so that no inconsiderable degree of security and liberty to the subject is almost inseparable from, and essential to, the subsistence and duration of a great monarchy. But it is easy for petty princes to practise an arbitrary and irregular exercise of power, by which their people are reduced to a condition of miserable slavery. Indeed, very few of them, in the course of ages, are capable of conceiving any other means of maintaining the ostentatious state, the luxurious and indolent pride, which they mistake for greatness. I heartily wish that this observation and censure may not, in some instances, be applicable to great landed proprietors in some parts of Britain." — Travelling Memorandums, vol. i. p. 123. 1792.

[24] The exciting effects of a poor soil, or climate, or of severe usage, on the productive powers of various vegetable species, have been long and often remarked. Flavel describes, in one of his ingenious emblems, illustrative of the influence of affliction on the Christian, an orchard tree, which had been beaten with sticks and stones, till it presented a sorely stunted and mutilated appearance; but which, while the fairer and more vigorous trees around it were rich in only leaves, was laden with fruit, — a direct consequence, it is shown, of the hard treatment to which it had been subjected. I have heard it told in a northern village, as a curious anecdote, that a large pear tree, which during a vigorous existence of nearly fifty years, had borne scarce a single pear, had, when in a state of decay, and for a few years previous to its death, borne immense crops of from two to three bolls each season. And the skilful gardener not unfrequently avails himself of the principle on which both phenomena seem to have occurred, — that exhibited in the beaten and that in the

decaying tree,—in rendering his barren plants fruitful. He has recourse to it even when merely desirous of ascertaining the variety of pear or apple which some thriving sapling, slow in bearing, is yet to produce. Selecting some bough which may be conveniently lopped away without destroying the symmetry of the tree, he draws his knife across the bark, and inflicts on it a wound, from which, though death may not ensue for some two or three twelvemonths, it cannot ultimately recover. Next spring the wounded branch is found to bear its bunches of blossoms; the blossoms set into fruit; and while in the other portions of the plant all is vigorous and barren as before, the dying part of it, as if sobered by the near prospect of dissolution, is found fulfilling the proper end of its existence. Soil and climate, too, exert, it has been often remarked, a similar influence. In the united parishes of Kirkmichael and Culicuden, in the immediate neighborhood of Cromarty, much of the soil is cold and poor, and the exposure ungenial; and "in most parts, where hardwood has been planted," says the Rev. Mr. Sage of Resolis, in his "Statistical Account," "it is stinted in its growth, and bark-bound. Comparatively young trees of ash," he shrewdly adds, "*are covered with seed,—an almost infallible sign that their natural growth is checked.* The leaves, too, fall off about the beginning of September."

of eminent Scientific Men; an Index of Important Papers in Scientific Journals, Reports, &c. Edited by David A. Wells, A.M. With a Portrait of Prof. O.M. Mitchell, 12mo, cloth, $1.25.

Volumes of the same Work for years 1850 to 1858 inclusive. With Portraits of Professors Agassiz, Silltman, Henry, Bache, Maury, Hitchcock, Richard M. Hoe, Profs. Jeffries Wyman, and H.D. Rogers. Nine volumes, 12mo, cloth, $1.25 per vol.

This work, issued annually, contains all important facts discovered or announced during the year.

☞Each volume is distinct in itself, and contains *entirely new matter*.

INFLUENCE OF THE HISTORY OF SCIENCE UPON INTELLECTUAL EDUCATION. By William Whewell, D.D., of Trinity College, Eng., and the alleged author of "Plurality of Worlds." 12mo, cloth, 25 cts.

THE NATURAL HISTORY OF THE HUMAN SPECIES; Its Typical Forms and Primeval Distribution. By Charles Hamilton Smith, With an Introduction containing an Abstract of the views of Blumenbach, Prichard, Bachman, Agassiz, and other writers of repute. By Samuel Kneeland, Jr., M.D. With elegant Illustrations. 12mo, cloth, $1.25.

"The marks of practical good sense, careful observation, and deep research, are displayed in every page. The introductory essay of some seventy or eighty pages forms a valuable addition to the work. It comprises an abstract of the opinions advocated by the most eminent writers on this subject. The statements are made with strict impartiality, and, without a comment, left to the judgment of the reader."—*Sartain's Magazine.*

KNOWLEDGE IS POWER. A View of the Productive Forces of Modern Society, and the Results of Labor, Capital, and Skill. By Charles Knight. With numerous Illustrations. American Edition. Revised, with Additions, by David A. Wells, Editor of the "Annual of Scientific Discovery." 12mo, cloth, $1.25.

☞This is emphatically a book *for the peopl*e. It contains an immense amount of important information, which everybody ought to be in possession of; and the volume should be placed in every family, and in every School and Public Library in the land. The facts and illustrations are drawn from almost every branch of skilful industry, and it is a work which the mechanic and artisan of every description will be sure to read with a RELISH.

IMPORTANT WORKS.

A TREATISE ON THE COMPARATIVE ANATOMY OF THE ANIMAL KINGDOM. By Profs. C. TH. Von Siebold and H. Stannius. Translated from the German, with Notes, Additions, &c. By Waldo I. Burnett, M.D., Boston. One elegant octavo volume, cloth, $3.00.

This is believed to be incomparably the best and most complete work on the subject extant; and its appearance in an English dress, with the additions of the American Translator, is everywhere welcomed by men of science in this country.

UNITED STATES EXPLORING EXPEDITION; during the years 1838, 1839, 1840, 1841, 1842, under Charles Wilkes, U.S.N. Vol. XII.

Mollusca and Shells. By Augustus A. Gould, M.D. Elegant quarto volume, cloth, $6.00.

THE LANDING AT CAPE ANNE; or, The Charter of the First Permanent Colony on the Territory of the Massachusetts Company. Now discovered, and first published from the ORIGINAL MANUSCRIPT, with an inquiry into its authority, and a History of the Colony, 1624–1628, Roger Conant, Governor. By J. Wingate Thornton. 8vo, cloth, $1.50.

☞"A rare contribution to the early history of New England."—*Mercantile Journal.*

LAKE SUPERIOR; Its Physical Character, Vegetation, and Animals. By L. Agassiz and others. One volume octavo, elegantly Illustrated, cloth, $3.50.

THE HALLIG; or, the Sheepfold in the Waters. A Tale of Humble Life on the Coast of Schleswig. Translated from the German of Biernatski, by Mrs. George P. Marsh. With a Biographical Sketch of the Author. 12mo, cloth, $1.00.

As a revelation of an entire new phase in human society, this work strongly reminds the reader of Miss Bremer's tales. In originality and brilliancy of imagination, it is not inferior to those:—its aim is far higher.

THE CRUISE OF THE NORTH STAR; A Narrative of the Excursion made by Mr. Vanderbilt's Party in the Steam Yacht, in her Voyage to England, Russia, Denmark, France, Spain, Italy, Malta, Turkey, Madeira, &c. By Rev. John Overton Choules, D.D. With elegant Illustrations, &c. 12mo, cloth, gilt backs and sides, $1.50; cloth, gilt, $2.00; Turkey, gilt, $3.00.

PILGRIMAGE TO EGYPT; embracing a Diary of Explorations on the Nile, with Observations Illustrative of the Manners, Customs, and Institutions of the People, and of the present condition of the Antiquities and Ruins. By Hon. J.V.C. Smith, late Mayor of the City of Boston. With numerous elegant Engravings. 12mo, cloth, $1.25.

POETICAL WORKS.

COMPLETE POETICAL WORKS OF WILLIAM COWPER; with a Life and Critical Notices of his Writings. Elegant Illustrations. 16mo, cloth, $1.00.

POETICAL WORKS OF SIR WALTER SCOTT. Life and Illustrations. 16mo, cloth, $1.00.

MILTON'S POETICAL WORKS. With a Life and elegant Illustrations. 16mo, cloth, $1.00.

☞The above Poetical Works, by standard authors, are all of uniform size and style, printed on fine paper from clear, distinct type, with new and elegant illustrations, richly bound in full gilt, and plain.

WORKS OF HUGH MILLER.

THE OLD RED SANDSTONE; or, New Walks in an Old Field. Illustrated with Plates and Geological Sections. New Edition, Revised and much Enlarged, by the addition of new matter and new Illustrations, etc. 12mo, cloth, $1.25.

This edition contains over *one hundred pages of entirely new matter*, from the pen of Hugh Miller. It contains, also, several additional new plates and cuts, the old plates re-engraved and improved, and an Appendix of new Notes.

"It is withal one of the most beautiful specimens of English composition to be found, conveying information on a most difficult and profound science, in a style at once novel, pleasing, and elegant."—Dr. Sprague—*Albany Spectator.*

THE FOOTPRINTS OF THE CREATOR; or, the Asterolepsis of Stromness, with numerous Illustrations. With a Memoir of the Author, by Louis Agassiz. 12mo, cloth, $1.00.

Dr. Buckland *said he would give his left hand to possess such power of description as this man.*

TESTIMONY OF THE ROCKS; or, Geology in its Bearings on the two Theologies, Natural and Revealed. "Thou shalt be in league with the stones of the field."—*Job.* With numerous elegant Illustrations. One volume, royal 12mo, cloth, $1.25.

This is the largest and most comprehensive Geological Work that the distinguished author has yet published. It exhibits the profound learning, the felicitous style, and the scientific perception, which characterize his former works, while it embraces the latest results of geological discovery. But the great charm of the book lies in those passages of glowing eloquence, in which, having spread out his facts, the author proceeds to make deductions from them of the most striking and exciting character. The work is profusely illustrated by engravings executed at Paris, in the highest style of French art.

THE CRUISE OF THE BETSEY; or, a Summer Ramble among the Fossiliferous Deposits of the Hebrides. With Rambles of a Geologist; or, Ten Thousand Miles over the Fossiliferous Deposits of Scotland. 12mo, cloth, $1.25.

Nothing need be said of it save that it possesses the same fascination for the reader that characterizes the author's other works.

MY SCHOOLS AND SCHOOLMASTERS; or, the Story of my Education. An Autobiography. With a full-length Portrait of the Author. 12mo, cloth, $1.25.

This is a personal narrative, of a deeply interesting and instructive character, concerning one of the most remarkable men of the age.

MY FIRST IMPRESSIONS OF ENGLAND AND ITS PEOPLE. With a fine Engraving of the author. 12mo, cloth, $1.00.

☞A very instructive book of travels, presenting the most perfectly life-like views of England and its people to be found in any language.

☞*The above six volumes are furnished in sets, printed and bound in uniform style*: viz.,

HUGH MILLER'S WORKS, Six Volumes. Elegant embossed cloth, $7.00, library sheep, $8.00; half calf, $12.00; antique, $12.00.

MACAULAY ON SCOTLAND. A Critique, from the "Witness." 16mo, flexible cloth, 25 cts.

GOULD AND LINCOLN,

59 WASHINGTON STREET, BOSTON.

Would call particular attention to the following valuable works described in their Catalogue of Publications, viz.:

Hugh Miller's Works.

Bayne's Works. Walker's Works. Miall's Works. Bungener's Work.

Annual of Scientific Discovery. Knight's Knowledge is Power.

Krummacher's Suffering Saviour.

Banvard's American Histories. The Aimwell Stories.

Newcomb's Works. Tweedie's Works. Chambers's Works. Harris' Works.

Kitto's Cyclopædia of Biblical Literature.

Mrs. Knight's Life of Montgomery. Kitto's History of Palestin.

Wheewell's Work. Wayland's Works. Agassiz's Works.

William's Works. Guyot's Works.

Thompson's Better Land. Kimball's Heaven. Valuable Works on Missions.

Haven's Mental Philosophy. Buchanan's Modern Atheism.

Cruden's Condensed Concordance. Eadie's Analytical Concordance.

The Psalmist: A Collection of Hymns.

Valuable School Books. Works for Sabbath Schools.

Memoir of Amos Lawrence.

Poetical Works of Milton, Cowpar, Scott. Elegant Miniature Volumes.

Arvine's Cyolopædia of Anecdotes.

Ripley's Notes on Gospels, Acts, and Romans.

Sprague's European Celebrities. Marsh's Camel and the Hallig.

Roget's Thesaurus of English Words.

Hackett's Notes on Acts. M'Whorter's Yahveh Christ.

Siebold and Stannius's Comparative Anatomy. Marco's Geological Map, U.S.

Religious and Miscellaneous Works.

Works in the various Departments of Literature, Science and Art.